Polymer Nanoparticles: Synthesis and Applications

Polymer Nanoparticles: Synthesis and Applications

Editor

Suguna Perumal

MDPI • Basel • Beijing • Wuhan • Barcelona • Belgrade • Manchester • Tokyo • Cluj • Tianjin

Editor
Suguna Perumal
Department of Chemistry
Sejong University
Seoul
Korea, South

Editorial Office
MDPI
St. Alban-Anlage 66
4052 Basel, Switzerland

This is a reprint of articles from the Special Issue published online in the open access journal *Polymers* (ISSN 2073-4360) (available at: www.mdpi.com/journal/polymers/special_issues/ polymer_nanoparticles_synthesis_applications).

For citation purposes, cite each article independently as indicated on the article page online and as indicated below:

LastName, A.A.; LastName, B.B.; LastName, C.C. Article Title. *Journal Name* **Year**, *Volume Number*, Page Range.

ISBN 978-3-0365-7071-6 (Hbk)
ISBN 978-3-0365-7070-9 (PDF)

© 2023 by the authors. Articles in this book are Open Access and distributed under the Creative Commons Attribution (CC BY) license, which allows users to download, copy and build upon published articles, as long as the author and publisher are properly credited, which ensures maximum dissemination and a wider impact of our publications.

The book as a whole is distributed by MDPI under the terms and conditions of the Creative Commons license CC BY-NC-ND.

Contents

About the Editor vii

Preface to "Polymer Nanoparticles: Synthesis and Applications" ix

Suguna Perumal
Polymer Nanoparticles: Synthesis and Applications
Reprinted from: *Polymers* **2022**, *14*, 5449, doi:10.3390/polym14245449 1

Herlina Marta, Dina Intan Rizki, Efri Mardawati, Mohamad Djali, Masita Mohammad and Yana Cahyana
Starch Nanoparticles: Preparation, Properties and Applications
Reprinted from: *Polymers* **2023**, *15*, 1167, doi:10.3390/polym15051167 5

Guangliang Liu and Kathleen McEnnis
Glass Transition Temperature of PLGA Particles and the Influence on Drug Delivery Applications
Reprinted from: *Polymers* **2022**, *14*, 993, doi:10.3390/polym14050993 35

Krissia Wilhelm Romero, María Isabel Quirós, Felipe Vargas Huertas, José Roberto Vega-Baudrit, Mirtha Navarro-Hoyos and Andrea Mariela Araya-Sibaja
Design of Hybrid Polymeric-Lipid Nanoparticles Using Curcumin as a Model: Preparation, Characterization, and In Vitro Evaluation of Demethoxycurcumin and Bisdemethoxycurcumin-Loaded Nanoparticles
Reprinted from: *Polymers* **2021**, *13*, 4207, doi:10.3390/polym13234207 53

Suguna Perumal, Raji Atchudan and Yong Rok Lee
Synthesis of Water-Dispersed Sulfobetaine Methacrylate–Iron Oxide Nanoparticle-Coated Graphene Composite by Free Radical Polymerization
Reprinted from: *Polymers* **2022**, *14*, 3885, doi:10.3390/polym14183885 71

Marta Ruiz-Bermejo, Pilar García-Armada, Pilar Valles and José L. de la Fuente
Semiconducting Soft Submicron Particles from the Microwave-Driven Polymerization of Diaminomaleonitrile
Reprinted from: *Polymers* **2022**, *14*, 3460, doi:10.3390/polym14173460 85

Maryam Wahab, Attya Bhatti and Peter John
Evaluation of Antidiabetic Activity of Biogenic Silver Nanoparticles Using *Thymus serpyllum* on Streptozotocin-Induced Diabetic BALB/c Mice
Reprinted from: *Polymers* **2022**, *14*, 3138, doi:10.3390/polym14153138 105

Srimala Sreekantan, Ang Xue Yong, Norfatehah Basiron, Fauziah Ahmad and Fatimah De'nan
Effect of Solvent on Superhydrophobicity Behavior of Tiles Coated with Epoxy/PDMS/SS
Reprinted from: *Polymers* **2022**, *14*, 2406, doi:10.3390/polym14122406 125

Ana Isabel Ribeiro, Vasyl Shvalya, Uroš Cvelbar, Renata Silva, Rita Marques-Oliveira and Fernando Remião et al.
Stabilization of Silver Nanoparticles on Polyester Fabric Using Organo-Matrices for Controlled Antimicrobial Performance
Reprinted from: *Polymers* **2022**, *14*, 1138, doi:10.3390/polym14061138 143

Jaime Bueno, Leire Virto, Manuel Toledano-Osorio, Elena Figuero, Manuel Toledano and Antonio L. Medina-Castillo et al.
Antibacterial Effect of Functionalized Polymeric Nanoparticles on Titanium Surfaces Using an In Vitro Subgingival Biofilm Model
Reprinted from: *Polymers* **2022**, *14*, 358, doi:10.3390/polym14030358 157

Rajangam Vinodh, Raji Atchudan, Hee-Je Kim and Moonsuk Yi
Recent Advancements in Polysulfone Based Membranes for Fuel Cell (PEMFCs, DMFCs and AMFCs) Applications: A Critical Review
Reprinted from: *Polymers* **2022**, *14*, 300, doi:10.3390/polym14020300 173

Rijuta Ganesh Saratale, Ganesh Dattatraya Saratale, Somin Ahn and Han-Seung Shin
Grape Pomace Extracted Tannin for Green Synthesis of Silver Nanoparticles: Assessment of Their Antidiabetic, Antioxidant Potential and Antimicrobial Activity
Reprinted from: *Polymers* **2021**, *13*, 4355, doi:10.3390/polym13244355 195

Mohd Shahrul Nizam Salleh, Roshafima Rasit Ali, Kamyar Shameli, Mohd Yusof Hamzah, Rafiziana Md Kasmani and Mohamed Mahmoud Nasef
Interaction Insight of Pullulan-Mediated Gamma-Irradiated Silver Nanoparticle Synthesis and Its Antibacterial Activity
Reprinted from: *Polymers* **2021**, *13*, 3578, doi:10.3390/polym13203578 211

Deivasigamani Ranjith Kumar, Kuppusamy Rajesh, Mostafa Saad Sayed, Ahamed Milton and Jae-Jin Shim
Development of Polydiphenylamine@Electrochemically Reduced Graphene Oxide Electrode for the D-Penicillamine Sensor from Human Blood Serum Samples Using Amperometry
Reprinted from: *Polymers* **2023**, *15*, 577, doi:10.3390/polym15030577 229

About the Editor

Suguna Perumal

Suguna Perumal has received a B.Sc. degree in Chemistry from Periyar University (TamilNadu, India) in 2002, M.Sc., in Chemistry from Bharathidasan University (TamilNadu, India) in 2004, and Ph.D. in Chemistry from Freie University Berlin (Germany). She had worked as Postdoctoral fellow and as a research professor with Prof. I. W. Cheong (2013-2019), Kyungpook National University, Daegu, South Korea. Additionally, she worked as a postdoctoral Research Associate and Research Professor at Yeungnam University (2020-2022) in Chemical Engineering Department. Currently, she is working as an Assistant Professor (March 2022) in Department of Chemistry at Sejong University, Seoul, South Korea. Her main research fields concerned with the synthesis of nanoparticles, block copolymers, and graphene-related works for different applications.

Preface to "Polymer Nanoparticles: Synthesis and Applications"

Polymer nanoparticles (PNPs) are formed by the spontaneous self-assembly of polymers that vary in size from 1 to 1000 nm. The self-assemblies of polymers are achieved by solvent evaporation, salting out, nanoprecipitation, desolvation, dialysis, ionic gelation, and spray-drying methods. In recent years, PNPs have been extensively employed as biomaterials because of their characteristic features. This includes biocompatibility, small size, high surface–volume ratio, and tunable surface and structure. Including biomaterials applications such as drug delivery, imaging, biosensors, and stimuli-responsive, PNPs are used in environmental and agricultural applications too. Their small size permits penetration through capillaries and is referred to as nanocarriers. PNPs protect the drug molecules and lead to controlled release and thus are used in drug delivery and diagnostics applications. Due to their high mechanical strength, optical and thermal properties, and conductivity, PNPs are used in imaging, sensors, catalysis, and water-treatment applications. Thus, the works related to polymer nanoparticles, their synthesis, and applications are covered in "Polymer Nanoparticles: Synthesis and Applications". The aim is to help the readers to expand their knowledge towards the development of novel PNPs for diverse applications.

Suguna Perumal
Editor

Editorial

Polymer Nanoparticles: Synthesis and Applications

Suguna Perumal

Department of Chemistry, Sejong University, Seoul 143747, Republic of Korea; suguna.perumal@gmail.com

Polymer nanoparticles (PNPs) are generally formed by the spontaneous self-assembly of polymers that vary size from 1 to 1000 nm [1]. Self-assembly of polymers or surfactant-directed polymers forms the PNPs. Self-assembly of polymers available at critical micelle concentration (CMC). CMC is a concentration above which PNPs are formed [2]. Typical PNPs are normal micelles and inverse micelles, as shown in Figure 1. The normal micelle will have a hydrophobic/oil core and hydrophilic/water shell, while the inverse micelle will have a hydrophilic/water core with a hydrophobic/oil shell [3]. In addition to normal or inverse structures, self-assembly nanostructures of PNPs include sphere, tubular, bottle-brush, rod-shaped, and so on [4].

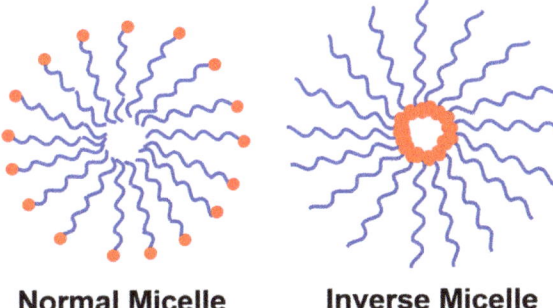

Figure 1. Schematic representation of normal and inverse micelles.

The preparation of PNPs are achieved by solvent evaporation, salting out, nanoprecipitation, desolvation, dialysis, ionic gelation, and spray drying methods [5,6]. Different types of polymers were employed for the preparation of PNPs, which includes natural polymers, for instance, gelatin, alginate, and albumin, and synthetic polymers such as random block copolymer, grafter polymer, block copolymer, and ionic polymers form PNPs [7,8]. PNPs show a wide range of applications which have been extensively employed as biomaterials in recent years because of their characteristic features. This includes biocompatibility, small size, high surface–volume ratio, and tunable surface and structure [1]. In addition to biomaterials applications such as drug delivery, imaging, biosensors, and stimuli-responsive systems, PNPs are used in environmental and agricultural applications [9–15]. The small size of PNPs permits penetration through capillaries, and thus, they are referred to as nanocarriers. PNPs protect the drug molecules, lead to controlled release, and are thus used in drug delivery and diagnostics applications [10,11]. Due to their high mechanical strength, optical and thermal properties, and conductivity, PNPs are used in imaging, sensors, catalysis, and water treatment applications [12–15].

Thus, this Special Issue was established to cover the exciting studies pertaining to polymeric materials and their applications. Romero et al. [16] reported about Pluronic F-127 stabilized polymeric lipid hybrid nanoparticles (PLHNs). Curcumin drugs, demethoxycurcumin (DMC) and bisdemethoycurcumin (BDM) were loaded in PLHNs. The prepared DMC-loaded PLHNs and BDM-loaded PLHNs were characterized by many techniques.

Overall, 88% of DMC and 68% of BDM were released from DMC- and BDM-loaded PLHNs at 180 min. The IC_{50} values for DMC- and BDM-loaded PLHNs were lower than the free ethanolic solutions of DMC and BDM. This confirms the improvement of antioxidant activity using DMC- and BDM-loaded PLHNs particles. Ruiz-Bermejo et al. [17] presented for the first time, the synthesis of submicron particles using diaminomaleonitrile polymers by microwave radiation. The reaction time was varied as follows: 16 min at 170 °C and 3.2 min at 190 °C. The structural, thermal, and electrochemical properties were studied carefully using various techniques. The obtained particles were ~230 nm with a long rice-like shape structure. The prepared polymers exhibited good semiconductor properties and can thus be a potential candidate for soft polymer materials.

Apart from the polymeric nanoparticles, the metal-incorporated polymer nanoparticles, and the preparation of metal nanoparticles from plant sources are also focused on. Salleh et al. [18] explained the synthesis of silver nanoparticles (AgNPs) using natural pullulan (AgNPs/PL) by the γ-irradiation process. The prepared AgNPs/PL was characterized by UV-Vis spectroscopy, X-ray powder diffraction (XRD), transmission electron microscopy (TEM), and Zeta potential analyses. Further, the AgNPs/PL was analyzed for antimicrobial activity against *Staphylococcus aureus* which showed high antibacterial activity with 11–15 nm as an average diameter of the inhibition zone at higher irradiation doses as 50 kGy. Grape pomace-extracted tannin was used as a reducing and stabilizing agent for AgNPs [19]. The prepared Ta-AgNPs showed a maximum at 420 nm in UV-Vis spectroscopy. The zeta potential measurement value of −28.48 suggests the stability of Ta-AgNPs. The surface morphology studies using TEM showed a size between 15 and 20 nm. Ta-AgNPs exhibited antidiabetic activity inhibition of α-amylase and α-glucosidase with IC_{50} values of 48.5 and 40.0 µg/mL, respectively. In addition, Ta-AgNPs were employed as a potent antioxidant and antibacterial agent. Polymeric nanoparticles were prepared using 2-hydroxyethyl methacrylate as a backbone monomer, ethylene glycol dimethacrylate as a cross-linker, and methacrylic acid as a functional monomer [20]. The prepared polymer was loaded with zinc and calcium nanoparticles, and their antibacterial effect was studied using an in vitro subgingival biofilm model. The prepared polyester-stabilized AgNPs and their antimicrobial performance against *Staphylococcus aureus* and *Escherichia coli* were systematically studied and reported [21]. The extract from the medicinal plant "Thymus serpyllum" was reported as a stabilizing and reducing agent in the preparation of AgNPs [22]. The antidiabetic activity on Streptozotocin-induced diabetic BALB/c mice was reported. Perumal et al. [23] synthesized the water-dispersible graphene composite. The graphene surface was functionalized with zwitterion polymer poly [2-(methacryloyloxy)ethyl]dimethyl-(3-sulfopropyl)ammonium hydroxide and iron oxide nanoparticle (FeNPs). The prepared composites were confirmed from various analyses such as XRD, Raman, SEM, TEM, X-ray photoelectron spectroscopy, and thermogravimetric analysis.

Moreover, Liu et al. [24] has summarized the glass transition temperature (T_g) of poly(lactic-co-glycolic acid) (PLGA) particles and their application towards drug delivery. The change in T_g of PLGA particles with a change in size, molecular weight, shape, and with ionic liquids was discussed in detail. The T_g of PLGA showed as an indicator for the controlled drug release. Vinodh et al. [25] reviewed and concisely reported on the polysulfone-based membrane for fuel cell application.

Thus, the articles that are published in this Special Issue will of particular interest for researchers who work with polymer materials. Additionally, these articles will be helpful in the further development of polymer materials for diverse applications.

Conflicts of Interest: The author declares no conflict of interest.

References

1. Zielińska, A.; Carreiró, F.; Oliveira, A.M.; Neves, A.; Pires, B.; Venkatesh, D.N.; Durazzo, A.; Lucarini, M.; Eder, P.; Silva, A.M.; et al. Polymeric Nanoparticles: Production, Characterization, Toxicology and Ecotoxicology. *Molecules* **2020**, *25*, 3731. [CrossRef] [PubMed]
2. Su, H.; Wang, F.; Ran, W.; Zhang, W.; Dai, W.; Wang, H.; Anderson, C.F.; Wang, Z.; Zheng, C.; Zhang, P.; et al. The role of critical micellization concentration in efficacy and toxicity of supramolecular polymers. *Proc. Natl. Acad. Sci. USA* **2020**, *117*, 4518–4526. [CrossRef] [PubMed]
3. Webber, S.E. Polymer Micelles: An Example of Self-Assembling Polymers. *J. Phys. Chem. B* **1998**, *102*, 2618–2626. [CrossRef]
4. Banik, B.L.; Fattahi, P.; Brown, J.L. Polymeric nanoparticles: The future of nanomedicine. *WIREs Nanomed. Nanobiotechnol.* **2016**, *8*, 271–299. [CrossRef] [PubMed]
5. Liu, J.; Lee, H.; Allen, C. Formulation of drugs in block copolymer micelles: Drug loading and release. *Curr. Pharm. Des.* **2006**, *12*, 4685–4701. [CrossRef] [PubMed]
6. Preethi, R.; Dutta, S.; Moses, J.A.; Anandharamakrishnan, C. Chapter 8—Green nanomaterials and nanotechnology for the food industry. In *Green Functionalized Nanomaterials for Environmental Applications*; Shanker, U., Hussain, C.M., Rani, M., Eds.; Elsevier: Amsterdam, The Netherlands, 2022; pp. 215–256.
7. Kamali, H.; Nosrati, R.; Malaekeh-Nikouei, B. Chapter 1—Nanostructures and their associated challenges for drug delivery. In *Hybrid Nanomaterials for Drug Delivery*; Kesharwani, P., Jain, N.K., Eds.; Woodhead Publishing: Cambridge, UK, 2022; pp. 1–26.
8. Peltonen, L.; Singhal, M.; Hirvonen, J. 1—Principles of nanosized drug delivery systems. In *Nanoengineered Biomaterials for Advanced Drug Delivery*; Mozafari, M., Ed.; Elsevier: Amsterdam, The Netherlands, 2020; pp. 3–25.
9. Adhikari, C. Polymer nanoparticles-preparations, applications and future insights: A concise review. *Polym.-Plast. Technol. Mater.* **2021**, *60*, 1996–2024. [CrossRef]
10. Singh, N.; Joshi, A.; Toor, A.P.; Verma, G. Chapter 27—Drug delivery: Advancements and challenges. In *Nanostructures for Drug Delivery*; Andronescu, E., Grumezescu, A.M., Eds.; Elsevier: Amsterdam, The Netherlands, 2017; pp. 865–886.
11. Goel, H.; Saini, K.; Razdan, K.; Khurana, R.K.; Elkordy, A.A.; Singh, K.K. Chapter 3—In vitro physicochemical characterization of nanocarriers: A road to optimization. In *Nanoparticle Therapeutics*; Kesharwani, P., Singh, K.K., Eds.; Academic Press: Cambridge, MA, USA, 2022; pp. 133–179.
12. Srikar, R.; Upendran, A.; Kannan, R. Polymeric nanoparticles for molecular imaging. *WIREs Nanomed. Nanobiotechnol.* **2014**, *6*, 245–267. [CrossRef] [PubMed]
13. Canfarotta, F.; Whitcombe, M.J.; Piletsky, S.A. Polymeric nanoparticles for optical sensing. *Biotechnol. Adv.* **2013**, *31*, 1585–1599. [CrossRef] [PubMed]
14. Shifrina, Z.B.; Matveeva, V.G.; Bronstein, L.M. Role of Polymer Structures in Catalysis by Transition Metal and Metal Oxide Nanoparticle Composites. *Chemical Reviews* **2020**, *120*, 1350–1396. [CrossRef] [PubMed]
15. Wen, Y.; Yuan, J.; Ma, X.; Wang, S.; Liu, Y. Polymeric nanocomposite membranes for water treatment: A review. *Environ. Chem. Lett.* **2019**, *17*, 1539–1551. [CrossRef]
16. Wilhelm Romero, K.; Quirós, M.I.; Vargas Huertas, F.; Vega-Baudrit, J.R.; Navarro-Hoyos, M.; Araya-Sibaja, A.M. Design of Hybrid Polymeric-Lipid Nanoparticles Using Curcumin as a Model: Preparation, Characterization, and In Vitro Evaluation of Demethoxycurcumin and Bisdemethoxycurcumin-Loaded Nanoparticles. *Polymers* **2021**, *13*, 4207. [CrossRef] [PubMed]
17. Ruiz-Bermejo, M.; García-Armada, P.; Valles, P.; de la Fuente, J.L. Semiconducting Soft Submicron Particles from the Microwave-Driven Polymerization of Diaminomaleonitrile. *Polymers* **2022**, *14*, 3460. [CrossRef] [PubMed]
18. Salleh, M.S.N.; Ali, R.R.; Shameli, K.; Hamzah, M.Y.; Kasmani, R.M.; Nasef, M.M. Interaction Insight of Pullulan-Mediated Gamma-Irradiated Silver Nanoparticle Synthesis and Its Antibacterial Activity. *Polymers* **2021**, *13*, 3578. [CrossRef]
19. Saratale, R.G.; Saratale, G.D.; Ahn, S.; Shin, H.-S. Grape Pomace Extracted Tannin for Green Synthesis of Silver Nanoparticles: Assessment of Their Antidiabetic, Antioxidant Potential and Antimicrobial Activity. *Polymers* **2021**, *13*, 4355. [CrossRef] [PubMed]
20. Bueno, J.; Virto, L.; Toledano-Osorio, M.; Figuero, E.; Toledano, M.; Medina-Castillo, A.L.; Osorio, R.; Sanz, M.; Herrera, D. Antibacterial Effect of Functionalized Polymeric Nanoparticles on Titanium Surfaces Using an In Vitro Subgingival Biofilm Model. *Polymers* **2022**, *14*, 358. [CrossRef]
21. Ribeiro, A.I.; Shvalya, V.; Cvelbar, U.; Silva, R.; Marques-Oliveira, R.; Remião, F.; Felgueiras, H.P.; Padrão, J.; Zille, A. Stabilization of Silver Nanoparticles on Polyester Fabric Using Organo-Matrices for Controlled Antimicrobial Performance. *Polymers* **2022**, *14*, 1138. [CrossRef]
22. Wahab, M.; Bhatti, A.; John, P. Evaluation of Antidiabetic Activity of Biogenic Silver Nanoparticles Using Thymus serpyllum on Streptozotocin-Induced Diabetic BALB/c Mice. *Polymers* **2022**, *14*, 3138. [CrossRef] [PubMed]
23. Perumal, S.; Atchudan, R.; Lee, Y.R. Synthesis of Water-Dispersed Sulfobetaine Methacrylate–Iron Oxide Nanoparticle-Coated Graphene Composite by Free Radical Polymerization. *Polymers* **2022**, *14*, 3885.
24. Liu, G.; McEnnis, K. Glass Transition Temperature of PLGA Particles and the Influence on Drug Delivery Applications. *Polymers* **2022**, *14*, 993. [CrossRef]
25. Vinodh, R.; Atchudan, R.; Kim, H.-J.; Yi, M. Recent Advancements in Polysulfone Based Membranes for Fuel Cell (PEMFCs, DMFCs and AMFCs) Applications: A Critical Review. *Polymers* **2022**, *14*, 300. [CrossRef] [PubMed]

Review

Starch Nanoparticles: Preparation, Properties and Applications

Herlina Marta [1,2,*], Dina Intan Rizki [1], Efri Mardawati [2,3], Mohamad Djali [1], Masita Mohammad [4] and Yana Cahyana [1]

1. Department of Food Technology, Universitas Padjadjaran, Bandung 45363, Indonesia
2. Research Collaboration Center for Biomass and Biorefinery between BRIN and Universitas Padjadjaran, Bandung 45363, Indonesia
3. Department of Agroindustrial Technology, Universitas Padjadjaran, Bandung 45363, Indonesia
4. Solar Energy Research Institute (SERI), Universitas Kebangsaan Malaysia, Bangi 43600, Selangor, Malaysia
* Correspondence: herlina.marta@unpad.ac.id

Abstract: Starch as a natural polymer is abundant and widely used in various industries around the world. In general, the preparation methods for starch nanoparticles (SNPs) can be classified into 'top-down' and 'bottom-up' methods. SNPs can be produced in smaller sizes and used to improve the functional properties of starch. Thus, they are considered for the various opportunities to improve the quality of product development with starch. This literature study presents information and reviews regarding SNPs, their general preparation methods, characteristics of the resulting SNPs and their applications, especially in food systems, such as Pickering emulsion, bioplastic filler, antimicrobial agent, fat replacer and encapsulating agent. The aspects related to the properties of SNPs and information on the extent of their utilisation are reviewed in this study. The findings can be utilised and encouraged by other researchers to develop and expand the applications of SNPs.

Keywords: starch nanoparticle; preparation; properties; applications

1. Introduction

Starch is a native polymer that is abundantly available and widely used worldwide [1,2]. Starch is a carbohydrate. Starch granules are a source of carbohydrates and renewable substances produced by plants in the form of granular constituents of plant parts, such as seeds, tubers and fruits, as sources of stored energy; the shape and nature of starch depend on the botanical source, climate and location from where it is isolated [3–7]. Globally, starch production is mainly based on four raw materials, namely corn, cassava, wheat and potatoes, with >75% of starch produced from corn [8,9].

Starch is a mixture of two macromolecules, namely amylose with linear chains of glucose molecules connected by α-1,4 glucosidic bonds, and amylopectin with branched chains consisting of short amylose groups connected by α-1,6 glucosidic bonds [10,11]. In general, starch contains about 20–30% amylose and 70–80% amylopectin, depending on the source [12]. The ratio of amylose/amylopectin content will affect the functional properties of starch, such as gelatinisation, viscosity and gel stability [13]. Starch has been widely used as an additive in the food industry and other industrial applications, such as pharmaceuticals, drug delivery and composites, because of its low cost and easy availability [14,15]. In the food industry, it is generally used as a thickener and an auxiliary to improve food texture and can be utilized to manufacture sauces, soups and puddings [16,17]. Native starch directly extracted from plants generally has limited industrial-scale applications. It has low thermal stability and resistance to external factors during storage, high brittleness and a hydrophilic nature [18]. Therefore, additional treatment, either physically, chemically or enzymatically, is needed to change the properties of native starch to overcome its limitations [16,19–25].

For decades, the food industry has attempted to improve the physicochemical properties of starch to ensure the final quality of food products [26]. Several starch studies have

focused on particle size and understanding the relationship between nano- and microscopic properties of different materials, where the resulting particles are small and have a large surface area [16,27]. This new alternative, the modification of starch macromolecules from micro- to nanoscale, produces nanostarches, which have attracted considerable attention due to their unique properties, and not only change the particle size, but also enhances their functional properties [6,26]. Nanostarches are of two types: starch nanocrystals (SNCs) are crystalline portions resulting from the disruption of amorphous domains in starch granules, and starch nanoparticles (SNPs) generated from gelatinised starch, which may include amorphous regions [27,28]. SNPs are considered promising biomaterials and offer various opportunities for quality improvement for the innovative development of competitive products [29,30]. SNPs have been widely used in active food packaging [31], bioactive compound encapsulation [32], nanocomposite film [33] and Pickering emulsion stabilisers [11,34].

This review focuses on SNPs, their various preparation methods, including physical, chemical and enzymatic treatments, and the resulting characteristic changes. Characterisation of SNPs in terms of size distribution, crystal structure and various properties compared with native starch and the extent to which SNPs have been utilised were also reviewed. The scheme in Figure 1 summarizes the preparation, properties and application of SNPs.

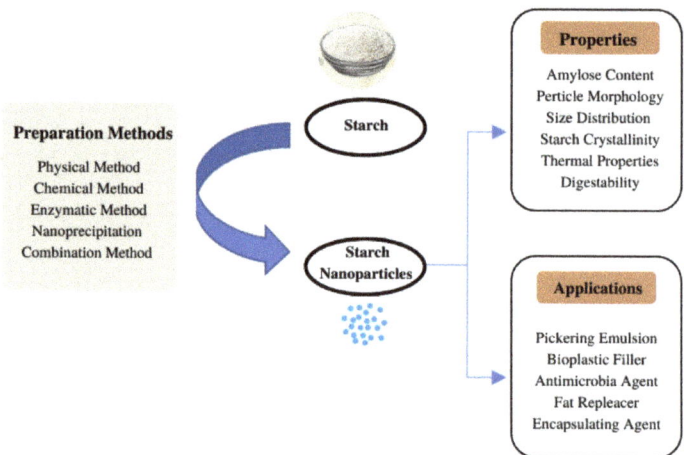

Figure 1. Scheme of SNP preparation, properties and applications.

2. Starch Nanoparticles

SNPs are produced through a nanotechnology process that produces nanoparticles with a size smaller than 1000 nm, but are larger than a single molecule [6,35]. Furthermore, Le Corre et al. [36] reported that the morphology and properties of the resulting nanoparticles depend on the botanical source. Some previous studies on SNPs, such as those from potatoes [6], bananas [37], water chestnuts [38], sago [39] and maize starches [40], have been widely carried out.

In general, the preparation of SNPs can be classified into 'top-down' and 'bottom-up' methods [41]. On the other hand, top-down methods, such as ultrasonication, homogenisation, gamma radiation, acid hydrolysis and others, involve the production of nanoparticles from the breakdown of large particles into small ones based on structural fragmentation using mechanical and chemical forces [30,42]. On the other hand, bottom-up methods produce nanoparticles using a thermodynamic process of controlled molecular assembly, such as nanoprecipitation or self-assembly [42]. Top-down approach technology is the most commonly used method to produce nanoparticles [10].

Top-down methods have the advantage of easy usage, but they are ineffective for the production of particles with the right size and shape [43]. Meanwhile, bottom-up methods produce nanoparticles with controllable shape and size, high yields and short duration times; they require specific chemical reagents and advanced equipment [26,44]. The main considerations in selecting the preparation method of nanoparticles may be related to the size of the resulting particles and scale of production. According to Pattekari et al. [45], top-down methods are more efficient on a larger scale and produce SNPs with a large particle size, whereas bottom-up methods can produce smaller particles, but are more suitable for use on a laboratory scale given the lower resulting yield.

SNPs have various distinctive properties. They are widely used compared to native starch because of their nanometric size; thus, starch has received considerable attention due to its large surface area per mass ratio and effectively increased interactions [38]. Based on several studies, SNPs are a promising alternative for the formation of stable emulsions, carriers of bioactive compounds and effective packaging developments for the production of food-grade films with great resistance to mechanical effects [35,38].

3. Preparation Method of Starch Nanoparticles

SNPs are obtained from the breakdown of granules produced by various methods. The preparation of nanoparticles can be classified based on the preparation method: physical, chemical, enzymatic methods or their combination. Table 1 presents the various methods for the preparation of SNPs from various starch sources.

Table 1. Preparation of SNPs by various methods.

Starch Source	Preparation Method	Preparation Condition	%Yield	Ref.
a. Top-down methods				
Cassava Waxy maize	Gamma irradiation	Doses 20 kGy (14 kG/h)	NR	[46]
Cassava	Gamma irradiation	Doses 20 kGy	NR	[47]
Green Sago	High-pressure homogenization	250 Mpa/5 passes 1 h (Refrigerated for 30 min/after time)	NR	[48]
High amylose maize	High-pressure homogenization	Starch was dispersed high-pressure homogenization was performed at 140, 200, and 250 MPa for 1–4 cycles	NR	[49]
Waxy maize	Ultrasonication	Ultrasonication (80% power, 8 °C, 20 kHz, 75 min	NR	[50]
Waxy maize	Ultrasonication	Sonication (80% power, 8 °C, 24 kHz, 75 min	NR	[51]
Cassava	Ultrasonication	Ultrasonication 8 °C, 24 kHz, 75 min	NR	[52]
Corn	Ultrasonication	mixture water-isopropanol (50/50 wt%) ultrasonication (100% power, 10 °C, 20 kHz, 75 min	NR	[53]
Quinoa Maize	Ultrasonication	The suspension is heated in solution NaOH (ultrasonication 20 kHz, 30 min)	NR	[54]
Waxy maize	Acid hydrolysis	3.16 M H_2SO_4, hydrolysis at 40 °C for 5 days	NR	[46]
Mung bean	Acid hydrolysis	3.16 M H_2SO_4, hydrolysis at 40 °C for 7 days	33.2	[55]
Waxy maize Normal maize High AM maize Potato Mungbean	Acid hydrolysis	3.16 M H_2SO_4, hydrolysis at 40 °C for 7 days	NR	[56]
Waxy maize High amylose maize	Acid hydrolysis	3.16 M H_2SO_4 hydrolysis at 40 °C for 6 days	NR	[57]
Water chesnut	Acid hydrolysis	3.16 M H_2SO_4 hydrolysis at 40 °C for 7 days	27.5	[38]
Andean potato	Acid hydrolysis	3.16 M H_2SO_4 hydrolysis at 40 °C for 5 days	NR	[58]
Waxy rice	Acid hydrolysis	2.2 M HCl hydrolysis at 35 °C for 7–10 days	NR	[59]

Table 1. Cont.

Starch Source	Preparation Method	Preparation Condition	%Yield	Ref.
Sago	Acid hydrolysis	2.2 M HCl hydrolysis 35 °C for 12–48 h	72–80	[39]
Sago	Combined acid hydrolysis and precipitation	HCl 2.2 M hydrolysis 35 °C for 12–48 h, then precipitation with ethanol HCl 2.2 M hydrolysis 35 °C for 12–48 h, then precipitation with butanol	20–25% 22–23%	[39]
Andean potato	Combined acid hydrolysis and –ultrasonication	3.16 M H_2SO_4 hydrolysis at 40 °C for 5 days, then sonication 4 °C, 26 kHz	NR	[58]
Waxy maize SNP	Combined cid hydrolysis and ultrasonication	3.16 M H_2SO_4 hydrolysis at 4/40 °C for 1–6 days, then ultrasonication 20 kHz, 3 min	78%	[60]
Tapioca	Combined nanoprecipitations and ultrasonication	Precipitation using aceton, then ultrasonication 60 min, 20 kHz, and 150 W	NR	[61]
Lotus seed	Combined enzymatic hydrolysis and ultrasonication	Hydrolysis with pullulanase enzyme (pH 4.6) at (30 ASPU/g of dry starch), 58 °C, 8 h, then ultrasonication 25 ± 1 kHz. Acid Stable Pullulanase Units (ASPU) is ASPU is defined as the amount of enzyme that liberates 1.0 mg glucose from starch in 1 min at pH 4.4 and 60 °C	NR	[62]
Waxy maize	Combined enzymatic hydrolysis and recrystallization	Hydrolysis with pullulanase enzyme (pH 5) at (30 ASPU/g of dry starch), 58 °C, 8 h. Followed by recrystallization at 4 °C 8 h	85%	[63]
Elephant foot yam	Combined enzymatic hydrolysis and recrystallization	Debranching by pullulanase, followed by recrystallization at 4 °C 12–24 h	56.66–61.33%	[64]
Waxy maize	Combined enzymatic hydrolysis and recrystallization	Hydrolysis with pullulanase enzyme at 58 °C 24 h, then recrystallized 5 °C	NR	[40]
b. Top-down methods				
Dry high amylose Corn Potato Tapioca Sweet potato Waxy corn	Nanoprecipitation	Absolute ethanol as a precipitate	NR	[6]
Green banana	Nanoprecipitation	Starch mixed in acetone and precipitated with water	NR	[37]
Tapioca	Nanoprecipitation	Produced with acetone	NR	[61]
Waxy maize	Nanoprecipitation	Starch mixed with ethanol	NR	[40]
Arrowroot	Nanoprecipitation	Produced by butanol	20.65–23.8	[65]
Potato	Nanoprecipitation	Produced by ethanol	NR	[66]

NR = Not reported.

3.1. Physical Methods

Various SNP preparations, such as gamma irradiation, high-pressure homogenisation (HPH) and ultrasonication, have been carried out. These methods are less complicated and less expensive, and can be used to reduce the use of chemicals to prevent leaving residues in the final product [48,67]. Furthermore, physical methods are less-time consuming than chemical methods. However, some physical methods have the disadvantage of being more energy consuming.

Physical preparation using gamma irradiation as an immediate modification technique that causes depolymerization by breaking glycosidic bonds and hydrolysing chemical bonds, and it results in the production of small starch fragments [41]. The preparation of SNPs with this method has been tested on various starch sources, such as cassava and waxy corn starches, with a general application of a 20 kGy dose and a resulting particle size of 20–50 nm [46,47].

Furthermore, HPH operates at high speed and shear rate of product flow [68]. Ahmad et al. [48] conducted HPH at a pressure of 250 MPa and reported that the repeated homogenisation process can result in significant size reduction. The preparation of nanoparticles using HPH has been tested on sago and high-amylose corn starches [48,49].

Ultrasonication has been widely used to prepare SNPs, such as those from corn, cassava, quinoa and corn starches [50–52,54]. Ultrasonic treatment is promising due to its high yield, being rapid and relatively simple without any purification steps. In the ultrasound modification, sound waves with frequencies higher than the threshold of the human hearing range (>16 kHz) are used. The sound waves generated by ultrasound generate mechanical energy disrupt the starch molecules, causing them to break apart into smaller particles. The mechanical energy also creates microscopic bubbles in the solution. These bubbles rapidly collapse, producing high pressure and temperature, which can also contribute to the breaking of starch molecules into smaller particles. In addition, the mechanical energy of the ultrasound also causes hydrodynamic stress in the solution, which can contribute to the formation of small particles [54,68–70]. SNPs preparation by ultrasonication uses a variety of powers, temperatures and times; in general, numerous studies have used 24 kHz power for 75 min, with the increased ultrasonication treatment time resulting in a decreased particle size [50,51]. Among various physical methods for obtaining SNPs, ultrasonication is more advantageous because of the more optimal yield produced than that obtained when using other physical methods. The yield produced using the ultrasonication method is close to 100%, higher than the acid hydrolysis process because the acid hydrolysis partially dissolves the material; of course, this method is very promising because of the high yields, it does not use chemical reagents, and it is faster and relatively simple, without purification steps [51,54].

3.2. Chemical Methods

The preparation of SNPs using chemical hydrolysis has been extensively studied. According to Wang and Copeland [71], the first acid hydrolysis, which was applied by Nageli, used sulfuric acid, whereas Lintner used hydrochloric acid as a solvent; at the end of the 19th Century, both methods were commercialised to treat starch granules and were carried out under gelatinisation temperature for a certain period to produce acid-hydrolysed starch. The preparation of SNPs by acid hydrolysis considers several factors, such as starch concentration, acid concentration, temperature, agitation and hydrolysis duration [72].

Table 1 shows the general chemical preparation of SNPs. Hydrolysis of SNPs generally uses optimal conditions, including 3.16 M H_2SO_4 or 2.2 M HCl and a 35–40 °C temperature for 12 h to several days. Dularia et al. [38] reported that the preparation of water chestnut SNPs by acid hydrolysis using 3.16 M H_2SO_4 for 7 days at 40 °C produced 27.5% SNPs. Another study reported that the use of sulfuric acid in mungbean starch resulted in a 32.2% yield of SNPs [55]. Maryam et al. [39] used 2.2 M HCl and produced a high yield of 80% in sago starch. Angellier et al. [73] reported that the hydrolysis yield was lower when using H_2SO_4 compared with HCl, with a lower time for SNP production; however, their method showed a more stable final suspension with H_2SO_4 due to the presence of sulphate groups on the surface of SNPs.

3.3. Enzymatic Methods

The enzymatic preparation of SNPs involves hydrolysis using enzymes. The commonly used enzymes include α-amylase, glucoamylase and pullulanase [72]. According to Qiu et al. [27], the enzymatic method is the most efficient for starch degradation. The most important thing in the conversion of starch by enzymatic hydrolysis is that the pullunase enzyme can rapidly hydrolyze α-1,6-glycosidic bonds, releasing a mixture of linear short-chain glucose units from the parent molecule amylopectin, and for hydrolysis treatment by α-amylase, which causes random cleavage of α-1,4-glycosidic bonds in amylose and amylopectin chains. This enzymatic hydrolysis results in cracks and erosion of the starch granules, resulting in a reduction in the size of the starch particles with the right degree of enzymatic hydrolysis. [63,74,75].

Kim et al. [74] reported that waxy rice starch hydrolysed using amylase had a large size of 500 nm and an irregular shape. Irregularly shaped SNPs produced by enzymatic hydrolysis are also found in waxy maize and lotus seed [62,75]. No study has reported the yield of SNPs produced by the enzymatic hydrolysis method, probably due to the limitations of research using this single procedure. However, SNPs preparation has been carried out using combined methods with enzymes, such as enzyme hydrolysis–ultrasonication [62] and enzyme hydrolysis–recrystallisation [40,64,76].

3.4. Nanoprecipitation

Nanoprecipitation, as a simple and the most commonly used technique to produce SNPs, is carried out by the gradual addition of aqueous polymer solutions or successive additions of nonsolvents to the polymer solution, which leads to the formation of nanoscale particles [41,61]. For SNP preparation by nanoprecipitation, the starch molecular chain must be completely dispersed in the solvent beforehand, and the process is mainly based on the deposition of the biopolymer interface and displacement of water-miscible semi-polar solvents from lipophilic solutions [61,77]. SNPs are usually prepared by the precipitation of a starch paste solution using ethanol, propanol, isopropanol or butanol [27]. According to Tan et al. [78], nanoprecipitation requires high levels of non-solvents, such as acetone, ethanol or isopropanol, which will inhibit the production and application of SNPs. Some previous studies have reported the nanoprecipitation preparation of SNPs from various starch sources (Table 1).

Wu et al. [77] reported that the particle size of SNPs produced is influenced by the proportion of non-solvent used in the nanoprecipitation method; that is, the particle size decreases when the proportion of non-solvent increases. Qin et al. [6] observed that the amylose–amylopectin ratio affects the characteristics of the resulting SNPs. Butanol can only form a complex and precipitate with amylose, but not with amylopectin. The processing of starch into shorter and more crystalline amylose via acid hydrolysis (lintnerisation) or acid-alcohol hydrolysis is required for the production of nano-sized particles by butanol complex [79,80]. The authors reported that the higher the amylose content of native starch, the higher the relative crystallinity of SNPs, and the resulting V-type diffraction pattern is derived from the single helical structure of the inclusion complex consisting of amylose and ethanol [6]. Winarti et al. [65] reported that SNPs from arrowroot starch produced by nanoprecipitation using butanol reached a yield of 20.65–23.8%.

3.5. Combined Methods

SNPs can also be prepared by combining various preparation methods. The combination of methods is intended to produce better SNP properties compared with those obtained using a single method. As shown in Table 1, the combined method of acid hydrolysis–nanoprecipitation using sago starch was reported by Maryam et al. [39]. This combined method can produce SNPs with smaller sizes compared with those obtained using acid hydrolysis alone. The authors reported that the addition of precipitation treatment for 12 h can provide a minimum particle size compared with that 24 h. This production was carried out based on the properties of amylose, which can form inclusion complexes with ethanol and n-butanol; the process triggered the formation of a single left helical structure as a result of the rearrangement of the starch structure, which was gelatinised. Acid hydrolysis–nanoprecipitation with ethanol produces a higher yield than precipitation with butanol [39].

Kim et al. [60] reported that the combined process of acid hydrolysis–ultrasonication using waxy corn starch and acid hydrolysis using H_2SO_4 for 6 days at low temperature (4 °C) can produce a hydrolysate that is resistant to ultrasonication treatment. Ultrasonication effectively broke down the starch hydrolysate produced. Previously turned into nanoparticles, the yield reached 78% with a particle size of 50–90 nm. The combined treatment of acid hydrolysis–ultrasonication can degrade lotus seed nanoparticles, which showed a significant effect on the resulting size; the increase in ultrasonic power produced high crystallinity with small particle sizes by weakening the interaction of starch molecules and destroying the amorphous regions [62]. Several studies have mentioned enzymatic hydrolysis–recrystallisation; as reported by Qin et al. [6], hydrolysis with pullulanase followed by recrystallisation at 4 °C on waxy corn starch increased crystallinity, produced SNPs with particle sizes of 60–120 nm and attained high yields above 85% compared with conventional hydrolysis. Enzymatic hydrolysis–recrystallisation of elephant foot yam starch yields 56.66–61.33% [64].

4. Properties of Starch Nanoparticles

4.1. Amylose Content

Amylose content can affect the physicochemical properties of starch, such as gel formation and adhesion. The amylose content may vary depending on the botanical source of starch granules [16,81]. The development of SNPs can result in differences in amylose content. Torres et al. [58] reported a drastic decrease in the amylose content of acid-hydrolysed nanoparticles in Andean potato starch by up to 80% compared with native starch; in addition, the combination treatment of hydrolysed SNPs with ultrasonication showed a decrease in amylose content. The decrease in amylose content can be attributed to hydrolysis, which eroded the amorphous regions consisting of amylose molecules, compared with crystalline regions of starch granules which are generally more resistant to hydrolysis [30]. The amylose and amylopectin ratio, the inter-chain organization and the type of crystallinity pattern play a significant role in NSPs properties. Furthermore, Bajer [82] reported that the influence of amylose content on nano-starch was crucial.

The preparation of arrowroot SNPs by nanoprecipitation with 5% butanol can increase the amylose content; the amylose content will increase by precipitation with alcohol because only amylose forms complexes with these alcohols. However, the results showed that the amylose content was not significantly different from that obtained with treatment using butanol at a concentration of 10% [65]. In addition, nanoparticles can be produced by physical preparation, such as HPH. This method triggers the destruction of the starch structure by reorganising the crystal region outside the granule, which can free the amylose, thereby increasing the amylose ratio [49]. Based on the work of Apostolidis and Mandala [49], the amylose content of corn starch obtained with this treatment can be influenced by two factors (pressure/cycle). The higher the pressure and the more cycles applied to this method, the greater the effect on the treated sample, where repeated treatments produce new structural domains from amylose leaching.

4.2. Particle Morphology and Size Distribution

The morphology, structure and size distribution of SNPs can be characterised through several testing techniques, such as atomic force microscopy, scanning electron microscopy and transmission electron microscopy by controlling the preparation conditions [68]. The morphology of SNPs may differ depending upon the botanical sources and preparation methods [81]. Table 2 presents a variety of starch sources, preparation methods and observations on the morphology and size of the tested SNPs.

Table 2. Morphological characteristics and size of SNPs.

Source	Preparation Method	Shape	Size (nm)	Ref.
Cassava	Gamma irradiation	Agglomerates	50	[47]
Cassava	Gamma irradiation	Laminar	20	[46]
Waxy maize		laminar Aggregates are formed	20–30	
High amylose maize	High-pressure homogenization	Aggregates and porous.	540	[49]
Green sago	High-pressure homogenization	Spherical	23.112	[48]

Table 2. *Cont.*

Source	Preparation Method	Shape	Size (nm)	Ref.
Cassava	Ultrasonication	Spherical	77.51	[52]
Quinoa Maize	Ultrasonication	flaky and porous flaky and porous	99 214	[54]
Waxy maize	Ultrasonication	Platelet-like	40	[53]
Waxy maize	Ultrasonication	Ellipsoidal	37	[50]
Waxy maize Normal maize High AM maize Potato Mungbean	Acid hydrolysis	Round or oval shapes	41.4 41.0 69.7 43.2 53.7	[56]
Andean potato	Acid hydrolysis	Elliptical-polyhedral shape	132.56–263.38	[58]
Waxy rice	Acid hydrolysis	Round but irregular	220–279.4	[59]
Unripe plantain fruits	Acid hydrolysis	Oval shape but fractured granules.	NR	[83]
Waxy maize High amylose maize	Acid hydrolysis	flat elliptical Round-polygonal	>500 268	[57]
Mungbean	Acid hydrolysis	slightly oval/irregular	141.772	[55]
Water chesnut	Acid hydrolysis	Irregular and rough surface	396	[38]
Sago	Acid hydrolysis	NR	789.30	[39]
Dry high-amylose corn Pea potato Corn Tapioca Sweet potato Waxy corn	Nanoprecipitation	Spherical and elliptical	20–80 30–150 50–225 15–80 30–110 40–100 20–200	[6]
Potato	Nanoprecipitation	Spherical and elliptical	50–150	[66]
Arrowroot	Nanoprecipitation	non-granular morphologies with porous	261.4	[65]
Green banana	Nanoprecipitation	NR	135.1	[37]
Waxy maize starch	Nanoprecipitation	Irregular	201.67	[40]
Tapioca	Nanoprecipitation	Spherical	219	[61]
Lotus seed	Enzymatic hydrolysis	irregular shapes	NR	[62]
Waxy rice	Enzymatic hydrolysis	Irregular shape	500	[74]
Waxy maize	Enzymatic hydrolysis	Irregular with erosion surface	NR	[75]
Sago	Combined acid hydrolysis and precipitation method with butanol	NR	7.57–178	[39]
	Combined acid hydrolysis and precipitation method with ethanol	NR	21.98–97.50	
Andean potato	Combined acid hydrolysis and ultrasonication	elliptical-polyhedral shape	153.63–366.76	[58]

Table 2. *Cont.*

Source	Preparation Method	Shape	Size (nm)	Ref.
Potato	Combined acid hydrolysis and ultrasonication	Spherical	40	[84]
Waxy maize	Combined acid hydrolysis and ultrasonication	Globular	40–90	[60]
Tapioca	Combined nanoprecipitation and ultrasonication	Spherical	163	[61]
lotus seed	Combined enzyme hydrolysis and ultrasonication	irregular shapes with the uneven surface	16.7–2420	[62]
Elephant foot yam	Combination enzyme and recrystallization	irregular to spherical shapes	182.07–198.1	[64]
Waxy maize	Combined enzyme hydrolysis and recrystallization	Spherical microscale coralloid aggregates	156	[40]
Waxy maize	Combined enzyme hydrolysis and recrystallization	Irregular	80–120	[63]

NR: Not Reported.

Based on various studies, SNPs can be round, flat, platelet, ellipsoidal or irregular with cracked and porous surfaces. The differences in morphology depend on the botanical source and preparation technique or modification of SNPs. The size distribution of different particles also varies depending on the starch source, where the smaller the starch granules, the smaller the nanoparticle scale produced [6,58,68]. A smaller nanoparticle size produces different functional characteristics from standard particle size, which has led to their use in various industrial developments [30]. Furthermore, amylose content can affect SNPs, with high amylose content producing large nanoparticles [30].

Physical preparation methods for SNPs can be carried out with various treatments. Numerous studies have used ultrasonication on cassava starch [52], quinoa and corn starch [54], corn starch [53] and waxy corn starch [50,51], which resulted in the size distribution of molecules and different morphological forms depending on the starch source and the time and frequency of sonication treatment. Remanan and Zhu [54] reported that ultrasonication significantly disrupted the starch granule structure and crystalline properties of starch. The reduction of granules to SNPs is influenced by the solvent composition used in this method, such as water content, which is a prerequisite for destroying starch size [53]. Other studies have used gamma irradiation at a dose of 20 kGy on cassava SNPs, and caused the formation of laminar aggregates with a large specific surface area; numerous OH groups on the surface were connected by hydrogen bonds, with a resulting particle size of 30–50 nm [46,47]. Physical treatment can be carried out using HPH. Apostolidis and Mandala [49] reported an increase in pressure, and the homogenisation cycle led to a more considerable size reduction, which was effective for starch breakdown. These results are also supported by research conducted on sago starch, where the treatment of five cycles of HPH resulted in a significant reduction in size [29].

Based on the research by Jeong and Shin [59], granule size decreased along with the increased days of acid hydrolysis treatment. Kim et al. [56] reported that the nanoparticles produced in hydrolysis using acid caused erosion of the amorphous lamellae and the release of nanocrystal components. During this treatment, the molecular distribution expanded as the hydrolysis time progressed due to surface erosion fragmentation during stirring, the hydrolysis process and agglomeration, which resulted in irregular shapes [83,85]. The resulting nano starch is diverse with different morphological characteristics of starch, which can be attributed to the biological origin and physiology of plant biochemistry [55].

Based on several studies that carried out the nanoprecipitation method on several starch sources, most of the SNPs are spherical/elliptical/irregular, and others exhibit slight aggression; SNPs occasionally show an uneven distribution with an average distribution range of 30 nm to >250 nm [6,37,40,65,66]. The resulting differences occur due to differences in starch sources, time and precipitation reagents used. Maryam et al. [39] combined the acid hydrolysis method with the precipitation of sago starch. They produced a substantially smaller molecular distribution than the acid hydrolysis treatment alone. In the precipitation process with hydrophobic components, such as butanol and ethanol, when amylose accommodates hydrophobic molecules, it will form a textured single helical crystal due to the rearrangement of the gelatinised starch structure.

According to Foresti et al. [75], hydrolysis using enzymes in starch occurs through three stages, namely, diffusion to the solid surface and adsorption until catalysis. The increase in hydrolysis time increases the fraction of fragmented particles. Grain fragmentation can be evidenced by a progressive reduction of the average grain diameter. Kim et al. [74] reported that enzymatic hydrolysis by amylase causes the starch surface to crack and become porous, which indicates that the enzyme penetrates the granules. They also concluded that the right degree of enzymatic hydrolysis can reduce starch particle size, but excessive hydrolysis can increase it.

4.3. Starch Crystallinity

The crystallinity level is the ratio between the mass of crystal domains and the total mass of whole SNPs, the crystallinity of which is mostly ascribed to amylopectin [86,87]. The crystalline structure of starch can be observed using X-ray diffraction (XRD). Using the X-ray diffraction pattern, starch can be classified into several types: A, B, C and V, whereas for the low quality x-ray diffraction pattern, it has about 70% of the starch polymer in an amorphous state [88,89]. The degree of crystallinity varies depending on the starch source and preparation method used: A-types have double helices tightly packed and are commonly found in cereal starches; B-types have a high amylose structure contained in tubers, stems and fruits, the crystalline part of which is formed of six left-handed parallel-stranded double helix packed in a relatively loosely packed hexagonal unit; C-types are considered a mixture of forms A and B and present in leguminous starches; and V-types can be observed during the formation of complexes between amylose and lipids [52,87]. Table 3 presents the crystallinity of SNPs from various preparation methods.

Table 3. Crystallinity of starch nanoparticles from various preparation methods.

Starch Source	Preparation Method	Crystallinity (%)	Crystalline Type		Ref.
			Native Starch	NSPs	
Cassava	Gamma irradiation	Decrease	NR	Amorphous	[46]
Waxy maize		Decrease	NR	Amorphous	
High amylose maize starch	High-pressure Homogenization	7.8	B-type	B-type	[49]
Cassava	Ultrasonication	Decrease	C-Type	Amorphous	[52]
Quinoa	Ultrasonication	Decrease	A-Type	Amorphous	[54]
Maize		Decrease	A-Type	Amorphous	
Waxy maize	Ultrasonication	-	A-Type	Amorphous	[53]
Waxy maize	Ultrasonication	Decrease	A-Type	Amorphous	[50]
High AM maize	Acid hydrolysis	61.4	A-type	B-type	[56]
Potato		89.4	B-type	B-type	

Table 3. Cont.

Starch Source	Preparation Method	Crystallinity (%)	Crystalline Type		Ref.
			Native Starch	NSPs	
Andean potato	Acid hydrolysis	42.2	B-type	B-type	[58]
Waxy rice	Acid hydrolysis	No change	A-type	A-type	[59]
Waxy maize High amylose maize	Acid hydrolysis	NR NR	A-type B-type	A-type A-type	[57]
Waxy maize	Acid hydrolysis	53	NR	A-type	[46]
Sago	Acid hydrolysis	36	NR	NR	[39]
Dry high amylose corn Pea Potato Corn Tapioca Sweet potato Waxy corn	Nanoprecipitation	39.8 31.5 26.3 23.2 19.3 20.7 7.1	B-type C-type B-type A-type A-type A-type A-type	V-type	[6]
Potato	Nanoprecipitation	23.5	B-type	V-type	[66]
Arrowroot	Nanoprecipitation	28.36–45.12	A-type	V-type	[65]
Waxy maize	Nanoprecipitation	NR	NR	V-type	[40]
Tapioca	Nanoprecipitation	12.53	A-type	V-type	[61]
lotus seed	Enzyme hydrolysis	65.07	B-type	B-type	[62]
Waxy rice	Enzymatic hydrolysis	NR	A-type	A-type	[74]
Waxy maize	Enzymatic hydrolysis	Increase	A-type	NR	[75]
Sago	Combined acid hydrolysis and precipitation method with ethanol Combined acid hydrolysis and precipitation method with butanol	41 34	NR NR	NR NR	[39]
Waxy maize	Combined cid hydrolysis andultrasonication	27.68	A-type	A-type	[60]
Tapioca	Combined nanoprecipitation and ultrasonication	6.49–15.21	A-type	V-type	[61]
lotus seed	Combined enzyme hydrolysis and ultrasonication	57.5–61.3	B-type	B-type	[62]
Elephant foot yam	Combined enzyme and recrystallization	41.30–43.22	C-type	B-type	[64]
Waxy maize starch	Combined enzyme hydrolysis and recrystallization	NR	NR	B +V-type	[40]
Waxy maize starch	Combined enzyme hydrolysis and recrystallization	45.28	A-type	B +V-type	[63]

NR: Not Reported.

Changes in the crystal structure of starch during the production of SNPs have been investigated, and studies have confirmed that different modifications applied to various starches will affect the category of starch crystallinity. SNPs made using physical treatments, such as gamma irradiation, ultrasonication and HPH, showed a decrease in crystallinity and an amorphous structure. Based on the research by Lamanna et al. [46], decreased crystallinity and amorphous patterns were observed in cassava and waxy maize starches. In the preparation using HPH, crystallinity also decreased due to the increase in homogenisation pressure applied to starch granules, but a crystalline B-type was maintained [90]. Moreover, da Silva et al. [52] reported that ultrasonication affects the crystal structure, which results in severe disruption of the crystal structure of amylopectin and the amorphous character of the resulting SNPs. This result was also observed in the ultrasonication treatment of waxy corn starch, which caused the loss of diffraction peaks; as a result, crystals were lost during the ultrasonication fragmentation process [50,53].

The XRD patterns of all SNPs generated from nanoprecipitation showed V-type crystallinity, which is unrelated to the native crystal type of starch [66]. The V-type crystallinity is derived from the single helical structure of amylose and ethanol. In addition, gelatinisation in this method destroys the crystallinity of A, B and C-types, with most of the relative crystallinity decreasing due to the weak crystallinity intensity of the single helix of SNPs and low number of single helices during the nanoprecipitation process. The number of single helices is low during the nanoprecipitation process [6,40]. Winarti et al. [65] reported that nanoprecipitation of arrowroot using a butanol complex caused an increase in the degree of crystallinity and a shift in the crystal from A to V-type. This shift in crystal type caused a loss in starch integrity during gelatinisation [65,91]. Qin et al. [6] investigated a positive correlation between the amylose content of starch and the relative crystallinity of starch with high amylose content, which resulted in an increased crystallinity in this nanoprecipitation treatment. The degree of crystallisation is influenced by the percentage of amylopectin, chain length, crystal size, orientation of the double helix in the crystal region and the degree of interaction between the double helix [64].

The SNPs obtained by acid hydrolysis of native starch granules have a higher crystallinity than the parent granules [59]. A-type starch is more resistant to acid hydrolysis than B-type crystals, where the crystal structure of B-type starch is more easily disturbed by acid hydrolysis than B-type crystals. In addition, the degree of hydrolysis is not positively related to the change in crystallinity as measured by X-ray analysis [56]. Furthermore, Maryam et al. [39] explained that the hydrolysis period can affect starch content. Starch crystallinity increases after hydrolysis, in which the hydrolysis process does not change the pattern of starch crystallinity, but only changes its crystallinity index. This condition indicates that hydrolysis only destroys the amorphous region to obtain more starch crystals. SNP crystallinity in enzyme-hydrolysed starch causes a relative increase in crystallinity associated with extensive degradation, especially in the amorphous region of starch granules (Foresti et al., 2014). In SNPs made from a combination of enzymatic hydrolysis followed by recrystallisation, the relative crystallinity increases, and a B + V-type crystalline pattern emerges [40,76].

4.4. Thermal Properties

The thermal transition behaviour of SNPs has been characterised using differential scanning calorimetry and thermogravimetric analysis. These thermal analyses are critical because they determine the conditions under which the use of SNPs is applied in industry. The resulting behaviour will determine the appropriate processing conditions to produce a final product that will remain stable [15,92]. Table 4 presents the thermal characteristics of SNPs from various studies.

Table 4. Thermal characteristics of starch nanoparticles.

Starch Source	Preparation Method	Technique	Result	Ref.
Cassava, waxy maize starch	Gamma irradiation	TGA	Degraded at a lower temperature than native starch and sudden decrease in weight loss	[46]
Green sago	High-Pressure homogenization	TGA	High degradation temperature	[48]
Cassava	Ultrasonication	TGA/DSC	SNPs are more thermally unstable and have low gelatinization temperature	[52]

Table 4. Cont.

Starch Source	Preparation Method	Technique	Result	Ref.
Quinoa Maize	Ultrasonication	DSC	T_o, T_p, T_c and ΔH decreased T_o, T_p and T_c decreased but ΔH increased	[54]
Waxy maize	Ultrasonication	DSC	ΔH decreased	[53]
Waxy maize	Ultrasonication	DSC	ΔH decreased	[50]
Waxy maize Normal maize	Acid hydrolysis	DSC	T_p and T_c and ΔH, but T_o decreased	[56]
Waxy rice	Acid hydrolysis	DSC	T_p and T_c increased as the hydrolysis time increased, but T_o and ΔT decreased	[59]
Unripe plantain fruit	Acid hydrolysis	DSC	T_p, T_c and ΔT increased as the hydrolysis time increased, but T_o decreased	[83]
Potato	Nanoprecipitation	TGA	Thermal degradation of SNPs started earlier than for native starch	[66]
Arrowroot	Nanoprecipitation	DSC	T_p decreased, T_o and ΔH increased	[65]
High amylose corn	Nanoprecipitation	DSC	T_o, T_p and T_c decreased, but ΔH increased	
Potato	Nanoprecipitation	DSC	T_o and ΔH decreased, but T_p increased	[6]
Pea corn Tapioca Sweet potato Waxy corn	Nanoprecipitation	DSC	T_o, T_p, T_c and ΔH decreased	
Tapioca	Combined nanoprecipitation and ultrasonication	DSC	T_o, T_p, T_c and ΔH decreased	[61]
Elephant foot yam	Combined enzyme and recrystallization	DSC	T_p, T_o, T_c increased, but T_c decreased at 24 h of hydrolysis	[64]
Waxy maize starch	Combined enzyme hydrolysis and recrystallization	TGA	Maximum degradation temperature decreased	[63]

SNPs prepared using physical methods, such as gamma irradiation, show a low drop in degradation temperature and a sudden weight loss. This finding is due to the initiation of SNP degradation at the surface, which has a high amount of hydroxyl groups [46]. da Silva et al. [52] reported a decrease in the degradation temperature of ultrasonicated cassava SNPs, which led to a lower thermal stability compared with that of native starch, in addition to a decrease in gelatinisation temperature caused by the weakening of hydrogen bonds in the amorphous region. According to Zhu [69], gelatinisation-associated SNPs can be affected by starch-type composition and ultrasonication experimental conditions. A decrease in gelatinisation enthalpy on ultrasonicated SNPs was also detected in quinone and waxy maize starches [50,53,54]. After ultrasonication, reductions in crystallinity and melting temperature were obtained due to disruption with starch particles. A significant increase in hydrogen and Van der Waals bonds occurred [48]. The modification of sago starch using high-pressure homogenisation caused a slight increase in the degradation temperature, which indicates an increase in thermal stability of modified starch after the treatment [48]. The combination method of enzyme hydrolysis followed by recrystallisation decreases thermal stability [63].

The thermal properties of SNPs obtained with acid hydrolysis exhibit increased T_p and T_c with the length of hydrolysis time, but a decrease in T_o after acid hydrolysis; the distribution of long-chain amylopectin in waxy rice starch changes to produce nanoparticles [59]. Increased values of T_p and T_c and decreased T_o were also reported in waxy maize, normal maize and unripe plantain fruit [56,83]. The decrease in T_o occurs due to the separation of the crystalline region from the unstable amorphous region. An increase in enthalpy was found in this treatment due to the rearrangement of the starch chain with the increase in crystallinity. However, the enthalpy can decrease at prolonged hydrolysis times, where differences in the susceptibility of the crystal region to hydrolysis may depend on the origin and crystal structure of starch [56]. According to Ding et al. [93], the enthalpy of an endothermic reaction reflects the number of crystals and double-helix chains that affect the amylose–amylopectin content and length distribution of amylopectin molecule. The thermal characteristics of various SNPs produced using the nanoprecipitation method, which were studied by Qin et al. [6], showed a decrease in the enthalpy of gelatinisation due to the single helical structure of nano starch being more susceptible to disintegration than native starch. The highest gelatinisation enthalpy and stability were found in high-amylose corn starch because it has a high amylose content and a high density of crystal structure [6].

4.5. Functional Properties

Several parameters have been used to examine the functional properties of SNPs. Winarti et al. [65] observed the functional properties of the swelling volume and solubility of arrowroot SNPs produced using the nanoprecipitation method with butanol; they showed increases in swelling volume (5.28–7.92 g/g) and solubility (9.43–16.89%) compared with the native ones. Other studies, such as those on potato and cassava SNPs obtained with mechanical treatment, showed a significant increase in swelling volume [94]. Jeong and Shin [59] reported that waxy rice SNPs prepared by acid hydrolysis method resulted in a decreased water binding capacity. However, no significant change was observed as a function of hydrolysis time.

4.6. Digestibility Properties

The digestion of starch granules is a complex process that includes the diffusion of enzymes to the substrate, which affects substrate porosity, the absorption of enzymes in starch-based materials and hydrolytic events. The in vitro digestibility of SNPs increases compared with that of native starch, generally due to the increased surface area of nano-sized starch [95]. Based on the research of Suriya et al. [64], the percentage of SNP digestibility was increased to 41.29–43.24% by debranching with pullulanase followed by recrystallisation for 12–24 h, with high starch digestibility resulting in shorter retrogradation time. According to Ding et al. [93], the digestibility of starch can be affected by the type of starch, particle size, crystallinity, amylose–amylopectin ratio and retrogradation conditions.

Meanwhile, studies on the enzymatic digestibility of arrowroot SNPs obtained by linearised method and butanol precipitation for 24 h caused a reduction in starch digestibility [65]. This finding was also observed in maize SNPs, which showed the lowest level of hydrolysis that can be attributed to the compact structure of SNPs formed during the recrystallisation of short-chain glucans; this condition resulted in an increased number of short chains, which made enzyme digestion more difficult [96]. Studies on the percentage of resistant starch resulting from hydrolysis showed maximal values until day 8, but the numbers drastically decreased on the next day in waxy rice starch [59]. Several studies have shown that a low hydrolysis rate of SNPs results in numerous non-hydrolysed SNPs. By contrast, acid hydrolysis digests recrystallise amorphous amylose to form a new double-helical structure, which is highly crystallised against enzymatic degradation; the formation of these SNPs is accompanied by crystallinity, which increases enzymatic resistance [65,96,97]. According to Oh et al. [98], when the concentration of SNP increases, digestion inhibition increases, in which SNPs change the secondary structure of and potentially inhibit α-amylase; thus, SNPs have the potential to reduce glucose absorption for diabetics.

5. Applications

5.1. Pickering Emulsion

Emulsion systems are applied in various fields, such as food, medical, pharmaceutical and cosmetic industries. However, the contact between water and oil is inherently unstable and will break down over time. Starch is considered a promising alternative for emulsion stabilisation because it is an abundant, cheap, non-allergenic material, and its biodegradability meets the increasing consumer demand for plant materials; emulsion stabilisers using starch have been applied in solid stable emulsions known as Pickering emulsion [99]. Pickering emulsion consists of solid particles which have been used to stabilize oil–water systems and monomer/polymers emulsions [100,101]. Pickering emulsions are often considered very stable due to the almost irreversible adsorption of the particulate stabiliser interface [100]. The use of particles can be a new strategy for emulsion stabilisers in food, such as bio-based particles using starch; numerous studies have shown that physical stability can be attributed to different stability mechanisms compared with conventional emulsifiers [102]. Solid particles can generally stabilise oil droplets by adsorption at the interface by accumulation to form a layer with high mechanical strength and overall functional properties, as well as stable Pickering emulsion depending on the oil–water interface [103]. The stabilisation of particles at the interface can affect barrier properties, thickness, charge and interfacial tension, which affect the behaviour of the resulting emulsion [104,105].

Nanostarch as a stabiliser of Pickering emulsion is promising because the emulsifying capability of nano starch can be significantly improved through hydrophobic modification [11]. In addition, the use of nano starch as a Pickering emulsion stabiliser has attracted widespread attention because of its small size, wide surface area, non-toxicity, biocompatibility, biodegradability, low cost and food-grade nature; thus, these starch particles can be used as attractive stabilisers [11,103,106]. Wei et al. [107] reported that the larger the size distribution of nanoparticles, the more stable the Pickering emulsion. Likewise, in a previous study [108], the Pickering emulsion was stable at a high concentration of total nanoparticles with a low oil ratio because a high number of particles were adsorbed to the interface and formed a physical barrier, which resulted in small droplets. Nanoparticles act as emulsion stabilisers by ensuring the homogeneous state of the emulsion. Nanostarch particles used as Pickering emulsion have been extensively studied [11,34,109–113]. Table 5 presents various applications of nano starch to stabilize Pickering emulsions.

Table 5. Application of SNPs compared with SNCs as Pickering emulsion.

Type Starch	Aqueous Phase	Oil Phase	Emulsion Type	Emulsification Method	Result	Ref.
Maize SNC	Water	Paraffin	o/w	Homogenization (10,000 rpm, 4 min)	- The emulsion is very stable up to 2 months of storage - Creaming is wholly inhibited at 6% SNC concentration	[114]
Taro SNP	NaCl	MCT oil	o/w	Homogenization (12,000 rpm, 2 min)	Emulsion with the best stability at an SNP concentration of 7% with an oil fraction of 0.5, up to 28 days	[115]
OSA amaranth and maize SNC	Phosphate buffer pH 7 (NaCl 0.2 M)	Canola oil	o/w	High shear mixer (22,000 rpm, 3 min)	the best emulsion stability on amaranth OSA starch nanocrystals (emulsion index 1.0 ± 0.02), for 10 days of storage	[116]
Waxy Maize SNC	Water	Parrafin	o/w and w/o	pH difference using HCl or NaOH, homogenization (10,000 rpm, 4 min)	- Decreased droplet rate and creaming as SNC increase - pH has no significant effect on creaming ability, but the emulsion stability significantly decreases at low pH	[117]

Table 5. *Cont.*

Type Starch	Aqueous Phase	Oil Phase	Emulsion Type	Emulsification Method	Result	Ref.
Tapioca, corn, and sweet potato SNC	NaCl	Soybean oil	o/w	High-speed homogenizer (10,000 rpm, 2 min)	- Creaming index 18–22% with emulsion drop size 29–32 m after 1 year of storage - Medium particle size produces the best emulsion stability (100–220 nm) - Best emulsion stability on corn starch suspension (2%) and oil fraction (0.5)	[34]
Breadfruit SNC	NaOH (0.1875 and 0.375 M)	MCT oil	o/w	Homogenization (10,000 rpm, 5 min)	Treatment of 5% starch concentration with 0.1875 M NaOH resulted in the best starch stability for 2 weeks of storage with the lowest cream index and the smallest droplets.	[118]
Maize SNC	Water	Corn oil	o/w	Homogenization (20,000 rpm, 3 min)	- Addition of SNP >9.1% can increase emulsion stability up to 95% - Emulsions using SNPs with a diameter of <30 nm produced the best increase in instability	[99]
Waxy maize SNC	Water	MCT oil	o/w	High-speed homogenizer (18,000 rpm, 4 min)	No o/w emulsion phase separation was detected during 30 days of storage	[119]
Corn SNC	Water	Sunflower oil	o/w	Homogenization (12,000 rpm, 5 min)	No cream was observed in the emulsion after storage for 6 months.	[120]
Oxidation of cassava, corn, and bean SNC	Water	Soybean oil	o/w	Homogenization 1 min	- Oxidated nanocrystals produce a stable suspension for up to 21 days - The addition of nanocrystals of cassava starch produces the best emulsion	[121]

o/w: oil in water, w/o: water in oil.

Li et al. [114] reported that Pickering emulsions made with the addition of 0.02 wt% nanocrystals can remain stable for more than 2 months without droplet variations and coalescences. Another study showed an increase in emulsion stability up to 28 days with the addition of taro SNPs at 7% concentration. In addition, emulsion stability does not always increase based on nanoparticle concentration because the resulting emulsion droplets can agglomerate and disrupt the structure [115]. Octenyl-succinic anhydride (OSA) SNC presents an increased emulsification capability compared with untreated starch due to the formation of a superficial charge that can increase the repulsion forces between oil droplets in the nanocrystals [116].

SNPs from acid-hydrolysed starch granules using H_2SO_4 or HCl can carry a surface charge and can be used as Pickering emulsion stabilisers; the repulsion that occurs in charged nanocrystals can play an important role in influencing the emulsion capability of SNPs in dispersions [117]. H_2SO_4-hydrolysed SNCs are negatively charged due to phosphate groups on the surface. When the pH is low, the electrostatic repulsion is reduced, which results in droplet aggregation and an increase in particle size in stabilised emulsion. However, the size of emulsion can be adjusted by changing the pH [117]. Haaj et al. [111] stated that HCl can give a better effect by causing polymer dispersion, which results in a smaller average size.

According to Miao et al. [122], high-branching SNPs play a crucial role in the physical stability of Pickering emulsions. A high-branching structure implies the stiffness of starch, which results in a thick adsorbed layer around the droplets with excellent barrier properties for a long time. Lu et al. [123] also reported that the stability capacity of SNPs depends on the starch source, where normal corn starch showed a better stability compared with high-amylose starch. The concentration of starch used in the emulsion can increase the stability of emulsion associated with particles in the continuous phase, which increases the emulsion viscosity and thereby inhibits the separation of the emulsion phase and aggregation of oil droplets; as a result, the emulsion becomes more stable [118,124,125]. In addition to the concentration of added nanoparticles, emulsion stability can be affected by SNP type, size and hydrophobicity, where a small hydrophobicity contributes to increased emulsion stability [34,99,126]. Figure 2 shows a graphical abstract of SNPs application as Pickering emulsion.

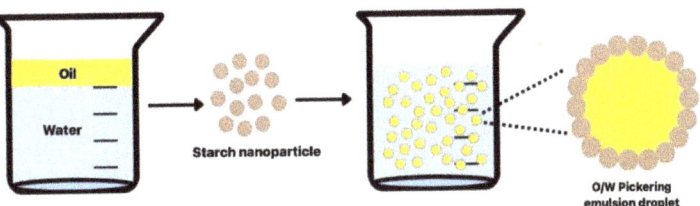

Figure 2. Application of starch nanoparticles as Pickering emulsion.

5.2. Bioplastic Filler

At present, plastics are increasingly and excessively used, which results in a negative impact on the environment and ecosystems. Therefore, efforts should be exerted to develop more eco-friendly plastics, such as bioplastics composed of biodegradable biopolymers. However, in general, bioplastic products have drawbacks, such as high permeability to water and oxygen, brittleness, low melting points, low mechanical strength, susceptibility to degradation during product storage or use and non-resistance to chemical compounds [127].

Bioplastics are produced from the fusion of biopolymers, plasticisers and fillers; additional components in the form of fillers, such as starch, can be used to improve the characteristics of bioplastic packaging [128,129]. The production of biopolymer nanocomposites containing fillers with nano-dimensions is one of the newest methods used to improving the functional properties of biopolymer films; the use of nanofillers in biofilms increases gas and thermal resistance [130]. The incorporation of nano starch from different botanical sources in packaging films as reinforcement components or polymer matrix fillers has been widely studied [96,131,132]. Starch has a good capability to make biodegradable films because of its suitable mechanical properties in bioplastic production [33]. The use of nano starch as a nanofiller can increase the modulus of elasticity and tensile strength and decrease elongation at break and water vapor permeability (WVP) [92]. Table 6 presents several applications of nano starch as fillers in bioplastics.

Table 6. Application of SNPs compared with SNCs as Fillers in Bioplastics.

Manufacturing Technique	Bioplastic Composition	Result	Ref.
Casting	Pea starch (5 g) + glycerol (1.5 g) + waxy maize acid hydrolysis SNC (5%)	There was a decrease in elongation, YM and TS by 57%, 305%, 73%, respectively. WVP reduction up to 62%	[133]
Casting	PU resin (polyurethane) + maize corn acid hydrolysis-ultrasonication SNP (20%)	The addition of 20% SNPs reduces WVP to 60% and oxygen permeability decreases to 75%. There was an increase in the value of T_g, T_m, and ΔH	[134]
Casting	Corn starch (7.5 g + glycerol (3 g) + taro enzymolysis SNP (10%)	Decrease in elongation and WVP by 24% and 56%, respectively, an increase in TS 161%. There was an increase in the value of T_g, T_m, and ΔH	[135]
Casting	Amanduble starch + glycerol + amadumbe acid hydrolysis SNC (2.5%)	There was an increase in TS of 62% and a decrease in WVP of 8.7%. There was an increase in the value of T_g, T_m, and ΔH	[136]
Casting	Potato starch + glycerol + amadumbe acid hydrolysis SNC (2.5%)	There was an increase in TS 288% and a decrease in WVP 11%	[136]
Casting	Polycaprolactone (PCL) + corn SNP (5%)	There was a decrease in the elongation value of 9%, an increase in YM 12% and TS 44%. T_m and ΔH decreased	[137]
Casting	Waterborne polyurethane (WPU) + pea SNC (10%)	The addition of SNCs by 10% showed a decrease in elongation by 27%, but there is an increase in TS and YM by 169% and 3733%, respectively	[138]
Casting	Cross-linked cassava starch + glycerol 2.5 g) + cassava SNC (6%)	Increase in young modulus and tensile strength, but decrease in elongation and water vapor permeability	[139]
Casting	Composite sago starch + sago SNP (6%)	Increase in elongation, YM and TS by 34%, 9% and 8%, respectively. WVP decrease up to 51%	[48]
Extrusion	PBA/TPS (70:30) + glycerol (7.5%), citric acid (0.6%), and stearic acid (0.3%) + cassava gamma irradiation SNP (0.6%)	There was an increase of about 20% in YM and TS. T_g: −34 °C and T_m: 117 °C	[47]
Extrusion	PBA/TPS (70:30) + glycerol (7.0%), citric acid (%) and stearic acid (0.3%) + cassava ultrasonication SNP 1%	The addition of 1% SNP can increase elongation by 35%, YM by 36%, TS by 27%, and decrease WVP up to 21.3%	[52]

YM: young modulus; TS: tensile strength; WVP: water vapor permeability; PBAT: polybutylen adipate-co-terephthalate; TPS: thermoplastic starch.

The addition of SNPs filler can be used as the main feature to obtain effective mechanical and thermal strength in the manufacture of bioplastics, where a structural change occurs in the resulting product. According to Al-Aseebee et al. [140], the mechanical properties of bioplastics are influenced by the interaction between the nanofiller and the matrix because nanoparticles have a large surface ratio. In general, in most cases, the modulus of elasticity and tensile strength increase, which is related to the decreased percentage of elongation in bioplastics produced by the addition of SNPs; similar results can be found in the manufacture of bioplastics using casting methods or other methods, such as extrusion [30,47]. The percentage of elongation related to the difference in stiffness decreases due to the interaction between the matrix and the processing agent [135]. The improvement of mechanical properties of the resulting nanocomposite films can be attributed to the structure and stiffness of nanoparticles, which limit the movement of starch chains [36,141].

In general, the increase in tensile strength with increased concentration of starch nanofillers is considered a result of the strong interactions between the filler and reinforcing matrices. According to Hakke et al. [134], the concentration of nanofillers increases the tensile strength resistance and flexural stress resistance of bioplastics, but the use of fillers decreases by more than 25% due to agglomeration of SNPs, which causes repulsion between the filler and the matrix. The authors also explained that the addition of nanofiller SNPs can induce interactions in the polymer matrix, such as hydrogen and non-covalent bonds; the increase in % elongation with the increase in SNP concentration also results in an increased Young's modulus of elasticity, which is higher until agglomeration begins, and nanofiller SNPs are distributed. The hydroxyl groups are uniformly connected to the surrounding polymer matrix, which provides additional strength to the applied force [134].

The interaction of filler and matrix in bioplastic packaging forms a strong bond, which causes difficulty for air and water vapor to penetrate through the film [142]. According to Hakke et al. [134], an increase in the concentration of added SNPs will increase film compactness; therefore, the more compact the resulting film, the longer the time required for water vapor and air to penetrate the matrix. The composition of bioplastic constituents used significantly determines the permeability of the resulting packaging; the higher the concentration of filler used, the denser the structure inside the packaging; thus, the packaging pores will be smaller, which causes difficulty for water vapor and air to penetrate the packaging walls [134,143].

Hakke et al. [134] reported a decrease in WVP along with the addition of SNP at a certain concentration; the addition of 20% SNP caused a 60% reduction in WVP; in addition to SNP concentration, temperature affected the WVP produced. According to Mukurumbira et al. [136], the presence of SNCs reduced the water affinity of the film, which formed between the starch matrix and nanocrystals, where the SNCs and starch film exhibited the same polarity; thus, the interaction of SNCs with the water matrix was minimal. In addition, the presence of dispersed SNCs created a winding path for the movement of molecules, and as a result, the longer diffusion path of water molecules affected the reduced permeability [133,135]. However, in the study of Li et al. [133], the addition of SNCs above 5% caused a slight increase in VWP, possibly due to aggregation; thus, SNCs failed to effectively prevent the migration of water molecules. Likewise, Ahmad et al. [48] showed an increase in WVP when the addition of nanoparticles was above 8%; this finding can occur when the use of filler exceeds the maximum concentration. Excess nanoparticles increase the water affinity of starch due to the abundance of hydroxyl groups and the high possibility of agglomeration in the filler [144,145].

According to Mukurumbira et al. [136], the thermal properties of nanocomposites are an important factor in determining suitable processing conditions. The increase in the composite film's thermal stability indicates the strong interaction between SNCs and the film [136]. An increase in melting temperature was observed with the addition of nanocrystals, and a decrease in film enthalpy was noted with the increased concentration of SNCs to the starch film; this finding may be due to SNCs inhibiting the lateral arrangement of starch chains and the crystallisation of starch films [133,136,143].

Basavegowda and Baek [31] reported that the mechanism of action of SNPs as a filler for increasing stability is the interaction between the filler and the matrix, which forms a barricade barrier that can inhibit the transfer of heat and energy. Hakke et al. [134] reported that the values of T_g, T_m and ΔH increased along with the addition of filler; the increase in ΔH can be associated with the interaction of active nanoparticle starch granules and binding to the packaging matrix; therefore, the increased changes in the polymer results in the increased energy requirement to change the bioplastic polymer; the investigation also showed that increasing the concentration of corn SNPs in the polyurethane solution can increase the cohesiveness of the resulting film. The glass transition temperature and nanocrystals increased, which was associated with the absorption of strong polymer chains on the nanocrystal surface; as a result, the polymer chain matrix bonds were formed [136].

According to Zou et al. [138], T_g and T_m can be affected in two opposite ways, including movement of the soft segment, which can be suppressed by steric resistance from nanocrystals, and hydrogen bonds on the surface of SNCs, which cause T_g and T_m to move to a higher temperature. On the other hand, the addition of SNCs may cleave the native interactions of soft and hard segments, which results in changes in the microphase matrix structure, where the soft segments can escape from the binding of hard segments, which causes a decrease in T_g and T_m [138].

5.3. Antimicrobial Agent

Raigond et al. [146] reported that incorporated nanoparticles can be used as antimicrobial agents that can improve food safety by minimising the growth of pathogenic microorganisms. Nanoparticles with a large surface area allow more microorganisms to adhere, which increases antimicrobial efficiency. SNPs can be developed for the compartmentalisation of active substances, such as stabilising antimicrobial compounds, which are known for their effectiveness [59,147]. Hakke et al. [134] reported that the addition of 5% SNPs to the manufacture of nanocomposites can attain the maximum reduction in bacterial resistance because the uniform distribution of SNPs in polyurethane solution can reduce the overall pore size of the film. In another study, the addition of SNPs to the stabilisation of potassium sorbate resulted in a retention capacity between 41.5–90 mg/g, which indicates that the added SNPs can be used as antimicrobial agents in food systems [148].

Qin et al. [149] showed that SNPs can significantly increase the antibacterial activity against *S. aureus* and *E. coli* from curcumin. Furthermore, Nieto-Suaza et al. [150] reported the preparation of films with banana starch and aloe vera with the addition of acetate SNPs and curcumin; the authors speculated that the resulting films could control microbial growth by increasing antibacterial activity in food products. Another study on nano starch loaded with carvacrol showed a good antimicrobial activity; that is, a 62% reduction in microbial growth of *E. coli*, 68.0% in *Salmonella typhimurium* and other tested bacteria [151]. Increased antibacterial activity was also found in SNPs loaded with polyherbal drugs [152].

Furthermore, Dai et al. [153] reported that the added nanoparticles can destroy bacterial cell walls and membranes, which results in the antimicrobial effect of bacterial apoptosis. The addition of 0.5 mg/mL starch to antibiotics increased the inhibition zone against *S. pyogenes*. Thus, SNPs can be used to increase the effectiveness of antibiotics [154]. Qin, et al. [155] concluded that branched SNPs obtained by ultrasonication can be used as an antibacterial enhancement factor in encapsulating epigallocatechin gallate (EGCG), especially against *E. coli*.

5.4. Fat Replacer

SNPs can be applied as an imitation or substitute for fat in food [92]. Fat substitutes act as imitators of triglycerides but do not replace fat on a gram-for-gram basis [30,41]. The particle size of starch is important in determining organoleptic tastes, such as the taste of fat in the mouth. Small-sized SNPs can be promising fat substitutes, and the mixing of SNPs with other components, such as smooth cream, produced properties similar to those of fats [41]. In addition, fat substitutes will decrease calorie levels [30].

Kaur et al. [156] reported that corn SNPs can replace fat in salad dressing products by up to 60% without reducing their quality characteristics. Another study revealed that using sweet potato SNPs as a substitute for fat in ice cream products allowed fat reduction, which is beneficial for low-fat ice cream production. The authors concluded that the application of SNPs can produce superior-quality products; that is, it significantly improved the texture of ice cream, which gained the approval of panellists [157]. Characteristics of fat substitutes can improve emulsion stabilisation, as shown in the study of Javidi et al. [120]. The authors revealed that fat replacement using corn SNCs resulted in a decreased droplet size and an increased zeta potential, which could be used to produce more hydrogen bonds; thus, the network between droplets formed was substantial. Nano starch as a fat substitute is useful as a stabiliser for oil–water emulsions. Considering its biodegradability, nano starch is promising in the food industry concerning public health [120].

5.5. Encapsulating Agent

SNPs can be applied in encapsulation systems, which are an attractive alternative for bioactive compounds [32,35]. The use of nano starch as a superior encapsulation material is due to its biocompatibility, low viscosity at high concentrations, large surface area, non-toxicity, low cost and ideal trapping of bioactive materials [26,86]. Several food ingredients and pharmaceutical application materials have been encapsulated using SNPs [158–162]. Table 7 presents several applications of SNPs as encapsulation agents.

Table 7. Application of native and modified SNPs as encapsulation agents.

	Type of Starch	Preparation Method	Encapsulation Compound	%Encapsulation Efficiency	Ref.
a.	Native starch				
	Banana starch	Nanoprecipitation	Curcumin	85.23	[37]
	Waxy maize	Nanoprecipitation	Polyphenols	60–70	[163]
	Quiona	Nanoprecipitation	Piroxicam	84	[164]
	Insoluble porous starch	Nanoprecipitation	Paclitaxel	73.92	[165]
	Soluble SNPs	Ethanol precipitate	Vitamin E	91.63	[166]
	Horse chestnut Water chestnut Lotus Stem	Acid hydrolysis	Catechin	59.09 48.30 55	[167]
	Quiona Maize starch	Ultasonication	Rutin	67.4 63.1	[54]
	Normal corn high-amylose Waxy corn	Ultrasonication	Anthocyanin	52.5 45.5 49.4	[168]
	Horse chestnut Lotus Stem Water chesnut	Ball milling	Resveratrol	81.46 75.83 73.37	[169]
b.	Modified starch				
	Acetylated Banana	Nanoprecipitation	Curcumin	82.23–92.12	[150]
	Acetylated Banana	Nanoprecipitation	Curcumin	90.63	[37]
	Acetylated corn	Nanoprecipitation	Ciprofloxacin	20.5–89.1	[170]
	OSA Waxy maize	Emulsion-diffusion	Conjugated linoleic acid	>97	[171]
	Debranched Waxy corn	Enzyme hydrolysis with pullulanase	Epigallocatechin gallate	84.4	[172]
	Debranched waxy maize SNPs	Ultrasonication combined with recrystallization	Epigallocatechin gallate	>80	[155]

According to Ahmad and Gani [169], encapsulation efficiency (EE) determines the amount of core material trapped in the carrier material, and the percentage depends on the number of compounds initially loaded during the encapsulation process. The highest percentage of EE in starch without modification treatment was shown in vitamin E (VE), with soluble SNPs reaching 91.63%; this finding also indicated that most of VE can be trapped in SNPs [166]. Based on Table 7, the highest proportion of EE up to above 97% was produced in conjugated linoleic acid (CLA) encapsulation using encapsulating agents from waxy corn OSA nanoparticles; CLA was effectively trapped in nanostructured particles and can be absorbed with the initial modification treatment, which will effectively increase the

EE [171]. Likewise, the acetylation of banana SNPs shows a better capacity for curcumin encapsulation than nanoparticles without acetylation [37].

The EEs of various SNPs are different [167]. The highest EE was found in horse chestnuts, and it was caused by the smallest diameter size; small particles have better EE and can form a better film around the core and retain encapsulated molecules [167,173]. Remanan and Zhu [54] reported that a difference in the efficiency of routine encapsulation of quinoa SNPs with corn starch. The smaller particle size of starch quinone nanoparticles with larger specific surface area and stronger adhesion may have contributed to the larger EE, which resulted in better retention of rutin in the encapsulated system [174]. According to Zhu [175], the routine encapsulation of nanoparticles can be caused by the formation of non-inclusion complexes by hydrogen bonds, hydrophobic interactions and electrostatic and ionic interactions. Molecular interactions occur between dissolved rutin and starch chains mostly due to the presence of hydrogen bonds (between the hydroxyl group of rutin with oxygen atoms from starch glycosidic bonds) [176].

Numerous researchers have argued that the difference in EE can be caused by differences in the type of starch and the degree of interaction between starch molecules and bioactive compounds, which can facilitate the incorporation of these compounds in starch networks [54,177]. EE also depends on several other factors, such as the concentration of core material, encapsulation reaction and synthesis process [178]. In general, the application of SNPs as an encapsulating agent can increase the percentage of EE. However, Ahmad and Gani [169] reported the encapsulation of resveratrol and SNPs at a ratio of 1:40, which resulted in a decrease in EE; this result is related to the high ratio of active ingredients, which may not be completely trapped; thus, a decrease in the EE of resveratrol in SNPs was observed. Qin et al. [155] reported that the length of ultrasonic irradiation of unbranched SNPs to be encapsulated caused a gradual decrease in the EE of EGCG, but the combination of irradiation and re-crystallisation techniques improved the EE. EGCG encapsulation using debranched waxy corn with different treatments also increased the EE [172].

6. Conclusions and Future Research

SNPs have been widely studied. They are found in various shapes and sizes based on the starch source and size reduction method used. In general, the preparation of SNPs can be classified into 'top-down' and 'bottom-up' methods. Starch-produced nanoparticles are used as Pickering emulsions, bioplastic fillers, antimicrobials, fat replacers and encapsulating agents. Thus, these nanoparticles have the potential to be produced on a large scale and further developed into food products.

The development of starch-based nanoparticles has attracted remarkable interest from researchers because of their biocompatibility, non-toxicity, low cost and use as disinfectants. SNPs have been extracted and tested from various botanical sources and developed using different preparation methods. SNPs have been used in various applications, such as reinforcement in polymer matrices, Pickering emulsions, antimicrobial agents, encapsulating agents, fat substitutes, etc., and caused increases in specific properties of the resulting product. In the future, to further expand their application field, we can use SNPs to provide new solutions, especially in food products, as a constituent component for the production of more innovative products from organoleptic and utilisation perspectives, such as the development of low-fat products and functional foods. Significant increases in the absorption of bioactive compounds increase their bioavailability and bioactivity. Thus, further research should optimise the production process of SNPs and determine the potential effects of functional products from SNPs.

Author Contributions: Conceptualization, H.M.; methodology, D.I.R.; validation, H.M., Y.C. and E.M.; investigation, E.M. and M.D.; writing—original draft preparation, H.M.; writing—review and editing, Y.C., M.M. and D.I.R.; visualization, D.I.R. and M.M.; supervision, H.M., E.M. and M.D.; project administration, H.M. All authors have read and agreed to the published version of the manuscript.

Funding: This research was funded by the Internal Research Grant of Universitas Padjadjaran, Bandung, Indonesia grant number: 2203/UN6.3.1/PT.00/2022 and the APC was funded by Universitas Padjadjaran.

Institutional Review Board Statement: Not applicable.

Data Availability Statement: Not applicable.

Conflicts of Interest: The authors declare no conflict of interest.

References

1. Olayil, R.; Arumuga Prabu, V.; DayaPrasad, S.; Naresh, K.; Rama Sreekanth, P.S. A Review on the Application of Bio-Nanocomposites for Food Packaging. *Mater. Today Proc.* **2022**, *56*, 1302–1306. [CrossRef]
2. Paulos, G.; Mrestani, Y.; Heyroth, F.; Gebre-Mariam, T.; Neubert, R.H.H. Fabrication of Acetylated Dioscorea Starch Nanoparticles: Optimization of Formulation and Process Variables. *J. Drug Deliv. Sci.* **2016**, *31*, 83–92. [CrossRef]
3. Bertoft, E. Understanding Starch Structure: Recent Progress. *Agronomy* **2017**, *7*, 56. [CrossRef]
4. Castro, L.M.G.; Alexandre, E.M.C.; Saraiva, J.A.; Pintado, M. Impact of High Pressure on Starch Properties: A Review. *Food Hydrocoll.* **2020**, *106*, 105877. [CrossRef]
5. Montero, B.; Rico, M.; Rodriguez-Llamazares, S.; Barral, L.; Bouza, R. Effect of Nanocellulose as a Filler on Biodegradable Thermoplastic Starch Films from Tuber, Cereal and Legume. *Carbohydr. Polym.* **2017**, *157*, 1094–1104. [CrossRef]
6. Qin, Y.; Liu, C.; Jiang, S.; Xiong, L.; Sun, Q. Characterization of Starch Nanoparticles Prepared by Nanoprecipitation: Influence of Amylose Content and Starch Type. *Ind. Crops Prod.* **2016**, *87*, 182–190. [CrossRef]
7. Marta, H.; Cahyana, Y.; Djali, M.; Arcot, J.; Tensiska, T. A Comparative Study on the Physicochemical and Pasting Properties of Starch and Flour from Different Banana (*Musa* Spp.) Cultivars Grown in Indonesia. *Int. J. Food Prop.* **2019**, *22*, 1562–1575. [CrossRef]
8. Vilpoux, O.F.; Brito, V.H.; Cereda, M.P. Starch Extracted from Corms, Roots, Rhizomes, and Tubers for Food Application. In *Starches for Food Application*; Academic Press: Cambridge, MA, USA, 2019; pp. 103–165.
9. Waterschoot, J.; Gomand, S.V.; Fierens, E.; Delcour, J.A. Production, Structure, Physicochemical and Functional Properties of Maize, Cassava, Wheat, Potato and Rice Starches. *Starch-Stärke* **2015**, *67*, 14–29. [CrossRef]
10. Ashfaq, A.; Khursheed, N.; Fatima, S.; Anjum, Z.; Younis, K. Application of Nanotechnology in Food Packaging: Pros and Cons. *J. Agric.Res.* **2022**, *7*, 100270. [CrossRef]
11. Lu, H.; Tian, Y. Nanostarch: Preparation, Modification, and Application in Pickering Emulsions. *J. Agric. Food Chem.* **2021**, *69*, 6929–6942. [CrossRef]
12. Colussi, R.; Pinto, V.Z.; El Halal, S.L.; Vanier, N.L.; Villanova, F.A.; Marques, E.S.R.; da Rosa Zavareze, E.; Dias, A.R. Structural, Morphological, and Physicochemical Properties of Acetylated High-, Medium-, and Low-Amylose Rice Starches. *Carbohydr. Polym.* **2014**, *103*, 405–413. [CrossRef] [PubMed]
13. Klaochanpong, N.; Puttanlek, C.; Rungsardthong, V.; Puncha-arnon, S.; Uttapap, D. Physicochemical and Structural Properties of Debranched Waxy Rice, Waxy Corn and Waxy Potato Starches. *Food Hydrocoll.* **2015**, *45*, 218–226. [CrossRef]
14. Ogunsona, E.; Ojogbo, E.; Mekonnen, T. Advanced Material Applications of Starch and Its Derivatives. *Eur. Polym. J.* **2018**, *108*, 570–581. [CrossRef]
15. Tagliapietra, B.L.; de Melo, B.G.; Sanches, E.A.; Plata-Oviedo, M.; Campelo, P.H.; Clerici, M.T.P.S. From Micro to Nanoscale: A Critical Review on the Concept, Production, Characterization, and Application of Starch Nanostructure. *Starch-Stärke* **2021**, *73*, 2100079. [CrossRef]
16. Chavan, P.; Sinhmar, A.; Nehra, M.; Thory, R.; Pathera, A.K.; Sundarraj, A.A.; Nain, V. Impact on Various Properties of Native Starch after Synthesis of Starch Nanoparticles: A Review. *Food. Chem.* **2021**, *364*, 130416. [CrossRef]
17. Suma, P.F.; Urooj, A. Isolation and Characterization of Starch from Pearl Millet (*Pennisetum typhoidium*) Flours. *Int. J. Food Prop.* **2015**, *18*, 2675–2687. [CrossRef]
18. Żarski, A.; Bajer, K.; Kapusniak, J. Review of the Most Important Methods of Improving the Processing Properties of Starch toward Non-Food Applications. *Polymers* **2021**, *13*, 832. [CrossRef]
19. Marta, H.; Cahyana, Y.; Djali, M. Densely Packed-Matrices of Heat Moisture Treated-Starch Determine the Digestion Rate Constant as Revealed by Logarithm of Slope Plots. *J. Food Sci. Technol.* **2021**, *58*, 2237–2245. [CrossRef]
20. Marta, H.; Cahyana, Y.; Djali, M.; Pramafisi, G. The Properties, Modification, and Application of Banana Starch. *Polymers* **2022**, *14*, 3092. [CrossRef]
21. Marta, H.; Cahyana, Y.; Bintang, S.; Soeherman, G.P.; Djali, M. Physicochemical and Pasting Properties of Corn Starch as Affected by Hydrothermal Modification by Various Methods. *Int. J. Food Prop.* **2022**, *25*, 792–812. [CrossRef]
22. Marta, H.; Cahyana, Y.; Arifin, H.R.; Khairani, L. Comparing the Effect of Four Different Thermal Modifications on Physicochemical and Pasting Properties of Breadfruit (*Artocarpus altilis*) Starch. *Int. Food Res J.* **2019**, *26*, 269–276.
23. Cahyana, Y.; Wijaya, E.; Halimah, T.S.; Marta, H.; Suryadi, E.; Kurniati, D. The Effect of Different Thermal Modifications on Slowly Digestible Starch and Physicochemical Properties of Green Banana Flour (*Musa acuminata colla*). *Food Chem.* **2019**, *274*, 274–280. [CrossRef]

24. Marta, H.; Hasya, H.N.L.; Lestari, Z.I.; Cahyana, Y.; Arifin, H.R.; Nurhasanah, S. Study of Changes in Crystallinity and Functional Properties of Modified Sago Starch (*Metroxylon* Sp.) Using Physical and Chemical Treatment. *Polymers* **2022**, *14*, 4845. [CrossRef]
25. Cahyana, Y.; Rangkuti, A.; Siti Halimah, T.; Marta, H.; Yuliana, T. Application of Heat-Moisture-Treated Banana Flour as Composite Material in Hard Biscuit. *CYTA J. Food* **2020**, *18*, 599–605. [CrossRef]
26. Alves, M.J.d.S.; Chacon, W.D.C.; Gagliardi, T.R.; Agudelo Henao, A.C.; Monteiro, A.R.; Ayala Valencia, G. Food Applications of Starch Nanomaterials: A Review. *Starch-Stärke* **2021**, *73*, 2100046. [CrossRef]
27. Qiu, C.; Hu, Y.; Jin, Z.; McClements, D.J.; Qin, Y.; Xu, X.; Wang, J. A Review of Green Techniques for the Synthesis of Size-Controlled Starch-Based Nanoparticles and Their Applications as Nanodelivery Systems. *Trends Food Sci. Technol.* **2019**, *92*, 138–151. [CrossRef]
28. Le Corre, D.; Angellier-Coussy, H. Preparation and Application of Starch Nanoparticles for Nanocomposites: A Review. *React. Funct. Polym.* **2014**, *85*, 97–120. [CrossRef]
29. Ahmad, M.; Gani, A.; Hassan, I.; Huang, Q.; Shabbir, H. Production and Characterization of Starch Nanoparticles by Mild Alkali Hydrolysis and Ultra-Sonication Process. *Sci. Rep.* **2020**, *10*, 3533. [CrossRef]
30. Kim, H.Y.; Park, S.S.; Lim, S.T. Preparation, Characterization and Utilization of Starch Nanoparticles. *Colloids. Surf. B Biointerfaces* **2015**, *126*, 607–620. [CrossRef]
31. Basavegowda, N.; Baek, K.H. Advances in Functional Biopolymer-Based Nanocomposites for Active Food Packaging Applications. *Polymers* **2021**, *13*, 4198. [CrossRef]
32. Morán, D.; Gutiérrez, G.; Blanco-López, M.C.; Marefati, A.; Rayner, M.; Matos, M. Synthesis of Starch Nanoparticles and Their Applications for Bioactive Compound Encapsulation. *Appl. Sci.* **2021**, *11*, 4547. [CrossRef]
33. Marta, H.; Wijaya, C.; Sukri, N.; Cahyana, Y.; Mohammad, M. A Comprehensive Study on Starch Nanoparticle Potential as a Reinforcing Material in Bioplastic. *Polymers* **2022**, *14*, 4875. [CrossRef] [PubMed]
34. Ge, S.; Xiong, L.; Li, M.; Liu, J.; Yang, J.; Chang, R.; Liang, C.; Sun, Q. Characterizations of Pickering Emulsions Stabilized by Starch Nanoparticles: Influence of Starch Variety and Particle Size. *Food. Chem.* **2017**, *234*, 339–347. [CrossRef] [PubMed]
35. Campelo, P.H.; Sant'Ana, A.S.; Pedrosa Silva Clerici, M.T. Starch Nanoparticles: Production Methods, Structure, and Properties for Food Applications. *Curr. Opin. Food Sci.* **2020**, *33*, 136–140. [CrossRef]
36. Le Corre, D.; Bras, J.; Dufresne, A. Starch Nanoparticles: A Review. *Biomacromolecules* **2010**, *11*, 1139–1153. [CrossRef]
37. Acevedo-Guevara, L.; Nieto-Suaza, L.; Sanchez, L.T.; Pinzon, M.I.; Villa, C.C. Development of Native and Modified Banana Starch Nanoparticles as Vehicles for Curcumin. *Int. J. Biol. Macromol.* **2018**, *111*, 498–504. [CrossRef]
38. Dularia, C.; Sinhmar, A.; Thory, R.; Pathera, A.K.; Nain, V. Development of Starch Nanoparticles Based Composite Films from Non-Conventional Source-Water Chestnut (*Trapa bispinosa*). *Int. J. Biol. Macromol.* **2019**, *136*, 1161–1168. [CrossRef]
39. Maryam; Kasim, A.; Novelina; Emriadi. Preparation and Characterization of Sago (*Metroxylon* Sp.) Starch Nanoparticles Using Hydrolysis-Precipitation Method. *J. Phys. Con. Ser.* **2020**, *1481*, 012021. [CrossRef]
40. Wang, F.; Chang, R.; Ma, R.; Tian, Y. Eco-Friendly and Superhydrophobic Nano-Starch Based Coatings for Self-Cleaning Application and Oil-Water Separation. *Carbohydr. Polym.* **2021**, *271*, 118410. [CrossRef]
41. Kaur, J.; Kaur, G.; Sharma, S.; Jeet, K. Cereal Starch Nanoparticles—A Prospective Food Additive: A Review. *Crit. Rev. Food Sci. Nutr.* **2018**, *58*, 1097–1107. [CrossRef]
42. Yu, M.; Ji, N.; Wang, Y.; Dai, L.; Xiong, L.; Sun, Q. Starch-Based Nanoparticles: Stimuli Responsiveness, Toxicity, and Interactions with Food Components. *Compr. Rev. Food. Sci. Food. Saf.* **2021**, *20*, 1075–1100. [CrossRef]
43. Abid, N.; Khan, A.M.; Shujait, S.; Chaudhary, K.; Ikram, M.; Imran, M.; Haider, J.; Khan, M.; Khan, Q.; Maqbool, M. Synthesis of Nanomaterials Using Various Top-Down and Bottom-up Approaches, Influencing Factors, Advantages, and Disadvantages: A Review. *Adv. Colloid Interface Sci.* **2022**, *300*, 102597. [CrossRef]
44. Wei, B.; Cai, C.; Tian, Y. Nano-Sized Starch: Preparations and Applications. In *Functional Starch and Applications in Food*; Springer: Berlin/Heidelberg, Germany, 2018; pp. 147–176.
45. Pattekari, P.; Zheng, Z.; Zhang, X.; Levchenko, T.; Torchilin, V.; Lvov, Y. Top-Down and Bottom-up Approaches in Production of Aqueous Nanocolloids of Low Solubility Drug Paclitaxel. *Phys. Chem. Chem. Phys.* **2011**, *13*, 9014–9019. [CrossRef]
46. Lamanna, M.; Morales, N.J.; Garcia, N.L.; Goyanes, S. Development and Characterization of Starch Nanoparticles by Gamma Radiation: Potential Application as Starch Matrix Filler. *Carbohydr. Polym.* **2013**, *97*, 90–97. [CrossRef]
47. González Seligra, P.; Eloy Moura, L.; Famá, L.; Druzian, J.I.; Goyanes, S. Influence of Incorporation of Starch Nanoparticles in Pbat/Tps Composite Films. *Polym. Int.* **2016**, *65*, 938–945. [CrossRef]
48. Ahmad, A.N.; Lim, S.A.; Navaranjan, N.; Hsu, Y.-I.; Uyama, H. Green Sago Starch Nanoparticles as Reinforcing Material for Green Composites. *Polymer* **2020**, *202*, 122646. [CrossRef]
49. Apostolidis, E.; Mandala, I. Modification of Resistant Starch Nanoparticles Using High-Pressure Homogenization Treatment. *Food Hydrocol.* **2020**, *103*, 105677. [CrossRef]
50. Bel Haaj, S.; Thielemans, W.; Magnin, A.; Boufi, S. Starch Nanocrystals and Starch Nanoparticles from Waxy Maize as Nanoreinforcement: A Comparative Study. *Carbohydr. Polym.* **2016**, *143*, 310–317. [CrossRef]
51. Bel Haaj, S.; Magnin, A.; Petrier, C.; Boufi, S. Starch Nanoparticles Formation Via High Power Ultrasonication. *Carbohydr. Polym.* **2013**, *92*, 1625–1632. [CrossRef]
52. Da Silva, N.M.C.; Correia, P.R.C.; Druzian, J.I.; Fakhouri, F.M.; Fialho, R.L.L.; de Albuquerque, E.C.M.C. Pbat/Tps Composite Films Reinforced with Starch Nanoparticles Produced by Ultrasound. *Int. J. Polym. Sci.* **2017**, *2017*, 1–10. [CrossRef]

53. Boufi, S.; Bel Haaj, S.; Magnin, A.; Pignon, F.; Imperor-Clerc, M.; Mortha, G. Ultrasonic Assisted Production of Starch Nanoparticles: Structural Characterization and Mechanism of Disintegration. *Ultrason. Sonochem.* **2018**, *41*, 327–336. [CrossRef] [PubMed]
54. Remanan, M.K.; Zhu, F. Encapsulation of Rutin Using Quinoa and Maize Starch Nanoparticles. *Food Chem* **2021**, *353*, 128534. [CrossRef] [PubMed]
55. Roy, K.; Thory, R.; Sinhmar, A.; Pathera, A.K.; Nain, V. Development and Characterization of Nano Starch-Based Composite Films from Mung Bean (*Vigna radiata*). *Int. J. Biol. Macromol.* **2020**, *144*, 242–251. [CrossRef] [PubMed]
56. Kim, H.-Y.; Lee, J.H.; Kim, J.-Y.; Lim, W.-J.; Lim, S.-T. Characterization of Nanoparticles Prepared by Acid Hydrolysis of Various Starches. *Starch-Stärke* **2012**, *64*, 367–373. [CrossRef]
57. Perez Herrera, M.; Vasanthan, T. Rheological Characterization of Gum and Starch Nanoparticle Blends. *Food Chem.* **2018**, *243*, 43–49. [CrossRef]
58. Torres, F.G.; Arroyo, J.; Tineo, C.; Troncoso, O. Tailoring the Properties of Native Andean Potato Starch Nanoparticles Using Acid and Alkaline Treatments. *Starch-Stärke* **2019**, *71*, 1800234. [CrossRef]
59. Jeong, O.; Shin, M. Preparation and Stability of Resistant Starch Nanoparticles, Using Acid Hydrolysis and Cross-Linking of Waxy Rice Starch. *Food Chem.* **2018**, *256*, 77–84. [CrossRef]
60. Kim, H.Y.; Park, D.J.; Kim, J.Y.; Lim, S.T. Preparation of Crystalline Starch Nanoparticles Using Cold Acid Hydrolysis and Ultrasonication. *Carbohydr. Polym* **2013**, *98*, 295–301. [CrossRef]
61. Hedayati, S.; Niakousari, M.; Mohsenpour, Z. Production of Tapioca Starch Nanoparticles by Nanoprecipitation-Sonication Treatment. *Int. J. Biol. Macromol.* **2020**, *143*, 136–142. [CrossRef]
62. Lin, X.; Sun, S.; Wang, B.; Zheng, B.; Guo, Z. Structural and Physicochemical Properties of Lotus Seed Starch Nanoparticles Prepared Using Ultrasonic-Assisted Enzymatic Hydrolysis. *Ultrason. Sonochem.* **2020**, *68*, 105199. [CrossRef]
63. Sun, Q.; Gong, M.; Li, Y.; Xiong, L. Effect of Retrogradation Time on Preparation and Characterization of Proso Millet Starch Nanoparticles. *Carbohydr. Polym.* **2014**, *111*, 133–138. [CrossRef]
64. Suriya, M.; Reddy, C.K.; Haripriya, S.; Harsha, N. Influence of Debranching and Retrogradation Time on Behavior Changes of Amorphophallus Paeoniifolius Nanostarch. *Int. J. Biol. Macromol.* **2018**, *120*, 230–236. [CrossRef]
65. Winarti, C.; Sunarti, T.C.; Mangunwidjaja, D.; Richana, N. Preparation of Arrowroot Starch Nanoparticles by Butanol-Complex Precipitation, and Its Application as Bioactive Encapsulation Matrix. *Int. Food Res. J.* **2014**, *21*, 2207–2213.
66. Caicedo Chacon, W.D.; Ayala Valencia, G.; Aparicio Rojas, G.M.; Agudelo Henao, A.C. Mathematical Models for Prediction of Water Evaporation and Thermal Degradation Kinetics of Potato Starch Nanoparticles Obtained by Nanoprecipitation. *Starch-Stärke* **2019**, *71*, 1800081. [CrossRef]
67. BeMiller, J.N. Physical Modification of Starch. In *Starch in Food*; Woodhead Publishing: Cambridge, UK, 2018; pp. 223–253.
68. Sun, Q. Starch Nanoparticles. In *Starch in Food*; Woodhead Publishing: Cambridge, UK, 2018; pp. 691–745.
69. Zhu, F. Impact of Ultrasound on Structure, Physicochemical Properties, Modifications, and Applications of Starch. *Trends Food Sci. Technol.* **2015**, *43*, 1–17. [CrossRef]
70. Minakawa, A.F.K.; Faria-Tischer, P.C.S.; Mali, S. Simple Ultrasound Method to Obtain Starch Micro- and Nanoparticles from Cassava, Corn and Yam Starches. *Food Chem.* **2019**, *283*, 11–18. [CrossRef]
71. Wang, S.; Copeland, L. Effect of Acid Hydrolysis on Starch Structure and Functionality: A Review. *Crit. Rev. Food Sci. Nutr.* **2015**, *55*, 1081–1097. [CrossRef]
72. Hassan, N.A.; Darwesh, O.M.; Smuda, S.S.; Altemimi, A.B.; Hu, A.; Cacciola, F.; Haoujar, I.; Abedelmaksoud, T.G. Recent Trends in the Preparation of Nano-Starch Particles. *Molecules* **2022**, *27*, 5497. [CrossRef]
73. Angellier, H.; Choisnard, L.; Molina-Boisseau, S.; Ozil, P.; Dufresne, A. Optimization of the Preparation of Aqueous Suspensions of Waxy Maize Starch Nanocrystals Using a Response Surface Methodology. *Biomacromolecules* **2004**, *5*, 1545–1551. [CrossRef]
74. Kim, J.Y.; Park, D.J.; Lim, S.T. Fragmentation of Waxy Rice Starch Granules by Enzymatic Hydrolysis. *Cer. Chem.* **2008**, *85*, 182–187. [CrossRef]
75. Foresti, M.L.; Williams Mdel, P.; Martinez-Garcia, R.; Vazquez, A. Analysis of a Preferential Action of Alpha-Amylase from B. Licheniformis Towards Amorphous Regions of Waxy Maize Starch. *Carbohydr. Polym.* **2014**, *102*, 80–87. [CrossRef] [PubMed]
76. Sun, Q.; Li, G.; Dai, L.; Ji, N.; Xiong, L. Green Preparation and Characterisation of Waxy Maize Starch Nanoparticles through Enzymolysis and Recrystallisation. *Food Chem.* **2014**, *162*, 223–228. [CrossRef] [PubMed]
77. Wu, X.; Chang, Y.; Fu, Y.; Ren, L.; Tong, J.; Zhou, J. Effects of Non-Solvent and Starch Solution on Formation of Starch Nanoparticles by Nanoprecipitation. *Starch-Stärke* **2016**, *68*, 258–263. [CrossRef]
78. Tan, Y.; Xu, K.; Li, L.; Liu, C.; Song, C.; Wang, P. Fabrication of Size-Controlled Starch-Based Nanospheres by Nanoprecipitation. *ACS Appl. Mater. Interfaces* **2009**, *1*, 956–959. [CrossRef]
79. Kim, J.-Y.; Lim, S.-T. Complex Formation between Amylomaize Dextrin and N-Butanol by Phase Separation System. *Carbohydr. Polym.* **2010**, *82*, 264–269. [CrossRef]
80. Kim, J.-Y.; Yoon, J.-W.; Lim, S.-T. Formation and Isolation of Nanocrystal Complexes between Dextrins and N-Butanol. *Carbohydr. Polym.* **2009**, *78*, 626–632. [CrossRef]
81. Alcázar-Alay, S.C.; Meireles, M.A.A. Physicochemical Properties, Modifications and Applications of Starches from Different Botanical Sources. *Food Sci Technol.* **2015**, *35*, 215–236. [CrossRef]
82. Bajer, D. Nano-Starch for Food Applications Obtained by Hydrolysis and Ultrasonication Methods. *Food Chem.* **2023**, *402*, 134489. [CrossRef]

83. Hernandez-Jaimes, C.; Bello-Perez, L.A.; Vernon-Carter, E.J.; Alvarez-Ramirez, J. Plantain Starch Granules Morphology, Crystallinity, Structure Transition, and Size Evolution Upon Acid Hydrolysis. *Carbohydr. Polym.* **2013**, *95*, 207–213. [CrossRef]
84. Shabana, S.; Prasansha, R.; Kalinina, I.; Potoroko, I.; Bagale, U.; Shirish, S.H. Ultrasound Assisted Acid Hydrolyzed Structure Modification and Loading of Antioxidants on Potato Starch Nanoparticles. *Ultrason. Sonochem.* **2019**, *51*, 444–450. [CrossRef]
85. LeCorre, D.; Bras, J.; Dufresne, A. Evidence of Micro- and Nanoscaled Particles During Starch Nanocrystals Preparation and Their Isolation. *Biomacromolecules* **2011**, *12*, 3039–3046. [CrossRef]
86. Kumari, S.; Yadav, B.S.; Yadav, R.B. Synthesis and Modification Approaches for Starch Nanoparticles for Their Emerging Food Industrial Applications: A Review. *Food Res. Int.* **2020**, *128*, 108765. [CrossRef]
87. Dufresne, A. Crystalline Starch Based Nanoparticles. *Curr. Opin.Colloid Interface Sci.* **2014**, *19*, 397–408. [CrossRef]
88. Liu, Y.; Xu, Y.; Yan, Y.; Hu, D.; Yang, L.; Shen, R. Application of Raman Spectroscopy in Structure Analysis and Crystallinity Calculation of Corn Starch. *Starch-Stärke* **2015**, *67*, 612–619. [CrossRef]
89. Pérez, S.; Bertoft, E. The Molecular Structures of Starch Components and Their Contribution to the Architecture of Starch Granules: A Comprehensive Review. *Starch-Stärke* **2010**, *62*, 389–420. [CrossRef]
90. Li, H.; Gidley, M.J.; Dhital, S. High-Amylose Starches to Bridge the "Fiber Gap": Development, Structure, and Nutritional Functionality. *Compr. Rev. Food Sci. Food Saf.* **2019**, *18*, 362–379. [CrossRef]
91. Ma, X.; Jian, R.; Chang, P.R.; Yu, J. Fabrication and Characterization of Citric Acid-Modified Starch Nanoparticles/Plasticized-Starch Composites. *Biomacromolecules* **2008**, *9*, 3314–3320. [CrossRef]
92. Ali Razavi, S.M.; Amini, A.M. Starch Nanomaterials: A state-of-the-Art Review and Future Trends. In *Novel Approaches of Nanotechnology in Food*; Academic Press: Cambridge, MA, USA, 2016; pp. 237–269.
93. Ding, Y.; Zheng, J.; Zhang, F.; Kan, J. Synthesis and Characterization of Retrograded Starch Nanoparticles through Homogenization and Miniemulsion Cross-Linking. *Carbohydr. Polym.* **2016**, *151*, 656–665. [CrossRef]
94. Szymońska, J.; Targosz-Korecka, M.; Krok, F. Characterization of Starch Nanoparticles. *J. Phys: Conf. Ser.* **2009**, *146*, 012027. [CrossRef]
95. Li, F.; Zhu, A.; Song, X.; Ji, L. Novel Surfactant for Preparation of Poly(L-Lactic Acid) Nanoparticles with Controllable Release Profile and Cytocompatibility for Drug Delivery. *Colloids Surf. B: Biointerfaces* **2014**, *115*, 377–383. [CrossRef]
96. Liu, C.; Jiang, S.; Han, Z.; Xiong, L.; Sun, Q. In vitro Digestion of Nanoscale Starch Particles and Evolution of Thermal, Morphological, and Structural Characteristics. *Food Hydrocoll.* **2016**, *61*, 344–350. [CrossRef]
97. Nagahata, Y.; Kobayashi, I.; Goto, M.; Nakaura, Y.; Inouchi, N. The Formation of Resistant Starch During Acid Hydrolysis of High-Amylose Corn Starch. *J. Appl. Glycosci.* **2013**, *60*, 123–130. [CrossRef]
98. Oh, S.M.; Lee, B.H.; Seo, D.H.; Choi, H.W.; Kim, B.Y.; Baik, M.Y. Starch Nanoparticles Prepared by Enzymatic Hydrolysis and Self-Assembly of Short-Chain Glucans. *Food Sci. Biotechnol.* **2020**, *29*, 585–598. [CrossRef] [PubMed]
99. Choi, H.D.; Hong, J.S.; Pyo, S.M.; Ko, E.; Shin, H.Y.; Kim, J.Y. Starch Nanoparticles Produced Via Acidic Dry Heat Treatment as a Stabilizer for a Pickering Emulsion: Influence of the Physical Properties of Particles. *Carbohydr. Polym.* **2020**, *239*, 116241. [CrossRef] [PubMed]
100. Jiang, H.; Sheng, Y.; Ngai, T. Pickering Emulsions: Versatility of Colloidal Particles and Recent Applications. *Curr. Opin Colloid Interface Sci.* **2020**, *49*, 1–15. [CrossRef]
101. Ben Ayed, E.; Magnin, A.; Putaux, J.-L.; Boufi, S. Vinyltriethoxysilane-Functionalized Starch Nanocrystals as Pickering Stabilizer in Emulsion Polymerization of Acrylic Monomers. Application in Nanocomposites and Pressure-Sensitive Adhesives. *J. Colloid Interface Sci.* **2020**, *578*, 533–546. [CrossRef]
102. Berton-Carabin, C.C.; Schroen, K. Pickering Emulsions for Food Applications: Background, Trends, and Challenges. *Ann.Rev. Food Sci. Technol.* **2015**, *6*, 263–297. [CrossRef]
103. Xu, T.; Yang, J.; Hua, S.; Hong, Y.; Gu, Z.; Cheng, L.; Li, Z.; Li, C. Characteristics of Starch-Based Pickering Emulsions from the Interface Perspective. *Trends Food Sci. Technol.* **2020**, *105*, 334–346. [CrossRef]
104. McClements, D.J.; Jafari, S.M. Improving Emulsion Formation, Stability and Performance Using Mixed Emulsifiers: A Review. *Adv. Colloid Interface Sci.* **2018**, *251*, 55–79. [CrossRef]
105. Sjoo, M.; Emek, S.C.; Hall, T.; Rayner, M.; Wahlgren, M. Barrier Properties of Heat Treated Starch Pickering Emulsions. *J. Colloid Interface Sci.* **2015**, *450*, 182–188. [CrossRef]
106. Tavernier, I.; Wijaya, W.; Van der Meeren, P.; Dewettinck, K.; Patel, A.R. Food-Grade Particles for Emulsion Stabilization. *Trends Food Sci. Technol.* **2016**, *50*, 159–174. [CrossRef]
107. Wei, B.; Xu, X.; Jin, Z.; Tian, Y. Surface Chemical Compositions and Dispersity of Starch Nanocrystals Formed by Sulfuric and Hydrochloric Acid Hydrolysis. *PLoS ONE* **2014**, *9*, e86024. [CrossRef]
108. Tao, S.; Jiang, H.; Gong, S.; Yin, S.; Li, Y.; Ngai, T. Pickering Emulsions Simultaneously Stabilized by Starch Nanocrystals and Zein Nanoparticles: Fabrication, Characterization, and Application. *Langmuir* **2021**, *37*, 8577–8584. [CrossRef]
109. Cazotti, J.C.; Smeltzer, S.E.; Smeets, N.M.B.; Dubé, M.A.; Cunningham, M.F. Starch Nanoparticles Modified with Styrene Oxide and Their Use as Pickering Stabilizers. *Polym. Chem.* **2020**, *11*, 2653–2665. [CrossRef]
110. Fonseca-Florido, H.A.; Vázquez-García, H.G.; Méndez-Montealvo, G.; Basilio-Cortés, U.A.; Navarro-Cortés, R.; Rodríguez-Marín, M.L.; Castro-Rosas, J.; Gómez-Aldapa, C.A. Effect of Acid Hydrolysis and Osa Esterification of Waxy Cassava Starch on Emulsifying Properties in Pickering-Type Emulsions. *LWT-Food Sci. Technol.* **2018**, *91*, 258–264. [CrossRef]

111. Haaj, S.B.; Thielemans, W.; Magnin, A.; Boufi, S. Starch Nanocrystal Stabilized Pickering Emulsion Polymerization for Nanocomposites with Improved Performance. *ACS Appl. Mater. Interfaces* **2014**, *6*, 8263–8273. [CrossRef]
112. Qi, L.; Luo, Z.; Lu, X. Facile Synthesis of Starch-Based Nanoparticle Stabilized Pickering Emulsion: Its Ph-Responsive Behavior and Application for Recyclable Catalysis. *Green Chem.* **2018**, *20*, 1538–1550. [CrossRef]
113. Qian, X.; Lu, Y.; Xie, W.; Wu, D. Viscoelasticity of Olive Oil/Water Pickering Emulsions Stabilized with Starch Nanocrystals. *Carbohydr. Polym.* **2020**, *230*, 115575. [CrossRef]
114. Li, C.; Sun, P.; Yang, C. Emulsion Stabilized by Starch Nanocrystals. *Starch-Stärke* **2012**, *64*, 497–502. [CrossRef]
115. Shao, P.; Zhang, H.; Niu, B.; Jin, W. Physical Stabilities of Taro Starch Nanoparticles Stabilized Pickering Emulsions and the Potential Application of Encapsulated Tea Polyphenols. *Int. J. Biol. Macromol.* **2018**, *118*, 2032–2039. [CrossRef]
116. Sánchez de la Concha, B.B.; Agama-Acevedo, E.; Agurirre-Cruz, A.; Bello-Pérez, L.A.; Alvarez-Ramírez, J. Osa Esterification of Amaranth and Maize Starch Nanocrystals and Their Use in "Pickering" Emulsions. *Starch-Stärke* **2020**, *72*, 1900271. [CrossRef]
117. Li, C.; Li, Y.; Sun, P.; Yang, C. Starch Nanocrystals as Particle Stabilisers of Oil-in-Water Emulsions. *J. Sci. Food. Agric.* **2014**, *94*, 1802–1807. [CrossRef] [PubMed]
118. Harsanto, B.W.; Pranoto, Y.; Supriyanto; Kartini, I. Breadfruit-Based Starch Nanoparticles Prepared Using Nanoprecipitation to Stabilize a Pickering Emulsion. *J. Southwest Jiaotong Univ.* **2021**, *56*, 372–383. [CrossRef]
119. Ye, F.; Miao, M.; Jiang, B.; Campanella, O.H.; Jin, Z.; Zhang, T. Elucidation of Stabilizing Oil-in-Water Pickering Emulsion with Different Modified Maize Starch-Based Nanoparticles. *Food Chem.* **2017**, *229*, 152–158. [CrossRef]
120. Javidi, F.; Razavi, S.M.A.; Mohammad Amini, A. Cornstarch Nanocrystals as a Potential Fat Replacer in Reduced Fat O/W Emulsions: A Rheological and Physical Study. *Food Hydrocoll.* **2019**, *90*, 172–181. [CrossRef]
121. Daniel, T.H.G.; Bedin, A.C.; Souza, É.C.F.; Lacerda, L.G.; Demiate, I.M. Pickering Emulsions Produced with Starch Nanocrystals from Cassava (*Manihot esculenta crantz*), Beans (*Phaseolus vulgaris* L.), and Corn (*Zea mays* L.). *Starch-Stärke* **2020**, *72*, 1900326. [CrossRef]
122. Miao, M.; Li, R.; Jiang, B.; Cui, S.W.; Zhang, T.; Jin, Z. Structure and Physicochemical Properties of Octenyl Succinic Esters of Sugary Maize Soluble Starch and Waxy Maize Starch. *Food Chem.* **2014**, *151*, 154–160. [CrossRef]
123. Lu, X.; Wang, Y.; Li, Y.; Huang, Q. Assembly of Pickering Emulsions Using Milled Starch Particles with Different Amylose/Amylopectin Ratios. *Food Hydrocoll.* **2018**, *84*, 47–57. [CrossRef]
124. Chang, S.; Chen, X.; Liu, S.; Wang, C. Novel Gel-Like Pickering Emulsions Stabilized Solely by Hydrophobic Starch Nanocrystals. *Int. J. Biol. Macromol.* **2020**, *152*, 703–708. [CrossRef]
125. Mao, L.; Miao, S. Structuring Food Emulsions to Improve Nutrient Delivery During Digestion. *Food Eng. Rev.* **2015**, *7*, 439–451. [CrossRef]
126. Ogunlaja, S.B.; Pal, R.; Sarikhani, K. Effects of Starch Nanoparticles on Phase Inversion of Pickering Emulsions. *Can. J. Chem. Eng.* **2018**, *96*, 1089–1097. [CrossRef]
127. Abe, M.M.; Martins, J.R.; Sanvezzo, P.B.; Macedo, J.V.; Branciforti, M.C.; Halley, P.; Botaro, V.R.; Brienzo, M. Advantages and Disadvantages of Bioplastics Production from Starch and Lignocellulosic Components. *Polymers* **2021**, *13*, 2484. [CrossRef]
128. Folino, A.; Karageorgiou, A.; Calabrò, P.S.; Komilis, D. Biodegradation of Wasted Bioplastics in Natural and Industrial Environments: A Review. *Sustainability* **2020**, *12*, 6030. [CrossRef]
129. Maulida; Kartika, T.; Harahap, M.B.; Ginting, M.H.S. Utilization of Mango Seed Starch in Manufacture of Bioplastic Reinforced with Microparticle Clay Using Glycerol as Plasticizer. *IOP Conf. Ser. Mater. Sci. Eng.* **2018**, *309*, 012068.
130. Hosseini, S.N.; Pirsa, S.; Farzi, J. Biodegradable Nano Composite Film Based on Modified Starch-Albumin/Mgo; Antibacterial, Antioxidant and Structural Properties. *Polym. Test.* **2021**, *97*, 107182. [CrossRef]
131. Fan, H.; Ji, N.; Zhao, M.; Xiong, L.; Sun, Q. Characterization of Starch Films Impregnated with Starch Nanoparticles Prepared by 2,2,6,6-Tetramethylpiperidine-1-Oxyl (Tempo)-Mediated Oxidation. *Food Chem.* **2016**, *192*, 865–872. [CrossRef]
132. Martinez, S.; Rivon, C.; Troncoso, O.P.; Torres, F.G. Botanical Origin as a Determinant for the Mechanical Properties of Starch Films with Nanoparticle Reinforcements. *Starch-Stärke* **2016**, *68*, 935–942. [CrossRef]
133. Li, X.; Qiu, C.; Ji, N.; Sun, C.; Xiong, L.; Sun, Q. Mechanical, Barrier and Morphological Properties of Starch Nanocrystals-Reinforced Pea Starch Films. *Carbohydr. Polym.* **2015**, *121*, 155–162. [CrossRef]
134. Hakke, V.S.; Landge, V.K.; Sonawane, S.H.; Babu, G.U.; Ashokkumar, M.; Flores, E.M. The Physical, Mechanical, Thermal and Barrier Properties of Starch Nanoparticle (Snp)/Polyurethane (Pu) Nanocomposite Films Synthesised by an Ultrasound-Assisted Process. *Ultrason. Sonochem.* **2022**, *88*, 106069. [CrossRef]
135. Dai, L.; Qiu, C.; Xiong, L.; Sun, Q. Characterisation of Corn Starch-Based Films Reinforced with Taro Starch Nanoparticles. *Food Chem.* **2015**, *174*, 82–88. [CrossRef]
136. Mukurumbira, A.R.; Mellem, J.J.; Amonsou, E.O. Effects of Amadumbe Starch Nanocrystals on the Physicochemical Properties of Starch Biocomposite Films. *Carbohydr. Polym.* **2017**, *165*, 142–148. [CrossRef] [PubMed]
137. Kong, J.; Yu, Y.; Pei, X.; Han, C.; Tan, Y.; Dong, L. Polycaprolactone Nanocomposite Reinforced by Bioresource Starch-Based Nanoparticles. *Int. J. Biol. Macromol.* **2017**, *102*, 1304–1311. [CrossRef] [PubMed]
138. Zou, J.; Zhang, F.; Huang, J.; Chang, P.R.; Su, Z.; Yu, J. Effects of Starch Nanocrystals on Structure and Properties of Waterborne Polyurethane-Based Composites. *Carbohydr. Polym.* **2011**, *85*, 824–831. [CrossRef]
139. Dai, L.; Zhang, J.; Cheng, F. Cross-Linked Starch-Based Edible Coating Reinforced by Starch Nanocrystals and Its Preservation Effect on Graded Huangguan Pears. *Food Chem.* **2019**, *311*, 125891. [CrossRef]

140. Al-Aseebee, M.D.F.; Rashid, A.H.; Naje, A.S.; Jayaraj, J.J.; Maridurai, T. Influence of Rice Starch Nanocrystals on the Film Properties of the Bionanocomposite Edible Films Produced from Native Rice Starch. *Dig. J. Nanomater. Biostructures* **2021**, *16*, 697–704. [CrossRef]
141. Jouyandeh, M.; Paran, S.M.R.; Jannesari, A.; Saeb, M.R. 'Cure Index' for Thermoset Composites. *Prog. Org. Coat.* **2019**, *127*, 429–434. [CrossRef]
142. Wahyuningtiyas, N.E.; Suryanto, H. Properties of Cassava Starch Based Bioplastic Reinforced by Nanoclay. *J. Mech. Eng. Sci. Technol.* **2018**, *2*, 20–26. [CrossRef]
143. Piyada, K.; Waranyou, S.; Thawien, W. Mechanical, Thermal and Structural Properties of Rice Starch Films Reinforced with Rice Starch Nanocrystals. *Int. Food Res. J.* **2013**, *20*, 439–449.
144. Balakrishnan, P.; Gopi, S.; MS, S.; Thomas, S. Uv Resistant Transparent Bionanocomposite Films Based on Potato Starch/Cellulose for Sustainable Packaging. *Starch-Stärke* **2018**, *70*, 1700139. [CrossRef]
145. Jiang, S.; Liu, C.; Wang, X.; Xiong, L.; Sun, Q. Physicochemical Properties of Starch Nanocomposite Films Enhanced by Self-Assembled Potato Starch Nanoparticles. *LWT-Food Sci. Technol.* **2016**, *69*, 251–257. [CrossRef]
146. Raigond, P.; Sood, A.; Kalia, A.; Joshi, A.; Kaundal, B.; Raigond, B.; Dutt, S.; Singh, B.; Chakrabarti, S.K. Antimicrobial Activity of Potato Starch-Based Active Biodegradable Nanocomposite Films. *Potato Res.* **2018**, *62*, 69–83. [CrossRef]
147. Alzate, P.; Zalduendo, M.M.; Gerschenson, L.; Flores, S.K. Micro and Nanoparticles of Native and Modified Cassava Starches as Carriers of the Antimicrobial Potassium Sorbate. *Starch-Stärke* **2016**, *68*, 1038–1047. [CrossRef]
148. Alzate, P.; Gerschenson, L.; Flores, S. Ultrasound Application for Production of Nano-Structured Particles from Esterified Starches to Retain Potassium Sorbate. *Carbohydr. Polym.* **2020**, *247*, 116759. [CrossRef]
149. Qin, Y.; Wang, J.; Qiu, C.; Hu, Y.; Xu, X.; Jin, Z. Self-Assembly of Metal–Phenolic Networks as Functional Coatings for Preparation of Antioxidant, Antimicrobial, and Ph-Sensitive-Modified Starch Nanoparticles. *ACS Sustain. Chem. Eng.* **2019**, *7*, 17379–17389. [CrossRef]
150. Nieto-Suaza, L.; Acevedo-Guevara, L.; Sánchez, L.T.; Pinzón, M.I.; Villa, C.C. Characterization of Aloe Vera-Banana Starch Composite Films Reinforced with Curcumin-Loaded Starch Nanoparticles. *Food Struct.* **2019**, *22*, 100131. [CrossRef]
151. Fonseca, L.M.; Cruxen, C.; Bruni, G.P.; Fiorentini, A.M.; Zavareze, E.D.R.; Lim, L.T.; Dias, A.R.G. Development of Antimicrobial and Antioxidant Electrospun Soluble Potato Starch Nanofibers Loaded with Carvacrol. *Int. J. Biol. Macromol.* **2019**, *139*, 1182–1190. [CrossRef]
152. Nallasamy, P.; Ramalingam, T.; Nooruddin, T.; Shanmuganathan, R.; Arivalagan, P.; Natarajan, S. Polyherbal Drug Loaded Starch Nanoparticles as Promising Drug Delivery System: Antimicrobial, Antibiofilm and Neuroprotective Studies. *Process Biochem.* **2020**, *92*, 355–364. [CrossRef]
153. Dai, X.; Guo, Q.; Zhao, Y.; Zhang, P.; Zhang, T.; Zhang, X.; Li, C. Functional Silver Nanoparticle as a Benign Antimicrobial Agent That Eradicates Antibiotic-Resistant Bacteria and Promotes Wound Healing. *ACS Appl. Mater. Interfaces* **2016**, *8*, 25798–25807. [CrossRef]
154. Ismail, N.S.; Gopinath, S.C.B. Enhanced Antibacterial Effect by Antibiotic Loaded Starch Nanoparticle. *J. Assoc. Arab Univ. Basic Appl. Sci.* **2018**, *24*, 136–140. [CrossRef]
155. Qin, Y.; Xue, L.; Hu, Y.; Qiu, C.; Jin, Z.; Xu, X.; Wang, J. Green Fabrication and Characterization of Debranched Starch Nanoparticles Via Ultrasonication Combined with Recrystallization. *Ultrason. Sonochem.* **2020**, *66*, 105074. [CrossRef]
156. Kaur, J.; Kaur, G.; Sharma, S. Corn Starch Nanoparticles: Preparation, Characterization, and Utilization as a Fat Replacer in Salad Dressing. *Acta Aliment.* **2019**, *48*, 204–212. [CrossRef]
157. Surendra Babu, A.; Parimalavalli, R.; Jagan Mohan, R. Effect of Modified Starch from Sweet Potato as a Fat Replacer on the Quality of Reduced Fat Ice Creams. *J. Food Meas. Charact.* **2018**, *12*, 2426–2434. [CrossRef]
158. Altan, A.; Aytac, Z.; Uyar, T. Carvacrol Loaded Electrospun Fibrous Films from Zein and Poly(Lactic Acid) for Active Food Packaging. *Food Hydrocoll.* **2018**, *81*, 48–59. [CrossRef]
159. Dandekar, P.; Jain, R.; Stauner, T.; Loretz, B.; Koch, M.; Wenz, G.; Lehr, C.M. A Hydrophobic Starch Polymer for Nanoparticle-Mediated Delivery of Docetaxel. *Macromol. Biosci.* **2012**, *12*, 184–194. [CrossRef] [PubMed]
160. Hasanvand, E.; Fathi, M.; Bassiri, A. Production and Characterization of Vitamin D3 Loaded Starch Nanoparticles: Effect of Amylose to Amylopectin Ratio and Sonication Parameters. *J. Food Sci. Technol.* **2018**, *55*, 1314–1324. [CrossRef]
161. Mwangi, W.W.; Lim, H.P.; Low, L.E.; Tey, B.T.; Chan, E.S. Food-Grade Pickering Emulsions for Encapsulation and Delivery of Bioactives. *Trends Food Sci. Technol.* **2020**, *100*, 320–332. [CrossRef]
162. Sharif, H.R.; Abbas, S.; Majeed, H.; Safdar, W.; Shamoon, M.; Khan, M.A.; Shoaib, M.; Raza, H.; Haider, J. Formulation, Characterization and Antimicrobial Properties of Black Cumin Essential Oil Nanoemulsions Stabilized by Osa Starch. *J. Food Sci. Technol.* **2017**, *54*, 3358–3365. [CrossRef]
163. Liu, C.; Ge, S.; Yang, J.; Xu, Y.; Zhao, M.; Xiong, L.; Sun, Q. Adsorption Mechanism of Polyphenols onto Starch Nanoparticles and Enhanced Antioxidant Activity under Adverse Conditions. *J. Funct. Foods* **2016**, *26*, 632–644. [CrossRef]
164. Bhatia, M.; Rohilla, S. Formulation and Optimization of Quinoa Starch Nanoparticles: Quality by Design Approach for Solubility Enhancement of Piroxicam. *Saudi Pharm. J.* **2020**, *28*, 927–935. [CrossRef]
165. Wang, L.; Zhao, X.; Yang, F.; Wu, W.; Wu, M.; Li, Y.; Zhang, X. Loading Paclitaxel into Porous Starch in the Form of Nanoparticles to Improve Its Dissolution and Bioavailability. *Int. J. Biol. Macromol.* **2019**, *138*, 207–214. [CrossRef]

166. Kheradvar, S.A.; Nourmohammadi, J.; Tabesh, H.; Bagheri, B. Starch Nanoparticle as a Vitamin E-Tpgs Carrier Loaded in Silk Fibroin-Poly(Vinyl Alcohol)-Aloe Vera Nanofibrous Dressing. *Colloids Surf. B Biointerfaces* **2018**, *166*, 9–16. [CrossRef]
167. Ahmad, M.; Mudgil, P.; Gani, A.; Hamed, F.; Masoodi, F.A.; Maqsood, S. Nano-Encapsulation of Catechin in Starch Nanoparticles: Characterization, Release Behavior and Bioactivity Retention During Simulated in-Vitro Digestion. *Food Chem.* **2019**, *270*, 95–104. [CrossRef]
168. Escobar-Puentes, A.A.; García-Gurrola, A.; Rincón, S.; Zepeda, A.; Martínez-Bustos, F. Effect of Amylose/Amylopectin Content and Succinylation on Properties of Corn Starch Nanoparticles as Encapsulants of Anthocyanins. *Carbohydr. Polym.* **2020**, *250*, 116972. [CrossRef]
169. Ahmad, M.; Gani, A. Ultrasonicated Resveratrol Loaded Starch Nanocapsules: Characterization, Bioactivity and Release Behaviour under in-Vitro Digestion. *Carbohydr. Polym.* **2020**, *251*, 117111. [CrossRef]
170. Mahmoudi Najafi, S.H.; Baghaie, M.; Ashori, A. Preparation and Characterization of Acetylated Starch Nanoparticles as Drug Carrier: Ciprofloxacin as a Model. *Int. J. Biol. Macromol.* **2016**, *87*, 48–54. [CrossRef]
171. Yang, J.; He, H.; Gu, Z.; Cheng, L.; Li, C.; Li, Z.; Hong, Y. Conjugated Linoleic Acid Loaded Starch-Based Emulsion Nanoparticles: In Vivo Gastrointestinal Controlled Release. *Food Hydrocoll.* **2020**, *101*, 105477. [CrossRef]
172. Liu, Q.; Cai, W.; Zhen, T.; Ji, N.; Dai, L.; Xiong, L.; Sun, Q. Preparation of Debranched Starch Nanoparticles by Ionic Gelation for Encapsulation of Epigallocatechin Gallate. *Int. J. Biol. Macromol.* **2020**, *161*, 481–491. [CrossRef]
173. Zhu, F. Encapsulation and Delivery of Food Ingredients Using Starch Based Systems. *Food Chem.* **2017**, *229*, 542–552. [CrossRef]
174. Qi, L.; Ji, G.; Luo, Z.; Xiao, Z.; Yang, Q. Characterization and Drug Delivery Properties of Osa Starch-Based Nanoparticles Prepared in [C3ohmim]Ac-in-Oil Microemulsions System. *ACS Sustain. Chem. Eng.* **2017**, *5*, 9517–9526. [CrossRef]
175. Zhu, F. Interactions between Starch and Phenolic Compound. *Trends Food Sci. Technol.* **2015**, *43*, 129–143. [CrossRef]
176. Karunaratne, R.; Zhu, F. Physicochemical Interactions of Maize Starch with Ferulic Acid. *Food Chem* **2016**, *199*, 372–379. [CrossRef] [PubMed]
177. Ceborska, M.; Zimnicka, M.; Wszelaka-Rylik, M.; Troć, A. Characterization of Folic Acid/Native Cyclodextrins Host–Guest Complexes in Solution. *J. Mol. Struct.* **2016**, *1109*, 114–118. [CrossRef]
178. Chin, S.F.; Mohd Yazid, S.N.A.; Pang, S.C. Preparation and Characterization of Starch Nanoparticles for Controlled Release of Curcumin. *Int. J. Polym. Sci.* **2014**, *2014*, 1–8. [CrossRef]

Disclaimer/Publisher's Note: The statements, opinions and data contained in all publications are solely those of the individual author(s) and contributor(s) and not of MDPI and/or the editor(s). MDPI and/or the editor(s) disclaim responsibility for any injury to people or property resulting from any ideas, methods, instructions or products referred to in the content.

Review

Glass Transition Temperature of PLGA Particles and the Influence on Drug Delivery Applications

Guangliang Liu and Kathleen McEnnis *

Otto H. York Department of Chemical and Materials Engineering, New Jersey Institute of Technology, Newark, NJ 07102, USA; gl242@njit.edu
* Correspondence: mcennis@njit.edu

Abstract: Over recent decades, poly(lactic-co-glycolic acid) (PLGA) based nano- and micro- drug delivery vehicles have been rapidly developed since PLGA was approved by the Food and Drug Administration (FDA). Common factors that influence PLGA particle properties have been extensively studied by researchers, such as particle size, polydispersity index (PDI), surface morphology, zeta potential, and drug loading efficiency. These properties have all been found to be key factors for determining the drug release kinetics of the drug delivery particles. For drug delivery applications the drug release behavior is a critical property, and PLGA drug delivery systems are still plagued with the issue of burst release when a large portion of the drug is suddenly released from the particle rather than the controlled release the particles are designed for. Other properties of the particles can play a role in the drug release behavior, such as the glass transition temperature (T_g). The T_g, however, is an underreported property of current PLGA based drug delivery systems. This review summarizes the basic knowledge of the glass transition temperature in PLGA particles, the factors that influence the T_g, the effect of T_g on drug release behavior, and presents the recent awareness of the influence of T_g on drug delivery applications.

Keywords: glass transition temperature; PLGA copolymers; drug delivery; nanoparticles

1. Introduction

The application of polymeric particles in drug delivery has been rapidly developed in the past several decades [1–9]. Particle-based therapeutics offer considerable benefits compared with traditional pharmaceuticals, such as controlling release rates, overcoming biological barriers, delivering hydrophobic drugs, and targeting specific sites [10–18]. Polymeric particles guard the encapsulated drugs from enzymatic reactions in order to prolong the half-life of the encapsulated drugs [19–21]. The tunable size of the polymeric particles enables the travel through cell membrane barriers [22–24]. Diverse manufacturing approaches and surface modifications offer opportunities for the polymeric particles to reach the desired organ, tissue, and cells, thus minimizing the toxicity at other sites [25–27]. All these benefits make polymeric particles a promising drug delivery strategy. Poly(lactic-co-glycolic acid) (PLGA) has been proven to be a successful polymeric drug carrier and widely used in drug delivery, tissue engineering, and cancer therapies [28,29] due to its biocompatibility and biodegradability. Currently, more than 20 different PLGA formulations have been approved by the U.S. FDA [30]. PLGA undergoes a hydrolysis process in body fluid and generates biodegradable metabolite substances, lactic acid and glycolic acid, which can be eliminated by the human body [31]. Additionally, the availability of different PLGA polymer degradation rates, ranging from days to months, can be used to design an appropriate release profile, keeping the drug concentration between maximum toxic concentration (MTC) and minimum effective concentration (MEC), and increasing patient compliance. In order to achieve promised pharmacodynamics, biodistribution, and toxicity levels, the key physicochemical properties need to be appropriately studied. Particle size,

size distribution, surface morphology, zeta potential, and loading efficiency are most commonly characterized because these parameters are the key factors that determine the drug release behaviors [32,33]. It has been proposed, however, that the glass transition temperature (T_g) will also impact the drug release behavior of polymeric nanoparticles [34–36]. During this transition, the increased polymer chain mobility allows the drug molecule to escape from the polymer chain entanglement, resulting in an increased release rate. Notably, during the preparation of PLGA particles, the nature of PLGA copolymer, reactant components, and manufacturing process will contribute to changes in the T_g (Figure 1) [37].

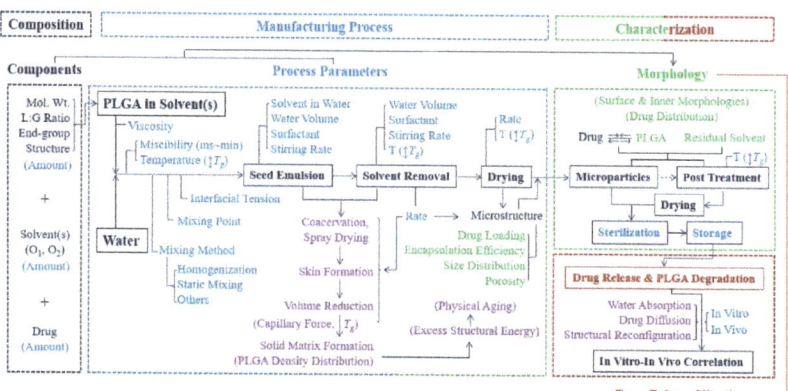

Figure 1. Flow chart of steps of manufacturing PLGA particles and their influence on particle properties [37]. Reprinted from Journal of Controlled Release, 329, Park, K.; Otte, A.; Sharifi, F.; Garner, J.; Skidmore, S.; Park, H.; Jhon, Y.K.; Qin, B.; Wang, Y., Formulation composition, manufacturing process, and characterization of poly(lactide-co-glycolide) microparticles, 1150–1161, Copyright (2021), with permission from Elsevier.

Glass transition temperature (T_g) is usually defined as the temperature range where the polymer transitions from a hard glassy state to a relative rubbery state and is normally detected by the rapid change in heat capacity, specific volume, or stiffness. The T_g is an important indicator of the physical properties of semi-crystalline and amorphous polymers. During the glass transition, the disordered chains in the amorphous portion start to escape the entanglements, increasing the polymer mobility on a macro scale which results in a soft rubbery substance. It has been well studied that during the glass transition, specific enthalpy, specific volume, thermal expansivity, motions of the polymer chain, and other parameters experience a dramatic change [38–42]. Notably, previous literature demonstrated that the T_g of drug loaded PLGA particles ranges from 30 °C to 60 °C [43,44], indicating that PLGA particles could undergo a glass transition in a 37 °C drug release environment. The substantial change in PLGA particle physicochemical properties could lead to a different drug release rate from the particle matrix and an uncontrolled release profile. Nevertheless, even though the drug release kinetics has been frequently investigated by researchers with respect to the physicochemical properties of PLGA copolymer, drug type, manufacturing process, and post-treatment, there is significantly less literature focusing on associating drug release kinetics with the T_g of PLGA particles.

Quite a few reviews have recently been reported about PLGA nanoparticles in drug delivery. Mir et al. summarized the application of PLGA nano carriers in cardiovascular diseases, inflammatory disease, neurodegenerative diseases as well as cancer therapy and theragnostic [45]. Xu et al. provides a review on experimental observations and theoretical models, inferring the relation between the manufacturing factors and drug release profiles (Figure 2). These include the inherent properties of the PLGA polymer, influence of the drug loaded into the particle, processing parameters, and release environment [46]. Ding et al. presented the approaches of PLGA particle preparation, which could be classified as emulsification-solvent evaporation, nanoprecipitation, microfluidics, spray-drying, and phase separation [47]. Rezvantalab et al. summarized passive targeting, active targeting, and magnetic targeting for PLGA drug delivery nanoparticles for cancer treatment. In addition, PLGA nanoparticles can be used with other therapies, such as magnetic hyperthermia, photodynamic and photothermal therapy, and gene therapy [48]. Ghitman et al. provided the comparison between the traditional approach of preparing PLGA-lipid nano vehicles and novel approaches, which were soft lithography and spray drying. Ghitman's review highlights the current challenges to fully understand the physicochemical properties of the nanocarriers and the interaction of targeting sites to determine the toxicity level and clinic safety [49]. Cunha et al. investigated the application of PLGA nanocarriers in neurodegenerative diseases, specifically the potential for PLGA nanocarriers to transport neuroprotective medicines across the blood-brain barrier [50]. Though many reviews exist exploring PLGA's role as a drug delivery vehicle, none exist that take the glass transition temperature of the particles into account. This review summaries the factors that influence the T_g of the PLGA copolymer, bare particles, and drug loaded particles. In addition, the connection of glass transition of PLGA particles and drug release behavior are discussed in terms of the mobility of PLGA particles, the physical ageing effect, and surface reconfiguration.

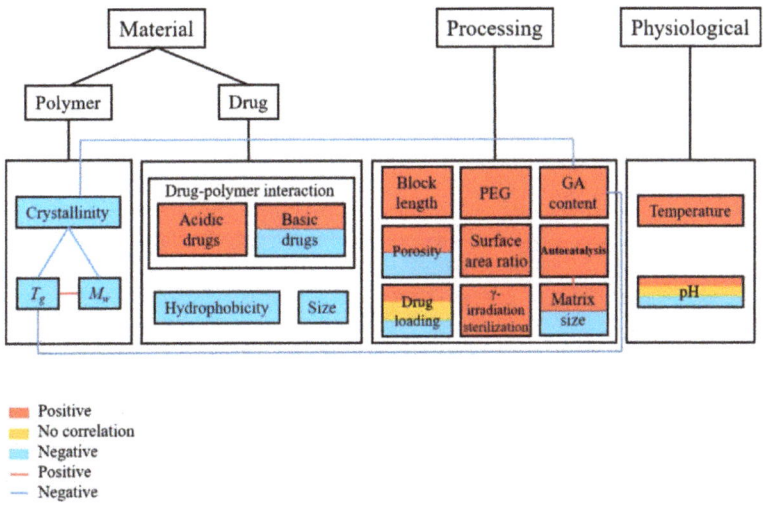

Figure 2. Link between the related parameters and the rate of drug release from PLGA carriers [46]. Reproduced with permission from Xu, Y. et al. Journal of Biomedical Materials Research Part B: Applied Biomaterials, published by John Wiley and Sons, Copyright 2017.

2. Glass Transition Temperature of PLGA Particles

2.1. PLGA Copolymer

PLGA is a linear random copolymer consisting of D,L-lactide and glycolide, which is usually prepared by polycondensation reaction and ring-opening polymerization of the two monomers (Figure 3) [51,52]. The physicochemical properties of the PLGA copolymer used in preparation are the determining factors of the properties of the PLGA particles, including monomer ratio, molecular weight, crystallinity, and end groups [35,53]. In addition, the final PLGA products are also affected by the approaches, reaction environment, and process parameters [54]. The molar ratio of the two monomers in the PLGA chains determines many physicochemical properties, such as the glass transition temperature, degradation rate, hydrophobicity, and degree of crystallinity [55,56]. In general, the T_g of PLGA increases when the copolymer has a rich content of PLA. Peter In Pyo et al. reported that, among four different ratio PLGAs, PLGA with a ratio of lactide to glycolide of 90:10 (PLGA90:10) had the highest T_g while PLGA with a ratio of 50:50 (PLGA50:50) had the lowest T_g of 35.7 °C [57]. Brostow et al. developed an equation to predict the glass transition temperature of physical mixtures of binary systems and copolymers [58],

$$T_g = x_1 T_{g1} + (1-x_1) T_{g2} + x_1(1-x_1) \times \left[a_0 + a_1(2x_1 - 1) + a_2(2x_1 - 1)^2 + a_3(2x_1 - 1)^3 \right] \quad (1)$$

where T_g is the glass transition temperature of the given sample, x_1 is the weight fraction of component 1, T_{g1} is the glass transition temperature of component 1, x_2 is the weight fraction of component 1, T_{g2} is the glass transition temperature of component 2, a_0, a_1, a_2, and a_3 are parameters for the given copolymer or binary system.

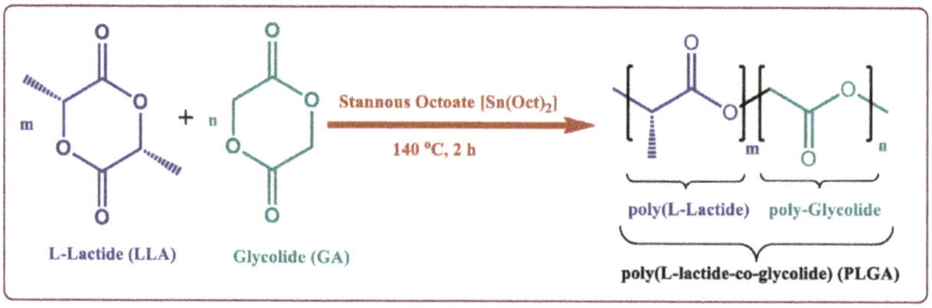

Figure 3. Copolymerization of PLGA by ring-opening method [52]. Reproduced from Butreddy, A. et al., International Journal of Molecular Sciences, 22, 2021, under Creative Commons Attribution 4.0 International License (http://creativecommons.org/licenses/by/4.0/ (accessed on 24 February 2022)).

It has also been reported that PLGA50:50 has the fastest degradation rate, which is due to the high percentage of hydrophilic glycolide, enabling water to penetrate the particle matrix and promote hydrolysis [59,60]. The Flory-Fox equation is a well-known empirical equation that describes the relationship between the number-average molecular weight and glass transition temperature, which was reported by Thomas et al. in 1950 [61]:

$$T_g = T_{g,\infty} - \frac{K}{M_n} \quad (2)$$

where $T_{g,\infty}$ is the highest glass transition temperature for a given polymer under the theoretical condition that the molecular weight is infinitely high, K is an empirical parameter for a given polymer sample which is related to the free volume, and M_n is the number-average molecular weight.

Briefly, T_g has a positive correlation with polymer molecular weight. As the polymer chains become longer, the concentration of chain ends decreases in a unit volume resulting in less free volume between chain ends, thus the T_g becomes higher [62]. Lee et al. illustrated that PLGA with molecular weight (MW) of 8000 g/mol has a T_g of 42.17 °C, and as the MW increased to 110,000 g/mol, the T_g rose to 52.62 °C [63]. Additionally, the crystallinity and the mobility of the polymer chain ends have a significant impact on free volume, and the T_g rises as the degree of crystallinity grows or as the density of end groups decreases [64–66].

2.2. Glass Transition Temperature of Polymeric Particles

When considering properties of substances at a nanoscale level, it is expected that the properties will be different from those of the bulk material, often because of the greater surface-to-volume ratio the nano substances have. Keddie's group reported the first systematic study on the size-dependent glass transition temperature of thin polystyrene films supported by silicon substrates. In their work, three different molecular weight polystyrenes were used to create thin films and their T_g was measured by ellipsometry. It was found that the T_g dropped substantially when the film became thinner [67]. Raegen et al. further investigated PS thin films on substrates under ambient, dry nitrogen, and vacuum environments (Figure 4) [68]. For all experiments, the T_g drop appeared with decreasing film thickness suggesting that the T_g reduction in PS thin films was an intrinsic property. The surface area to volume was greatly reduced in these thin films, however, the two surfaces were different: one was the supporting substrate, and the other was a free surface. In order to reduce the inequality from the free surface and interface of the thin film and to better understand the interfacial effect on a polymer's T_g, spherical nanoparticles have been investigated by several research groups because a 3-dimensional geometry reduces the interface to one and the increase of the surface area to volume ratio will exhibit more obvious interfacial effects. Zhang et al. prepared polystyrene nanoparticles (PS NPs) of different sizes and the T_g of the PS NPs suspended in water was measured by MDSC [69]. The results agreed with the trend of PS thin films, in which the PS nanoparticles of extremely small size will show a significant reduction in T_g. It was well accepted that the T_g reduction was caused by an enhanced mobile layer on the surface [67]. To further prove the interfacial effects on T_g shift, they synthesized PS/silica core-shell structural nanoparticles, for which the silica on the surface was defined as a hard shell. It was observed that the silica capped nanoparticle samples did not have a size dependent T_g reduction, which reinforced the conclusion that the mobile layer formed on the free surface will cause the T_g shift.

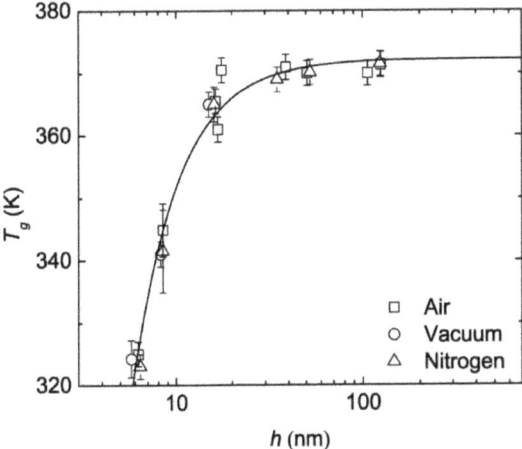

Figure 4. T_g of PS thin films under different measuring environment [68]. Reprinted by permission from Springer Nature Customer Service Centre GmbH: Springer, The European Physical Journal E, Effect of atmosphere on reductions in the glass transition of thin polystyrene films, Raegen, A.N.; Massa, M.V.; Forrest, J.A.; Dalnoki-Veress, K., 2008.

Christie et al. investigated the effects of the measurement environment on T_g by measuring the T_g of PS nanoparticles suspended in three different liquids: glycerol, ionic liquid (1-butyl-3-methylimidazolium trifluoromethanesulfonate, [BMIM][CF$_3$SO$_3$]), and water [70]. As shown in Figure 5, the T_g reduction of PS nanoparticles suspended in water, ionic liquid and glycerol will have a strong, independent, and weak correlation with the size. Also, the T_g reduction from particles suspended in water will be similar to those measured in air, because the interfaces of water-PS and air-PS are considered "soft" due to their low viscosities compared with the polymer. The higher viscosity of glycerol, however, will inhibit the mobility of the glycerol-PS interface, resulting in a relatively inert polymer chain in the mobile layer. When considering the suspension in ionic liquid, ionic interactions dominate the mobility in the mobile layer because the positively charged [BMIM] molecule will anchor onto the negatively charged PS surface, inhibiting the mobility of the polymer chains at the interface.

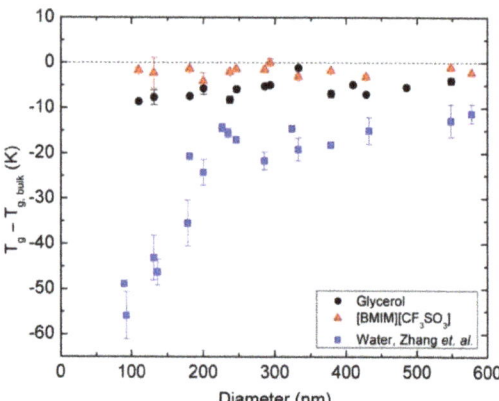

Figure 5. PS nanoparticles suspended in various liquids [70]. Reproduced with permission from Christie, D. et al. Journal of Polymer Science Part B: Polymer Physics, published by John Wiley and Sons, Copyright 2016.

Feng et al. proposed investigation in aqueous environments by preparing PS nanoparticles from a nonionic surfactant (Brij 98) and an anionic surfactant (sodium dodecyl benzene sulfonate (SDBS)) as well as surfactant-free particles [71]. A substantial reduction in T_g with decreasing size of surfactant-free particles was observed which corresponded to previous studies. Nanoparticle surface softness is critical in T_g characterization because of the high surface area to volume ratio in nano-size materials.

These studies demonstrate that polymeric particles under confinement exhibit variations in T_g as a result of surface and interfacial effects. Due to the challenges associated with residual surfactant, size distributions, and other factors, the size-T_g correlation has not been conducted explicitly on PLGA particles; nonetheless, similar trends are expected to occur in PLGA particles.

2.3. Drug Effect

PLGA particles are extensively employed for a broad range of drugs, including hydrophobic and hydrophilic drugs. Drugs that are hydrophobic are easier to encapsulate in PLGA than those that are hydrophilic. Hydrophilic medicines often have lower drug loading efficiencies because the drug molecules enter the aqueous phase before the PLGA chains form into particles [72]. For loading hydrophobic and hydrophilic drugs into PLGA microparticles, the most extensively utilized methods are emulsion-evaporation technique (oil/water or water/oil/water) (Figure 6) [73–75]. The single emulsion technique involves an organic phase which contains PLGA polymer and the hydrophobic drug in a suitable organic solvent and an aqueous phase which contains a stabilizer. Mechanical force provided by ultrasonication is utilized to form an oil in water emulsion, and the organic solvent is then extracted to solidify PLGA particles [76]. On the other hand, hydrophilic drugs are usually encapsulated by a double emulsion technique to prevent diffusion of the drug into the aqueous phase. The inner water phase containing the hydrophilic drug is added into the PLGA solution to form a primary water-in-oil emulsion. Then the primary emulsion is injected into the outer water phase with the presence of a stabilizer to create a double emulsion. The final step is similar to the single emulsion process, which is to evaporate the organic solvent to obtain PLGA particles [77]. Another conventional preparation of PLGA particles is nanoprecipitation, which involves mixing a miscible solvent and stabilizer in water [77]. During the diffusion of organic solvent into the aqueous phase, nucleation, nuclei growth, and aggregation are expected to occur in order to form final particles [78]. Microfluidic technology is a novel method to produce narrow size distribution PLGA particles with high drug encapsulation. A microfluidic chip is made up of micro size channels which ensures the mixing of inlet flows to be completed within milliseconds [33]. Electrospray jetting can also be utilized to prepare PLGA particles. Electrospray jetting usually consists of a high voltage source, syringe pump, and collector. By adjusting the voltage, distance between collector and syringe, and flow rate, a Taylor-cone forms at the needle, which results in a stable spray. Solvent in the small droplets experience evaporation and solid particles reach the collector [79].

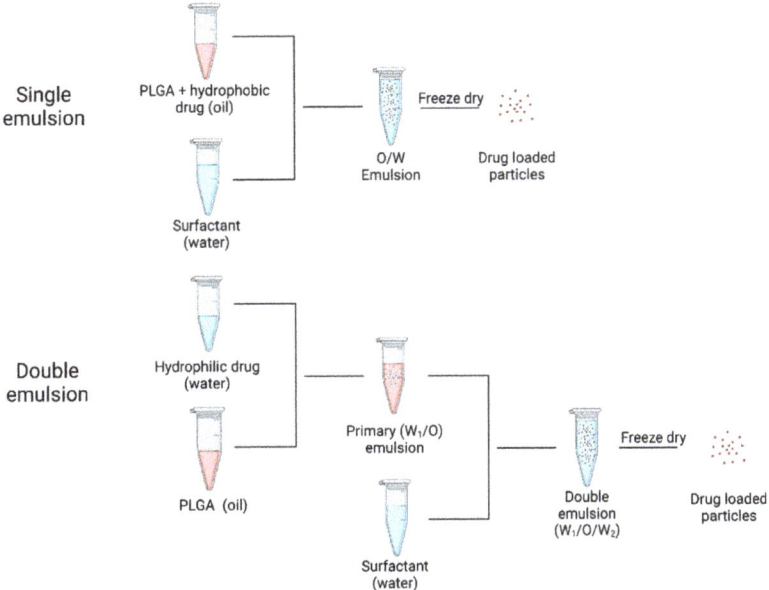

Figure 6. Processes of loading hydrophobic drug (single emulsion) and hydrophilic drug (double emulsion). Created in Biorender.com (accessed on 24 February 2022).

The drug type is tightly associated with drug release behavior, as hydrophobic molecules have a significantly lower degree of initial burst release than other pharmaceuticals due to their poor water solubility [80–82]. Steven et al. demonstrated the influence of a hydrophilic drug (aspirin) and a hydrophobic drug (haloperidol) on PLGA matrices release behavior, and the results showed that the drug with higher water solubility (aspirin) would give a relatively higher diffusion efficiency [83]. Apart from the influence of drug release behavior, the drug type will also determine the glass transition temperature of PLGA particles. Svenja et al. encapsulated flurbiprofen into 200 nm PLGA nanoparticles, and found that as the encapsulation efficiency increased, the T_g of PLGA nanoparticles decreased from 28.8 °C to 19.9 °C, and that the overall mobility was improved by a higher flurbiprofen loading efficiency. Furthermore, they prepared mTHPP-loaded PLGA nanoparticles and measured the T_g, which turned out to have no effect compared with the unloaded PLGA particles. Due to the higher molecular weight of mTHPP molecules, the rigid chemical structure, and relatively hydrophobic compounds, the mTHPP is unable to form a tight association with the polymer, preventing the polymer chain from becoming more mobile [34]. In order to investigate the plasticizing effect of different drugs in polymeric system, Siepmann et al. prepared thin films with metoprolol tartrate, chlorpheniramine maleate, and ibuprofen [84]. The experimental data illustrated that T_g of the thin films decreased with the increased drug loading efficiency, which demonstrated that three drugs acted as plasticizers in the polymeric system, where ibuprofen contributed the most plasticizing effect (Figure 7). It is expected that drug molecules penetrate in the spaces between the polymer chains, which increases the free volume and decrease the T_g of the polymeric system.

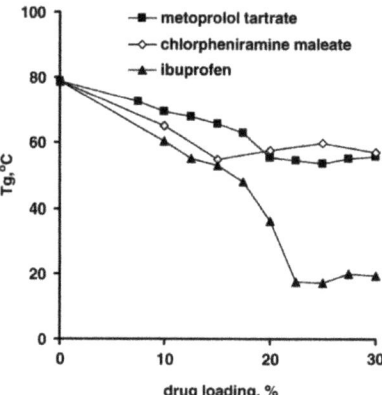

Figure 7. Plasticizing effects of drug types and drug loading efficiency [84]. Reprinted from Journal of Controlled Release, 115, Siepmann, F.; Le Brun, V.; Siepmann, J., Drugs acting as plasticizers in polymeric systems: A quantitative treatment, 298–306, Copyright (2006), with permission from Elsevier.

Recently, it has become more common to report the measured T_g of particles used for drug delivery applications. In this review, the T_g of PLGA nanoparticles from recent literature loaded with a drug are listed, along with the size, preparation, measurement, and heating rate, as shown in Table 1.

2.4. Water Content

Water is well established as a plasticizer in polymeric systems [91] and it lowers the T_g of many polymers. Passerini et al. reported that undried PLGA particles contain approximate 4.47% of moisture content and have a T_g of 27.7 °C, which is about 15 °C lower than that of bulk PLGA polymer Notably, the dried particles which undergo 3 days' lyophilization still contain 3.5% of residual moisture, and the T_g is 33.1 °C [92]. Susan et al. further investigated the influence of water uptake on the T_g of PLGA polymers. After incubation in 0.5% PVA solution, the dried PLGA polymer had a T_g approximately 15 °C higher than the wet polymer, and T_g recovery was achieved upon removing the moisture content (Figure 8). After 14 days' incubation, lower PLGA chains were found, indicating that degradation of PLGA occurred [93].

2.5. Residual Surfactant

Amphiphilic compounds are widely utilized in PLGA particle manufacturing processes to generate monodisperse particles, reduce surface tension of the particles, and avoid aggregation among the particles [94]. The interaction between PLGA chains and other substances such as encapsulated drugs, trapped stabilizer, and residual solvent, will eventually influence the mobility of the PLGA matrix [95]. Sahoo et al. reported that it was quite challenging to remove the remaining poly (vinyl alcohol) (PVA) from PLGA particles before freeze-drying. A logical explanation would be that PVA filled in the inner pockets and coated the surface of PLGA particles [96]. The remining surfactant had an effect on the particle parameters of PLGA particles, including particle size, zeta potential, size distribution, surface hydrophobicity, and protein loading, and also had a modest effect on the encapsulated protein's in vitro release. According to their report, the weight percentage of residual PVA could be up to 5% of PLGA particles prepared by emulsion-solvent evaporation technique [96]. Spek et al. illustrated the residual PVA present in PLGA particles by ^{1}H nuclear magnetic resonance spectroscopy (NMR), which turned out to be 9.9 wt.% from the bulk of the particles. For the PEG-PLGA particles, PVA content could be as high as 35 wt.% on the particles surface based on the X-ray photoelectron spectroscopy (XPS) results [97]. In order to demonstrate the surfactant effect on the T_g of polymeric particles, Feng et al. prepared PS nanoparticle with Brij 98 (nonionic surfactant) and sodium dodecyl benzene sulfonate (anionic surfactant) as well as surfactant-free particles. A substantial reduction in T_g with decreasing size of surfactant-free particles was observed which corresponded to previous studies. On the other hand, the PS nanoparticles prepared with nonionic surfactant showed a weak correlation between size and T_g reduction while anionic PS latex nanoparticles showed no correlation. The authors suggested that the incorporation of surfactant into the mobile layer influenced the free volume, thus affecting the T_g of the particles [71].

Table 1. Glass transition temperature of drug loaded PLGA nanoparticles.

PLGA LA:GA Mol wt. (g/mol)	Diameter (nm)	Model Drug	Preparation	T_g (°C)	Measurement Heating Rate	Ref.
50:50 7000–17,000	Around 200 Around 180 Around 170 Around 190	None Atorvastatin None Atorvastatin	STM [1] STM SUM [2] SUM	39.35 42.49 30.24 35.02	DSC 10 °C/min	[85]
50:50 54,000–69,000	Around 240 Around 230 Around 225 Around 180	None Atorvastatin None Atorvastatin	STM STM SUM SUM	47.66 47.62 25.98 28.00	DSC 10 °C/min	[85]
85:15 Unknown	391+/−160	Menthol	W/O/W	48.0	DSC 10 °C/min	[86]
75:25 14,000 [3]	162+/−3	None	Emulsion-evaporation	32.7+/−0.2	DSC 5 °C/min	[53]
75:25 32,000 [3]	155+/−5	None	Emulsion-evaporation	37.6+/−0.2	DSC 5 °C/min	[53]
75:25 32,000 [4]	213+/−18	None	Emulsion-evaporation	37.2+/−0.4	DSC 5 °C/min	[53]
75:25 14,000 [4]	238+/−18	None	Emulsion-evaporation	24.8+/−0.6	DSC 5 °C/min	[53]
50:50 38,000–54,000	Unknown Unknown	Enrofloxacin None	Emulsification-diffusion	32.9+/−0.8 31.26	MDSC 5 °C/min	[87]
62:38 18,400	282+/−43	Insulin	Emulsification-diffusion	43.14	DSC 10 °C/min	[88]
50:50 Unknown	211.9+/−2 170.9+/−2.1 256.3+/−9.4	Abiraterone acetate Docetaxel Abiraterone acetate/Docetaxel	Modified single emulsion	45.64 45.93 46.61	DSC 5 °C/min	[89]
50:50 Unknown	179+/−13 123+/−4	Rutin Rutin	Single solvent evaporation Microfluidics	46.19 44.03	DSC 5 °C/min	[33]
75:25 Unknown	Unknown	Simvastatin	Emulsion solvent evaporation	51.5	DSC 10 °C/min	[90]
Unknown	226.8+/−6.8 224.2+/−5.3 222.8+/−4.8 216.0+/−3.8 223.3+/−11.7 237.4+/−9.1	Flurbiprofen Flurbiprofen Flurbiprofen Flurbiprofen Flurbiprofen mTHPP	Emulsion diffusion	28.8+/−0.6 26.9+/−0.5 25.3+/−1.1 22.4+/−1.5 19.9+/−1.6 32.4+/−1.1	DSC 20 °C/min	[34]

[1] Standard method: modified emulsion diffusion evaporation method. [2] Sustainable method: modified solvent displacement method. [3] PLGA polymer has acid terminal functional groups. [4] PLGA polymer has ester terminal functional groups.

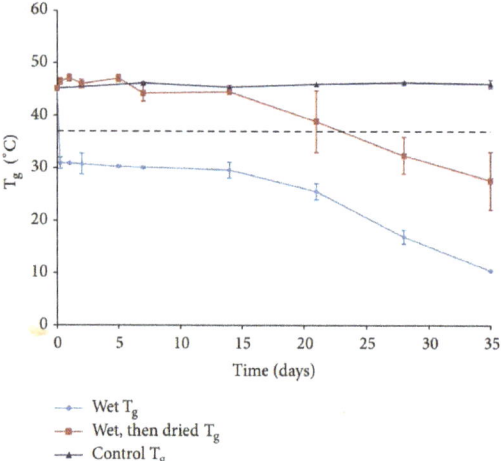

Figure 8. T_g of PLGA under different post-treatment [93]. Reproduced from D'Souza, S. et al., Advances in Biomaterials 2014, 2014, under Creative Commons Attribution 4.0 International License (http://creativecommons.org/licenses/by/4.0/ (accessed on 13 February 2022)).

3. Influence of T_g on Drug Delivery

3.1. Particle Mobility

Takeuchi et al. examined the effects of the glass transition temperature on drug release behavior of drug loaded PLGA nanoparticles (Figure 9). 200 nm PLGA and PLLGA nanoparticles were prepared with a T_g of 40.6 °C and 47.7 °C, respectively. The in vitro drug release study was carried out by dispersing the drug-loaded nanoparticles into PBS at 37 °C. More than 90% of the drug was released from the PLGA nanoparticles in the first two hours whereas only around 65% of the drug was released from PLLGA. The T_g of PLLGA was 7 °C higher than PLGA and the crystallinity of these two samples was similar, which demonstrated that the T_g strongly influenced the initial burst release [35].

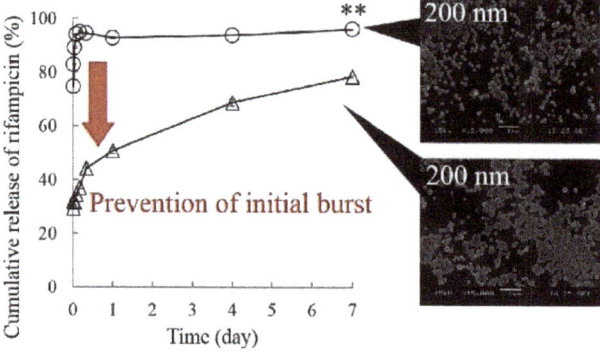

○: Poly (DL-lactide-co-glycolide) nanoparticles
△: Poly (L-lactide-co-glycolide) nanoparticles

In vitro release of rifampicin from nanoparticles at 37°C.

Figure 9. Effect of T_g on drug release profiles [35]. Reprinted from Colloids and Surfaces A: Physicochemical and Engineering Aspects, 520, Takeuchi, I.; Tomoda, K.; Hamano, A.; Makino, K., Effects of physicochemical properties of poly(lactide-co-glycolide) on drug release behavior of hydrophobic drug-loaded nanoparticles, 771–778, Copyright (2017), with permission from Elsevier.

Lappe et al. studied the correlation between T_g and the release profile kinetics by comparing the release of two model drugs from PLGA nanoparticles at different temperatures (Figure 10). At the initial incubation temperature of 37 °C, the FBP-NPs reached around 93% drug release within a short time. Even after shifting the release medium temperature to 10 °C, the released amount of drug was constant at 19%. When the starting temperature was 10°C, only 70% of the drug was released after the first 24 h of incubation time, and an addition of 23% drug release was observed upon the changing the temperature to 37 °C. mTHPP-NPs had the same release behavior while the total amount of released drug was lower than FBP particles, which was mainly caused by the drug type. This study demonstrated that when the release medium temperature is lower than the T_g of the nanoparticles, only the drug absorbed on the particle surface led to burst release, while at a higher temperature, the entrapped drug would also contribute to the burst release [34].

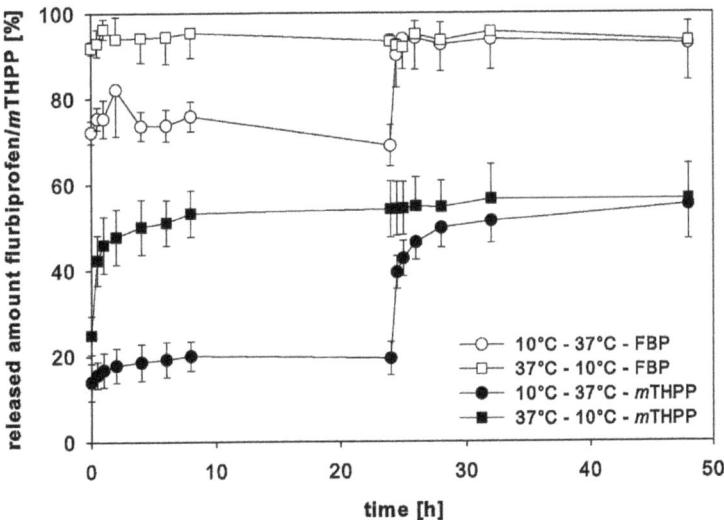

Figure 10. Release behaviors of drug loaded PLGA nanoparticles at different release temperatures [34]. Reprinted from International Journal of Pharmaceutics, 517, Lappe, S.; Mulac, D.; Langer, K., Polymeric nanoparticles—Influence of the glass transition temperature on drug release, 338–347, Copyright (2017), with permission from Elsevier.

3.2. Physical Ageing of Particles

Polymers are naturally non-equilibrium substances when they are in glassy states. During the process of cooling or solidification, polymer chains will reach a threshold where the thermal energy is inadequate for the polymer chains to rearrange on the given time scales [98]. As a result, the system loses its equilibrium and becomes arrested. The temperature where the glassy state develops is a cooling rate dependent parameter. Theoretically, the equilibrium state of a polymer can be achieved with an infinitely low cooling rate. Polymer in the glassy state, a non-equilibrium state, experiences a gradual relaxing process in order to achieve an equilibrium state, which is referred to as physical ageing or structural relaxation [99]. According to Figure 11, free volume decreases along with structural relaxation of the polymer (path A to B). As shown in Figure 12, in response to this phenomenon, the overall polymer matrix experiences shrinkage and micro spaces are created by rearranging of the local chains, which results in wide distribution of local density and helps the diffusion of water into the polymer matrix. Therefore, ageing time is very important to determine the penetration of water into polymeric particles. The time (t_∞) needed to achieve thermodynamic equilibrium can be described in following equation [100]:

$$t_\infty \sim 100 \times 10^{T_g - \frac{T}{3}} = 100 \times e^{1.77(T_g - T)} \qquad (3)$$

According to the equation, ageing time is associated with the difference between T_g of a given polymer and ageing temperature (T).

The extraction of solvent and solidification of PLGA chains during particle preparation is comparable to the process of quenching PLGA polymers [101,102]. Faster extraction of solvent leads to quicker rearrangement of PLGA chains; thus, particles will have more internal energy compared with those with slower extraction processes. The extra internal energy of PLGA particles is the driving force to relax the system toward thermodynamic equilibrium [103]. As mentioned before, when PLGA particles are placed in human body fluids, structural relaxation can be completed within a short time when the T_g is close to 37 °C, which indicates that the micro spaces are created immediately after the administration of drug loaded PLGA particles. The improved water penetration will carry more drug molecules and enhance the diffusion, which leads to the initial burst release. Therefore, Kinam et al. stated that glass transition temperature and time it takes to complete structural relaxation significantly influence drug release behavior at the early stage.

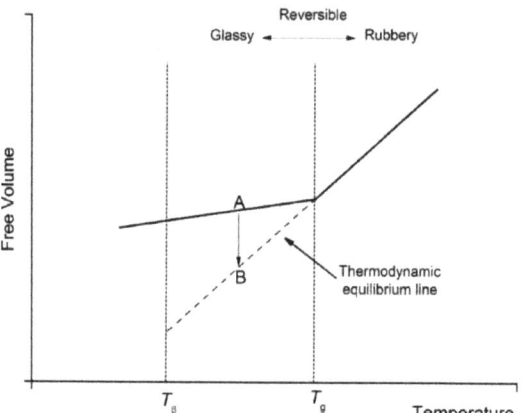

Figure 11. Explanation of structure relaxation in terms of free volume [104]. Reproduced from Motta Dias, M.H. et al., Mechanics of Time-Dependent Materials, 20, 2016, under Creative Commons Attribution 4.0 International License (http://creativecommons.org/licenses/by/4.0/ (accessed on 13 February 2022)).

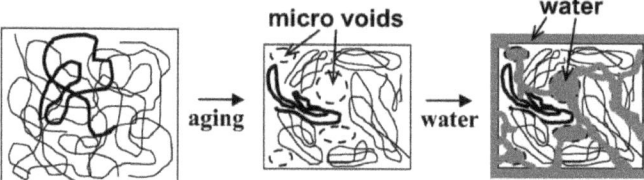

Figure 12. Appearance of micro spaces during polymer ageing process [102]. Reproduced with permission from Yoshioka, T. et al. Macromolecular Materials and Engineering, published by John Wiley and Sons, Copyright 2011.

3.3. Surface Reconfiguration

Hydration is the first and most important phase in the drug release process since water is required for drug disintegration and diffusion via the drug delivery system, whether in the form of biologic fluid or in vitro release medium [105]. Studies have shown that water penetration into PLGA particles can be completed within seconds through the porous structure [106]. The appearance of micro spaces during physical ageing of PLGA particles improves water absorption, resulting in an initial burst release of the drug. Water, on the other hand, acts as a plasticizer which decreases the glass transition temperature of PLGA particles, thus PLGA particles would be softer than the dry polymer state [92]. In addition, upon placement in release medium, a mobile layer forms on the particle surface, further softening the structure. All these plasticizing effects lead to surface reconfiguration, which closes the surface channels and inhibits the diffusion of drug and water penetration. This could explain the observed relative slow release rate following the initial burst [107].

4. Conclusions

PLGA-based nanoparticles have received a great deal of interest as drug delivery vehicles for a variety of therapeutic purposes. It has been shown that the T_g of polymeric nanoparticles has an impact on the drug release behavior, despite the fact that this physicochemical feature is often absent from many pharmaceutical research investigations. This review provides a comprehensive summary of variables affecting the T_g of the PLGA copolymer, including molecular weight and monomer ratio. Additionally, research with PS particles was highlighted to demonstrate the size effect on the T_g of polymeric particles and how that could affect PLGA particles. Drug type, moisture content, and residual surfactant are considerable parameters in altering T_g during drug release processes. Finally, the connection of drug release and glass transition temperature are illustrated by three different aspects, which are the mobility of PLGA particles, ageing time on structural relaxation, and surface reconfiguration. The investigation into the effect of T_g on drug release reveals that the T_g of PLGA particles may account for the majority of the drug release profiles observed. In summary, the glass transition temperature, as an excellent indicator of drug release profiles, could be utilized in manufacturing PLGA particles for designed controlled drug release (Figure 13).

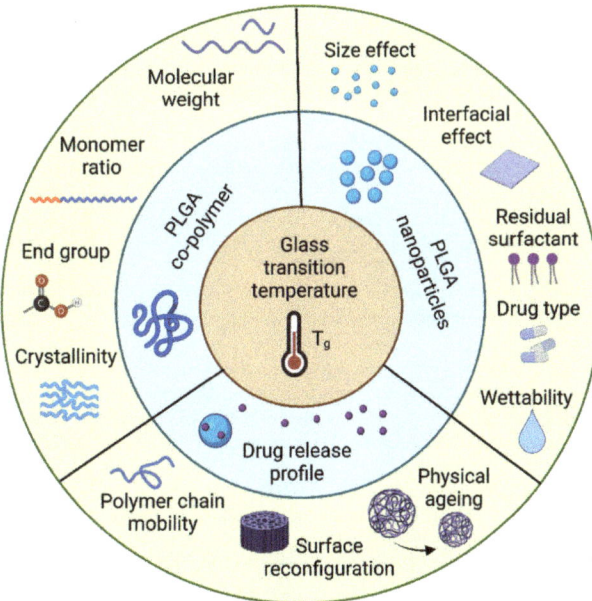

Figure 13. T_g is an inherent property of PLGA nanoparticles that can predict the drug release profiles. Created with Biorender.com.

Looking forward in the field, the T_g of particles for drug delivery should be reported in literature as it is critical to the behavior of the particles. Additionally, the factors affecting the T_g should also be routinely reported to bring reproducibility to the particle synthesis process and consistent behavior in drug delivery applications.

Author Contributions: Writing original draft preparation G.L., review editing and supervision K.M. All authors have read and agreed to the published version of the manuscript.

Funding: This research received no external funding.

Conflicts of Interest: The authors declare no conflict of interest.

References

1. Zeng, X.; Tao, W.; Meiab, L.; Huangab, L.; Tanab, C.; Fengabc, S.S. Cholic acid-functionalized nanoparticles of star-shaped PLGA-vitamin E TPGS copolymer for docetaxel delivery to cervical cancer. *Biomaterials* **2013**, *34*, 6058–6067. [CrossRef] [PubMed]
2. Cheng, C.-Y.; Pho, Q.-H.; Wu, X.-Y.; Chin, T.-Y.; Chen, C.-M.; Fang, P.-H.; Lin, Y.-C.; Hsieh, M.-F. PLGA Microspheres Loaded with β-Cyclodextrin Complexes of Epigallocatechin-3-Gallate for the Anti-Inflammatory Properties in Activated Microglial Cells. *Polymers* **2018**, *10*, 519. [CrossRef] [PubMed]
3. Shi, N.-Q.; Zhou, J.; Walker, J.; Li, L.; Hong, J.K.; Olsen, K.F.; Tang, J.; Ackermann, R.; Wang, Y.; Qin, B.; et al. Microencapsulation of luteinizing hormone-releasing hormone agonist in poly (lactic-co-glycolic acid) microspheres by spray-drying. *J. Control Release* **2020**, *321*, 756–772. [CrossRef] [PubMed]
4. Xu, K.; An, N.; Zhang, H.; Zhang, Q.; Zhang, K.; Hu, X.; Wu, Y.; Wu, F.; Xiao, J.; Zhang, H.; et al. Sustained-release of PDGF from PLGA microsphere embedded thermo-sensitive hydrogel promoting wound healing by inhibiting autophagy. *J. Drug Deliv. Sci. Technol.* **2019**, *55*, 101405. [CrossRef]
5. Zhai, J.; Wang, Y.-E.; Zhou, X.; Ma, Y.; Guan, S. Long-term sustained release Poly(lactic-co-glycolic acid) microspheres of asenapine maleate with improved bioavailability for chronic neuropsychiatric diseases. *Drug Deliv.* **2020**, *27*, 1283–1291. [CrossRef] [PubMed]
6. Zhang, C.; Yang, L.; Wan, F.; Bera, H.; Cun, D.; Rantanen, J.; Yang, M. Quality by design thinking in the development of long-acting injectable PLGA/PLA-based microspheres for peptide and protein drug delivery. *Int. J. Pharm.* **2020**, *585*, 119441. [CrossRef]
7. Lu, Y.; Wu, F.; Duan, W.; Mu, X.; Fang, S.; Lu, N.; Zhou, X.; Kong, W. Engineering a "PEG-g-PEI/DNA nanoparticle-in- PLGA microsphere" hybrid controlled release system to enhance immunogenicity of DNA vaccine. *Mater. Sci. Eng. C* **2019**, *106*, 110294. [CrossRef]
8. Lee, P.W.; Pokorski, J.K. Poly(lactic-co-glycolic acid) devices: Production and applications for sustained protein delivery. *Wiley Interdiscip. Rev. Nanomed. Nanobiotechnol.* **2018**, *10*, e1516. [CrossRef]
9. Shakeri, S.; Ashrafizadeh, M.; Zarrabi, A.; Roghanian, R.; Afshar, E.G.; Pardakhty, A.; Mohammadinejad, R.; Kumar, A.; Thakur, V.K. Multifunctional Polymeric Nanoplatforms for Brain Diseases Diagnosis, Therapy and Theranostics. *Biomedicines* **2020**, *8*, 13. [CrossRef]
10. Desai, N. Challenges in Development of Nanoparticle-Based Therapeutics. *AAPS J.* **2012**, *14*, 282–295. [CrossRef]
11. Zhai, P.; Chen, X.; Schreyer, D.J. PLGA/alginate composite microspheres for hydrophilic protein delivery. *Mater. Sci. Eng. C* **2015**, *56*, 251–259. [CrossRef]
12. Ozeki, T.; Kaneko, D.; Hashizawa, K.; Imai, Y.; Tagami, T.; Okada, H. Improvement of survival in C6 rat glioma model by a sustained drug release from localized PLGA microspheres in a thermoreversible hydrogel. *Int. J. Pharm.* **2012**, *427*, 299–304. [CrossRef] [PubMed]
13. Herrmann, V.L.; Hartmayer, C.; Planz, O.; Groettrup, M. Cytotoxic T cell vaccination with PLGA microspheres interferes with influenza A virus replication in the lung and suppresses the infectious disease. *J. Control. Release* **2015**, *216*, 121–131. [CrossRef] [PubMed]
14. Alange, V.V.; Birajdar, R.P.; Kulkarni, R.V. Functionally modified polyacrylamide-graft-gum karaya pH-sensitive spray dried microspheres for colon targeting of an anti-cancer drug. *Int. J. Biol. Macromol.* **2017**, *102*, 829–839. [CrossRef] [PubMed]
15. Terry, T.L.; Givens, B.E.; Rodgers, V.; Salem, A.K. Tunable Properties of Poly-DL-Lactide-Monomethoxypolyethylene Glycol Porous Microparticles for Sustained Release of Polyethylenimine-DNA Polyplexes. *AAPS PharmSciTech* **2019**, *20*, 23. [CrossRef] [PubMed]
16. Ni, G.; Yang, G.; He, Y.; Li, X.; Du, T.; Xu, L.; Zhou, S. Uniformly sized hollow microspheres loaded with polydopamine nanoparticles and doxorubicin for local chemo-photothermal combination therapy. *Chem. Eng. J.* **2020**, *379*. [CrossRef]
17. Yang, H.; Yang, Y.; Li, B.-Z.; Adhikari, B.; Wang, Y.; Huang, H.-L.; Chen, D. Production of protein-loaded starch microspheres using water-in-water emulsion method. *Carbohydr. Polym.* **2019**, *231*, 115692. [CrossRef]
18. Zhou, X.; Hou, C.; Chang, T.-L.; Zhang, Q.; Liang, J.F. Controlled released of drug from doubled-walled PVA hydrogel/PCL microspheres prepared by single needle electrospraying method. *Colloids Surf. B Biointerfaces* **2019**, *187*, 110645. [CrossRef]
19. Pack, D.W.; Hoffman, A.S.; Pun, S.; Stayton, P. Design and development of polymers for gene delivery. *Nat. Rev. Drug Discov.* **2005**, *4*, 581–593. [CrossRef]
20. Grigsby, C.; Leong, K.W. Balancing protection and release of DNA: Tools to address a bottleneck of non-viral gene delivery. *J. R. Soc. Interface* **2009**, *7*, S67–S82. [CrossRef]

21. Putney, S.D.; Burke, P.A. Improving protein therapeutics with sustained-release formulations. *Nat. Biotechnol.* **1998**, *16*, 153–157. [CrossRef] [PubMed]
22. Lerch, S.; Dass, M.; Musyanovych, A.; Landfester, K.; Mailänder, V. Polymeric nanoparticles of different sizes overcome the cell membrane barrier. *Eur. J. Pharm. Biopharm.* **2013**, *84*, 265–274. [CrossRef] [PubMed]
23. Wohlfart, S.; Gelperina, S.; Kreuter, J. Transport of drugs across the blood–brain barrier by nanoparticles. *J. Control. Release* **2012**, *161*, 264–273. [CrossRef]
24. Wang, D.; Wu, L.-P. Nanomaterials for delivery of nucleic acid to the central nervous system (CNS). *Mater. Sci. Eng. C* **2017**, *70*, 1039–1046. [CrossRef] [PubMed]
25. Moradian, H.; Keshvari, H.; Fasehee, H.; Dinarvand, R.; Faghihi, S. Combining NT3-overexpressing MSCs and PLGA microcarriers for brain tissue engineering: A potential tool for treatment of Parkinson's disease. *Mater. Sci. Eng. C* **2017**, *76*, 934–943. [CrossRef] [PubMed]
26. Li, Z.; Liu, X.; Chen, X.; Chua, M.X.; Wu, Y.-L. Targeted delivery of Bcl-2 conversion gene by MPEG-PCL-PEI-FA cationic copolymer to combat therapeutic resistant cancer. *Mater. Sci. Eng. C* **2017**, *76*, 66–72. [CrossRef]
27. Cheng, H.; Wu, Z.; Wu, C.; Wang, X.; Liow, S.S.; Li, Z.; Wu, Y.-L.; Cheng, H.; Wu, Z.; Wu, C.; et al. Overcoming STC2 mediated drug resistance through drug and gene co-delivery by PHB-PDMAEMA cationic polyester in liver cancer cells. *Mater. Sci. Eng. C* **2018**, *83*, 210–217. [CrossRef] [PubMed]
28. Acharya, S.; Sahoo, S.K. PLGA nanoparticles containing various anticancer agents and tumour delivery by EPR effect. *Adv. Drug Deliv. Rev.* **2011**, *63*, 170–183. [CrossRef]
29. Peres, C.; de Matos, A.I.N.; Conniot, J.; Sainz, V.; Zupančič, E.; Silva, J.M.; Graca, L.; Gaspar, R.; Préat, V.; Florindo, H.F. Poly(lactic acid)-based particulate systems are promising tools for immune modulation. *Acta Biomater.* **2017**, *48*, 41–57. [CrossRef]
30. Su, Y.; Zhang, B.; Sun, R.; Liu, W.; Zhu, Q.; Zhang, X.; Wang, R.; Chen, C. PLGA-based biodegradable microspheres in drug delivery: Recent advances in research and application. *Drug Deliv.* **2021**, *28*, 1397–1418. [CrossRef]
31. Kumari, A.; Yadav, S.K.; Yadav, S.C. Biodegradable polymeric nanoparticles based drug delivery systems. *Colloids Surf. B Biointerfaces* **2010**, *75*, 1–18. [CrossRef] [PubMed]
32. Karnik, R.; Gu, F.; Basto, P.; Cannizzaro, C.; Dean, L.; Kyei-Manu, W.; Langer, R.; Farokhzad, O.C. Microfluidic Platform for Controlled Synthesis of Polymeric Nanoparticles. *Nano Lett.* **2008**, *8*, 2906–2912. [CrossRef] [PubMed]
33. Vu, H.T.; Streck, S.; Hook, S.M.; McDowell, A. Utilization of Microfluidics for the Preparation of Polymeric Nanoparticles for the Antioxidant Rutin: A Comparison with Bulk Production. *Pharm. Nanotechnol.* **2019**, *7*, 469–483. [CrossRef] [PubMed]
34. Lappe, S.; Mulac, D.; Langer, K. Polymeric nanoparticles—Influence of the glass transition temperature on drug release. *Int. J. Pharm.* **2017**, *517*, 338–347. [CrossRef] [PubMed]
35. Takeuchi, I.; Tomoda, K.; Hamano, A.; Makino, K. Effects of physicochemical properties of poly(lactide-co-glycolide) on drug release behavior of hydrophobic drug-loaded nanoparticles. *Colloids Surf. A Physicochem. Eng. Asp.* **2017**, *520*, 771–778. [CrossRef]
36. Takeuchi, I.; Yamaguchi, S.; Goto, S.; Makino, K. Drug release behavior of hydrophobic drug-loaded poly(lactide-co-glycolide) nanoparticles: Effects of glass transition temperature. *Colloids Surf. A Physicochem. Eng. Asp.* **2017**, *529*, 328–333. [CrossRef]
37. Park, K.; Otte, A.; Sharifi, F.; Garner, J.; Skidmore, S.; Park, H.; Jhon, Y.K.; Qin, B.; Wang, Y. Formulation composition, manufacturing process, and characterization of poly(lactide-co-glycolide) microparticles. *J. Control Release* **2020**, *329*, 1150–1161. [CrossRef]
38. Djemour, A.; Sanctuary, R.; Baller, J. Mobility restrictions and glass transition behaviour of an epoxy resin under confinement. *Soft Matter* **2015**, *11*, 2683–2690. [CrossRef]
39. Huang, M.; Tunnicliffe, L.B.; Thomas, A.G.; Busfield, J.J. The glass transition, segmental relaxations and viscoelastic behaviour of particulate-reinforced natural rubber. *Eur. Polym. J.* **2015**, *67*, 232–241. [CrossRef]
40. Samith, V.; Ramos-Moore, E. Study of glass transition in functionalized poly(itaconate)s by differential scanning calorimetry, Raman spectroscopy and thermogravimetric analysis. *J. Non Cryst. Solids* **2015**, *408*, 37–42. [CrossRef]
41. Raczkowska, J.; Stetsyshyn, Y.; Awsiuk, K.; Lekka, M.; Marzec, M.; Harhay, K.; Ohar, H.; Ostapiv, D.; Sharan, M.; Yaremchuk, I.; et al. Temperature-responsive grafted polymer brushes obtained from renewable sources with potential application as substrates for tissue engineering. *Appl. Surf. Sci.* **2017**, *407*, 546–554. [CrossRef]
42. Stetsyshyn, Y.; Raczkowska, J.; Lishchynskyi, O.; Awsiuk, K.; Zemla, J.; Dąbczyński, P.; Kostruba, A.; Harhay, K.; Ohar, H.; Orzechowska, B. Glass transition in temperature-responsive poly(butyl methacrylate) grafted polymer brushes. Impact of thickness and temperature on wetting, morphology, and cell growth. *J. Mater. Chem. B* **2018**, *6*, 1613–1621. [CrossRef] [PubMed]
43. Fernández-Carballido, A.; Puebla, P.; Herrero-Vanrell, R.; Pastoriza, P. Radiosterilisation of indomethacin PLGA/PEG-derivative microspheres: Protective effects of low temperature during gamma-irradiation. *Int. J. Pharm.* **2006**, *313*, 129–135. [CrossRef] [PubMed]
44. Carrascosa, C.; Espejo, L.; Torrado, S.; Torrado, J.J. Effect of c-Sterilization Process on PLGA Microspheres Loaded with Insulin-Like Growth Factor-I (IGF-I). *J. Biomater. Appl.* **2003**, *18*, 95–108. [CrossRef]
45. Mir, M.; Ahmed, N.; Rehman, A.U. Recent applications of PLGA based nanostructures in drug delivery. *Colloids Surf. B Biointerfaces* **2017**, *159*, 217–231. [CrossRef]
46. Xu, Y.; Kim, C.-S.; Saylor, D.M.; Koo, D. Polymer degradation and drug delivery in PLGA-based drug-polymer applications: A review of experiments and theories. *J. Biomed. Mater. Res. Part B Appl. Biomater.* **2016**, *105*, 1692–1716. [CrossRef]

47. Ding, D.; Zhu, Q. Recent advances of PLGA micro/nanoparticles for the delivery of biomacromolecular therapeutics. *Mater. Sci. Eng. C* **2018**, *92*, 1041–1060. [CrossRef]
48. Rezvantalab, S.; Drude, N.; Moraveji, M.K.; Güvener, N.; Koons, E.K.; Shi, Y.; Lammers, T.; Kiessling, F. PLGA-Based Nanoparticles in Cancer Treatment. *Front. Pharmacol.* **2018**, *9*, 1260. [CrossRef]
49. Ghitman, J.; Biru, E.I.; Stan, R.; Iovu, H. Review of hybrid PLGA nanoparticles: Future of smart drug delivery and theranostics medicine. *Mater. Des.* **2020**, *193*, 108805. [CrossRef]
50. Cunha, A.; Gaubert, A.; Latxague, L.; Dehay, B. PLGA-Based Nanoparticles for Neuroprotective Drug Delivery in Neurodegenerative Diseases. *Pharmaceutics* **2021**, *13*, 1042. [CrossRef]
51. Erbetta, C.D.C. Synthesis and Characterization of Poly(D,L-Lactide-co-Glycolide) Copolymer. *J. Biomater. Nanobiotechnol.* **2012**, *3*, 208–225. [CrossRef]
52. Butreddy, A.; Gaddam, R.P.; Kommineni, N.; Dudhipala, N.; Voshavar, C. PLGA/PLA-Based Long-Acting Injectable Depot Microspheres in Clinical Use: Production and Characterization Overview for Protein/Peptide Delivery. *Int. J. Mol. Sci.* **2021**, *22*, 8884. [CrossRef] [PubMed]
53. Robin, B.; Albert, C.; Beladjine, M.; Legrand, F.-X.; Geiger, S.; Moine, L.; Nicolas, V.; Canette, A.; Trichet, M.; Tsapis, N.; et al. Tuning morphology of Pickering emulsions stabilised by biodegradable PLGA nanoparticles: How PLGA characteristics influence emulsion properties. *J. Colloid Interface Sci.* **2021**, *595*, 202–211. [CrossRef] [PubMed]
54. Streck, S.; Neumann, H.; Nielsen, H.M.; Rades, T.; McDowell, A. Comparison of bulk and microfluidics methods for the formulation of poly-lactic-co-glycolic acid (PLGA) nanoparticles modified with cell-penetrating peptides of different architectures. *Int. J. Pharm. X* **2019**, *1*, 100030. [CrossRef] [PubMed]
55. Makadia, H.K.; Siegel, S.J. Poly lactic-co-glycolic acid (PLGA) As biodegradable controlled drug delivery carrier. *Polymers* **2011**, *3*, 1377–1397. [CrossRef] [PubMed]
56. Xie, S.; Wang, S.; Zhu, L.; Wang, F.; Zhou, W. The effect of glycolic acid monomer ratio on the emulsifying activity of PLGA in preparation of protein-loaded SLN. *Colloids Surf. B Biointerfaces* **2009**, *74*, 358–361. [CrossRef]
57. Park, P.I.P.; Jonnalagadda, S. Predictors of glass transition in the biodegradable poly-lactide and poly-lactide-co-glycolide polymers. *J. Appl. Polym. Sci.* **2006**, *100*, 1983–1987. [CrossRef]
58. Brostow, W.; Chiu, R.; Kalogeras, I.M.; Vassilikou-Dova, A. Prediction of glass transition temperatures: Binary blends and copolymers. *Mater. Lett.* **2008**, *62*, 3152–3155. [CrossRef]
59. Koerner, J.; Horvath, D.; Groettrup, M. Harnessing Dendritic Cells for Poly (D,L-lactide-co-glycolide) Microspheres (PLGA MS)—Mediated Anti-tumor Therapy. *Front. Immunol.* **2019**, *10*, 707. [CrossRef]
60. Hsu, M.-Y.; Feng, C.-H.; Liu, Y.-W.; Liu, S.-J. An Orthogonal Model to Study the Effect of Electrospraying Parameters on the Morphology of poly (d,l)-lactide-co-glycolide (PLGA) Particles. *Appl. Sci.* **2019**, *9*, 1077. [CrossRef]
61. Fox, T.G., Jr.; Flory, P.J. Second-order transition temperatures and related properties of polystyrene. I. Influence of molecular weight. *J. Appl. Phys.* **1950**, *21*, 581–591. [CrossRef]
62. Zeng, X.M.; Martin, G.P.; Marriott, C. Effects of molecular weight of polyvinylpyrrolidone on the glass transition and crystallization of co-lyophilized sucrose. *Int. J. Pharm.* **2001**, *218*, 63–73. [CrossRef]
63. Lee, J.S.; Chae, G.S.; Khang, G.; Kim, M.S.; Cho, S.H.; Lee, H.B. The effect of gamma irradiation on PLGA and release behavior of BCNU from PLGA wafer. *Macromol. Res.* **2003**, *11*, 352–356. [CrossRef]
64. Mizuno, A.; Mitsuiki, M.; Motoki, M. Effect of Crystallinity on the Glass Transition Temperature of Starch. *J. Agric. Food Chem.* **1998**, *46*, 98–103. [CrossRef]
65. Kawai, K.; Fukami, K.; Thanatuksorn, P.; Viriyarattanasak, C.; Kajiwara, K. Effects of moisture content, molecular weight, and crystallinity on the glass transition temperature of inulin. *Carbohydr. Polym.* **2011**, *83*, 934–939. [CrossRef]
66. Jiang, X.; Yang, C.Z.; Tanaka, K.; Takahara, A.; Kajiyama, T. Effect of chain end group on surface glass transition temperature of thin polymer film. *Phys. Lett. A* **2001**, *281*, 363–367. [CrossRef]
67. Keddie, J.L.; Jones, R.A.; Cory, R.A. Size-Dependent Depression of the Glass Transition Temperature in Polymer Films. *Eur. Lett.* **1994**, *27*, 59–64. [CrossRef]
68. Raegen, A.; Massa, M.V.; Forrest, J.A.; Dalnoki-Veress, K. Effect of atmosphere on reductions in the glass transition of thin polystyrene films. *Eur. Phys. J. E* **2008**, *27*, 375–377. [CrossRef] [PubMed]
69. Zhang, C.; Guo, Y.; Priestley, R.D. Glass Transition Temperature of Polymer Nanoparticles under Soft and Hard Confinement. *Macromolecules* **2011**, *44*, 4001–4006. [CrossRef]
70. Christie, D.; Zhang, C.; Fu, J.; Koel, B.; Priestley, R.D. Glass transition temperature of colloidal polystyrene dispersed in various liquids. *J. Polym. Sci. Part B Polym. Phys.* **2016**, *54*, 1776–1783. [CrossRef]
71. Feng, S.; Li, Z.; Liu, R.; Mai, B.; Wu, Q.; Liang, G.; Gao, H.; Zhu, F. Glass transition of polystyrene nanospheres under different confined environments in aqueous dispersions. *Soft Matter* **2013**, *9*, 4614–4620. [CrossRef]
72. Ramazani, F.; Chen, W.; Van Nostrum, C.F.; Storm, G.; Kiessling, F.; Lammers, T.; Hennink, W.E.; Kok, R.J. Strategies for encapsulation of small hydrophilic and amphiphilic drugs in PLGA microspheres: State-of-the-art and challenges. *Int. J. Pharm.* **2016**, *499*, 358–367. [CrossRef] [PubMed]
73. Ali, M.; Walboomers, X.F.; Jansen, J.A.; Yang, F. Influence of formulation parameters on encapsulation of doxycycline in PLGA microspheres prepared by double emulsion technique for the treatment of periodontitis. *J. Drug Deliv. Sci. Technol.* **2019**, *52*, 263–271. [CrossRef]

74. Zhao, J.; Wang, L.; Fan, C.; Yu, K.; Liu, X.; Zhao, X.; Wang, D.; Liu, W.; Su, Z.; Sun, F.; et al. Development of near zero-order release PLGA-based microspheres of a novel antipsychotic. *Int. J. Pharm.* **2017**, *516*, 32–38. [CrossRef] [PubMed]
75. Yang, Z.; Liu, L.; Su, L.; Wu, X.; Wang, Y.; Liu, L.; Lin, X. Design of a zero-order sustained release PLGA microspheres for palonosetron hydrochloride with high encapsulation efficiency. *Int. J. Pharm.* **2019**, *575*, 119006. [CrossRef] [PubMed]
76. Wischke, C.; Schwendeman, S.P. Principles of encapsulating hydrophobic drugs in PLA/PLGA microparticles. *Int. J. Pharm.* **2008**, *364*, 298–327. [CrossRef]
77. Ansary, R.H.; Awang, M.B.; Rahman, M.M. Biodegradable Poly(D,L-lactic-co-glycolic acid)-Based Micro/Nanoparticles for Sustained Release of Protein Drugs—A Review. *Trop. J. Pharm. Res.* **2014**, *13*, 1179. [CrossRef]
78. Barreras-Urbina, C.G.; Ramírez-Wong, B.; López-Ahumada, G.A.; Ibarra, S.E.B.; Martínez-Cruz, O.; Tapia-Hernández, J.A.; Félix, F.R. Nano- and Micro-Particles by Nanoprecipitation: Possible Application in the Food and Agricultural Industries. *Int. J. Food Prop.* **2016**, *19*, 1912–1923. [CrossRef]
79. Morais, A.S.; Vieira, E.G.; Afewerki, S.; Sousa, R.B.; Honorio, L.; Cambrussi, A.N.C.O.; Santos, J.A.; Bezerra, R.D.S.; Furtini, J.A.O.; Silva-Filho, E.C.; et al. Fabrication of Polymeric Microparticles by Electrospray: The Impact of Experimental Parameters. *J. Funct. Biomater.* **2020**, *11*, 4. [CrossRef]
80. Cabezas, L.I.; Gracia, I.; De Lucas, A.; Rodriguez, J.F. Validation of a Mathematical Model for the Description of Hydrophilic and Hydrophobic Drug Delivery from Biodegradable Foams: Experimental and Comparison Using Indomethacin as Released Drug. *Ind. Eng. Chem. Res.* **2014**, *53*, 8866–8873. [CrossRef]
81. Hans, M.; Lowman, A. Biodegradable nanoparticles for drug delivery and targeting. *Curr. Opin. Solid State Mater. Sci.* **2002**, *6*, 319–327. [CrossRef]
82. Lee, L.Y.; Ranganath, S.H.; Fu, Y.; Zheng, J.L.; Lee, H.S.; Wang, C.-H.; Smith, K.A. Paclitaxel release from micro-porous PLGA disks. *Chem. Eng. Sci.* **2009**, *64*, 4341–4349. [CrossRef]
83. Siegel, S.; Kahn, J.B.; Metzger, K.; Winey, K.I.; Werner, K.; Dan, N. Effect of drug type on the degradation rate of PLGA matrices. *Eur. J. Pharm. Biopharm.* **2006**, *64*, 287–293. [CrossRef] [PubMed]
84. Siepmann, F.; Le Brun, V.; Siepmann, J. Drugs acting as plasticizers in polymeric systems: A quantitative treatment. *J. Control. Release* **2006**, *115*, 298–306. [CrossRef] [PubMed]
85. Grune, C.; Zens, C.; Czapka, A.; Scheuer, K.; Thamm, J.; Hoeppener, S.; Jandt, K.D.; Werz, O.; Neugebauer, U.; Fischer, D. Sustainable preparation of anti-inflammatory atorvastatin PLGA nanoparticles. *Int. J. Pharm.* **2021**, *599*, 120404. [CrossRef]
86. Holz, J.P.; Bottene, M.K.; Jahno, V.; Einloft, S.; Ligabue, R. Menthol-loaded PLGA Micro and Nanospheres: Synthesis, Characterization and Degradation in Artificial Saliva. *Mater. Res.* **2018**, *21*, 488. [CrossRef]
87. Shakiba, S.; Astete, C.E.; Cueto, R.; Rodrigues, D.F.; Sabliov, C.M.; Louie, S.M. Asymmetric flow field-flow fractionation (AF4) with fluorescence and multi-detector analysis for direct, real-time, size-resolved measurements of drug release from polymeric nanoparticles. *J. Control Release* **2021**, *338*, 410–421. [CrossRef]
88. Farid, E.A.; Davachi, S.M.; Pezeshki-Modaress, M.; Taranejoo, S.; Seyfi, J.; Hejazi, I.; Hakim, M.T.; Najafi, F.; D'Amico, C.; Abbaspourrad, A. Preparation and characterization of polylactic-co-glycolic acid/insulin nanoparticles encapsulated in methacrylate coated gelatin with sustained release for specific medical applications. *J. Biomater. Sci. Polym. Ed.* **2020**, *31*, 910–937. [CrossRef]
89. Sokol, M.B.; Nikolskaya, E.; Yabbarov, N.; Zenin, V.A.; Faustova, M.R.; Belov, A.V.; Zhunina, O.A.; Mollaev, M.D.; Zabolotsky, A.I.; Tereshchenko, O.G.; et al. Development of novel PLGA nanoparticles with co-encapsulation of docetaxel and abiraterone acetate for a highly efficient delivery into tumor cells. *J. Biomed. Mater. Res. Part B Appl. Biomater.* **2018**, *107*, 1150–1158. [CrossRef]
90. Sedki, M.; Khalil, I.; El-Sherbiny, I.M. Hybrid nanocarrier system for guiding and augmenting simvastatin cytotoxic activity against prostate cancer. *Artif. Cells Nanomed. Biotechnol.* **2018**, *46*, S641–S650. [CrossRef]
91. Levine, H.; Slade, L. Water as a plasticizer: Physico-chemical aspects of low-moisture polymeric systems. In *Water Science Reviews 3: Water Dynamics*; Cambridge University Press: Cambridge, UK, 1988; pp. 79–185. [CrossRef]
92. Passerini, N.; Craig, D. An investigation into the effects of residual water on the glass transition temperature of polylactide microspheres using modulated temperature DSC. *J. Control Release* **2001**, *73*, 111–115. [CrossRef]
93. D'Souza, S.; Dorati, R.; DeLuca, P.P. Effect of Hydration on Physicochemical Properties of End-Capped PLGA. *Adv. Biomater.* **2014**, *2014*, 834942. [CrossRef]
94. Anton, N.; Benoit, J.-P.; Saulnier, P. Design and production of nanoparticles formulated from nano-emulsion templates—A review. *J. Control Release* **2008**, *128*, 185–199. [CrossRef] [PubMed]
95. Bouissou, C.; Rouse, J.J.; Price, R.; Van Der Walle, C.F. The Influence of Surfactant on PLGA Microsphere Glass Transition and Water Sorption: Remodeling the Surface Morphology to Attenuate the Burst Release. *Pharm. Res.* **2006**, *23*, 1295–1305. [CrossRef] [PubMed]
96. Sahoo, S.K.; Panyam, J.; Prabha, S.; Labhasetwar, V. Residual polyvinyl alcohol associated with poly (d,l-lactide-co-glycolide) nanoparticles affects their physical properties and cellular uptake. *J. Control Release* **2002**, *82*, 105–114. [CrossRef]
97. Spek, S.; Haeuser, M.; Schaefer, M.; Langer, K. Characterisation of PEGylated PLGA nanoparticles comparing the nanoparticle bulk to the particle surface using UV/vis spectroscopy, SEC, 1H NMR spectroscopy, and X-ray photoelectron spectroscopy. *Appl. Surf. Sci.* **2015**, *347*, 378–385. [CrossRef]
98. Roth, C.B. *Polymer Glasses*, 1st ed.; CRC Press: Boca Raton, FL, USA, 2016; pp. 4–5.
99. Priestley, R.D.; Ellison, C.J.; Broadbelt, L.J.; Torkelson, J.M. Structural Relaxation of Polymer Glasses at Surfaces, Interfaces, and in between. *Science* **2005**, *309*, 456–459. [CrossRef]

100. Struik, L. *Physical Aging in Amorphous Polymers and Other Materials*; Elsevier: Amsterdam, The Netherlands, 1978.
101. Allison, S.D. Effect Of Structural Relaxation On The Preparation And Drug Release Behavior Of Poly(lactic-co-glycolic)acid Microparticle Drug Delivery Systems. *J. Pharm. Sci.* **2008**, *97*, 2022–2035. [CrossRef]
102. Yoshioka, T.; Kawazoe, N.; Tateishi, T.; Chen, G. Effects of Structural Change Induced by Physical Aging on the Biodegradation Behavior of PLGA Films at Physiological Temperature. *Macromol. Mater. Eng.* **2011**, *296*, 1028–1034. [CrossRef]
103. Brunacci, A.; Cowie, J.; Ferguson, R.; McEwen, I. Enthalpy relaxation in glassy polystyrenes: 1. *Polymer* **1997**, *38*, 865–870. [CrossRef]
104. Dias, M.H.M.; Jansen, K.M.B.; Luinge, J.W.; Bersee, H.E.N.; Benedictus, R. Effect of fiber-matrix adhesion on the creep behavior of CF/PPS composites: Temperature and physical aging characterization. *Mech. Time Depend. Mater.* **2016**, *20*, 245–262. [CrossRef] [PubMed]
105. Blasi, P.; D'Souza, S.S.; Selmin, F.; DeLuca, P.P. Plasticizing effect of water on poly(lactide-co-glycolide). *J. Control Release* **2005**, *108*, 1–9. [CrossRef] [PubMed]
106. Park, K.; Skidmore, S.; Hadar, J.; Garner, J.; Park, H.; Otte, A.; Soh, B.K.; Yoon, G.; Yu, D.; Yun, Y.; et al. Injectable, long-acting PLGA formulations: Analyzing PLGA and understanding microparticle formation. *J. Control Release* **2019**, *304*, 125–134. [CrossRef] [PubMed]
107. Park, K.; Otte, A.; Sharifi, F.; Garner, J.; Skidmore, S.; Park, H.; Jhon, Y.K.; Qin, B.; Wang, Y. Potential Roles of the Glass Transition Temperature of PLGA Microparticles in Drug Release Kinetics. *Mol. Pharm.* **2020**, *18*, 18–32. [CrossRef] [PubMed]

Article

Design of Hybrid Polymeric-Lipid Nanoparticles Using Curcumin as a Model: Preparation, Characterization, and In Vitro Evaluation of Demethoxycurcumin and Bisdemethoxycurcumin-Loaded Nanoparticles

Krissia Wilhelm Romero [1,2], María Isabel Quirós [1], Felipe Vargas Huertas [1], José Roberto Vega-Baudrit [2,3], Mirtha Navarro-Hoyos [1] and Andrea Mariela Araya-Sibaja [2,*]

[1] Laboratorio BIODESS, Escuela de Química, Universidad de Costa Rica, San Pedro de Montes de Oca, San José 2060, Costa Rica; krissia.wilhelm@ucr.ac.cr (K.W.R.); maria.quirosfallas@ucr.ac.cr (M.I.Q.); luis.vargashuertas@ucr.ac.cr (F.V.H.); mnavarro@codeti.org (M.N.-H.)
[2] Laboratorio Nacional de Nanotecnología LANOTEC-CeNAT-CONARE, Pavas, San José 1174-1200, Costa Rica; jvegab@gmail.com
[3] Laboratorio de Investigación y Tecnología de Polímeros POLIUNA, Escuela de Química, Universidad Nacional de Costa Rica, Heredia 86-3000, Costa Rica
* Correspondence: aaraya@cenat.ac.cr; Tel.: +506-2519-5700 (ext. 6016)

Abstract: Polymeric lipid hybrid nanoparticles (PLHNs) are the new generation of drug delivery systems that has emerged as a combination of a polymeric core and lipid shell. We designed and optimized a simple method for the preparation of Pluronic F-127-based PLHNs able to load separately demethoxycurcumin (DMC) and bisdemethoycurcumin (BDM). CUR was used as a model compound due to its greater availability from turmeric and its structure similarity with DMC and BDM. The developed method produced DMC and BDM-loaded PLHNs with a size average of 75.55 ± 0.51 and 15.13 ± 0.014 nm for DMC and BDM, respectively. An FT-IR analysis confirmed the encapsulation and TEM images showed their spherical shape. Both formulations achieved an encapsulation efficiency $\geq 92\%$ and an exhibited significantly increased release from the PLHN compared with free compounds in water. The antioxidant activity was enhanced as well, in agreement with the improvement in water dissolution; obtaining IC_{50} values of 12.74 ± 0.09 and 16.03 ± 0.55 for DMC and BDM-loaded PLHNs, respectively, while free curcuminoids exhibited considerably lower antioxidant values in an aqueous solution. Hence, the optimized PHLN synthesis method using CUR as a model and then successfully applied to obtain DMC and BDM-loaded PLHNs can be extended to curcuminoids and molecules with a similar backbone structure to improve their bioactivities.

Keywords: drug delivery; polymer-lipid hybrid nanoparticles; curcumin; demethoxycurcumin; bisdemethoxycurcumin; Pluronic F-127

1. Introduction

In the last years, polymeric nanoparticles (PNs) have been one of the most studied nanocarriers to achieve drug delivery challenges [1]. This nanosystem possess high structural integrity afforded by the rigidity of the polymer matrix [2]; therefore, presenting advantages such as simple preparation and design, good biocompatibility, broad-structure variety, and notable bio-imitative properties [3]. However, the main drawback of PN is their low encapsulation efficiency of water-soluble drugs due to the fast leakage of the drug from the nanoparticles during the high-energy emulsification step commonly employed in their preparation [4,5]. In this scenario, the new generation of PN, the polymeric-lipid hybrid nanoparticles (PLHNs) emerged as a combination of PN with lipid-derived nanoparticles. It possesses the characteristics and the advantages of both polymer and lipid-based particles [5,6] and because of its hydrophobic matrix and hydrophilic core, it is possible to load both lipid soluble and water-soluble drugs within the same particle [5]. PLHNs consist of a central polymer core that is surrounded by single or multiple lipids [7], in which the therapeutic substances are encapsulated, an inner lipid layer enveloping the polymer core, whose main function is to confer biocompatibility to the polymer core, and an outer lipid-polymeric layer [5,6]. The outer coating acts not only as a barrier toward diffusion but also prolongs the in vivo circulation time of PLHNs in systemic circulation [5]. In addition, the inner lipid layer slows down the polymer degradation rate, and also maintains functions as a molecular fence minimizing leakage of the encapsulated content during the preparation and increasing drug loading efficiency [8]. The typical structure representation of a PLHN is shown in Figure 1.

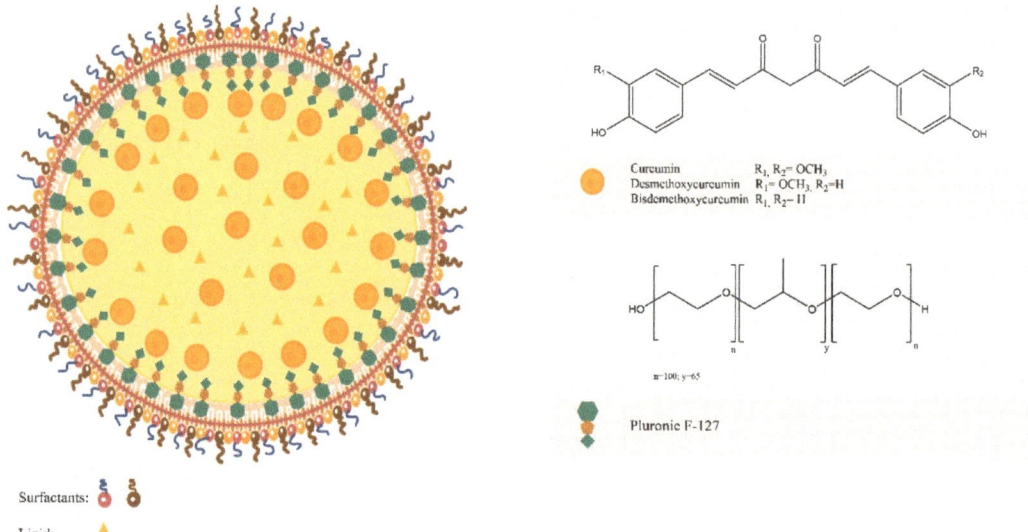

Figure 1. Schematic representation of PLHNs and chemical structure of Pluronic F-127, CUR, DMC, and BDM.

For drug delivery and PN preparation, poloxamers, a polyethylene oxide-polypropylene oxide-polyethylene oxide (PEO-PPO-PEO)-based triblock, commercialized under the Pluronics® trademark is one of the most used copolymers [9,10]. Its monomers can render an amphiphilic character in an aqueous solution based on the PEO solubility in water and the PPO insolubility. The PEO blocks are thus hydrophilic, while the PPO block is hydrophobic [11,12]. Besides leading to surface active properties, the block segregation also gives rise to self-assembly useful nanostructures [11]. Pluronic F-127 (PEO_{100}–PPO_{65}–PEO_{100}) is relatively non-toxic to cells and widely investigated because its great hydrophobic region favors micellization [13]. It is the preferred one among pluronics for drug delivery applications due its high biocompatibility [14] and is approved by FDA as excipient in oral, ophthalmic, and topical medicinal formulations [15]. In terms of formulation of nanoparticles, Pluronic 127 possesses different functional attributes that subsequently influence the core-shell lipid-polymer nanoparticulate system and directly affect the efficiency of the delivery system [9]. Recent studies have shown the positive influence of Pluronic F-127 on the loading enhancement of curcumin [15,16].

Turmeric has a great variety of curcuminoids which has generated interest in the last years due to their recognized bioactivities including antioxidant, anti-inflammatory, anti-microbial, anti-diabetic, and immunomodulatory [17–19]. Among them, bis-demethoxycurcumin (BDM), demethoxycurcumin (DMC) and curcumin (CUR) are the most representative curcuminoids in turmeric. CUR is the major component making up 80%, DMC 17%, and BDM 3% in commercially available crude products [20,21]. Hence, CUR has been the most extensively studied curcuminoid and has been attributed a broad range of biological activities [22–25]. In addition, different types of nanosystems for drug delivery with several therapeutic applications have been applied to CUR [26–29], mainly in an effort to bypass the glucuronidation pathway [17] but also to improve its low solubility and chemical instability [30]. The chemical structures of the three curcuminoids are shown in Figure 1. The similarity in their structures suggests that they may exhibit similar bioactivities. Indeed, there are reports studying the anti-inflammatory and anticancer properties of DMC and BDM [31,32]. In turn, evidence indicates DMC and BDM possess the same drawbacks compromising CUR's bioavailability [33,34]. Hence, these two curcuminoids are promising molecules for the improvement of their limited bioavailability through PLHNs.

In this study, we designed and optimized a simple method for the preparation of Pluronic F-127-based PLHNs able to load DMC and BDM individually using CUR as a model molecule because of its major presence in turmeric. Processing parameters including, ultrasonic probe, high-speed homogenization, mixing phases and homogenization speed were evaluated. Organic and aqueous phase's composition: lipid, drug loading, lipid-drug ratio, organic solvent, surfactant, and polymer amount were investigated. The developed method was applied to obtain DMC and BDM-loaded PLHNs. In addition, the characterization of the nanoparticles using the FT-IR spectroscopy, DLS TEM and DSC, and the evaluation of their in vitro release and antioxidant activity, were performed.

2. Materials and Methods

2.1. Materials

Curcumin (CUR), demethoxycurcumin (DMC), and bisdemethoxycurcumin (BDM) were obtained and isolated from Curcuma longa by BIODESS Laboratory (Costa Rica). CUR, DMC, and BDM analytical standards used in the UHPLC and UV quantification studies as well as Cholesterol (Chol) poloxamer 407 (Pluronic F-127), 2,2-diphenyl-1-picrylhidrazyl (DPPH), dichloromethane (CH_2Cl_2), phosphoric acid (H_3PO_4) and disodium hydrogen phosphate were purchased from Sigma-Aldrich. Sodium dihydrogen phosphate monohydrate was acquired from Merck. Polysorbatum 80 (Tween 80) was purchased from Sonntag and Rote S.A., and sorbitan monooleate (Span® 80) was supplied by LABQUIMAR S.A. Chloroform ($CHCl_3$), methanol (MeOH), and acetonitrile (MeCN) were purchased from JTBaker. All solvents were HPLC/UV grade or highly pure, and the water was purified using a Millipore system filtered through a Millipore membrane 0.22 µm Millipak 40.

2.2. Design and Optimization of the Polymer-Lipid Hybrid Nanoparticles (PLHNs) Method Using CUR as Model

The strategy for the design of the PLHN was to combine one method for preparing solid lipid nanoparticles with another to prepare polymeric nanoparticles. Therefore, the emulsion method reported by Rompicharla et al. 2017 [35] and the emulsification solvent diffusion method reported by Udompornmongkol et al. 2015 [36], respectively, were used as starting methods. The following were the initial parameters: the aqueous solution composed of 5 mg/mL of Pluronic F-127 prepared in acetic acid 0.1% and Tween 80: Span80 1:1 4% was added dropwise into the organic one containing a CUR-lipid ratio of 1:24 in a 1:1 mixture of MeOH: $CHCl_3$. Mixed phases were homogenized at 10,000 rpm for 10 min.

The optimization of the PLHN preparation method consisted of using different components of organic and aqueous phases as well as parameters related to mixing the phases and homogenization techniques for visually obtaining an emulsion and efficient formulations in terms of drug encapsulation. Hence, at each change, if the emulsion formation was visually confirmed, the next step was to evaluate the encapsulation efficiency (EE) both direct (EE_D) and indirect (EE_I) according to Equations (1) and (2), respectively, showed in the Section 2.3.1. Hence, the emulsion was placed in magnetic stirring for 10 min to eliminate the traces of the organic solvents. The particles were collected by ultracentrifugation and the CUR content was determined to calculate EE_I. Then, the emulsion was washed three times with ultrapure water to remove the remaining unencapsulated molecule and unreacted substances. Further, EE_D was calculated by determining the CUR content in 100 µL of the final formulation dissolved in 900 µL of MeOH. A blank of PLHN was prepared using the same parameters tested without adding the CUR in the organic phase.

2.2.1. Homogenization Technique

Two homogenization techniques were evaluated, first, a Cole Palmer Gex 30 Ultrasonic Processor (Cole Palmer, Illinois, IL, USA) operated at 130 watts and 20 KHertz provided with a 3 mm titanium probe. The aqueous and organic phases were mixed in a beaker in a one-time addition, and the solution was sonicated during two different periods of 30 and 45 min at intervals of 5 min. The second one consisted of a high speed ULTRA-TURRAX® T25 homogenizer (IKA, Staufen, Germany). The aqueous solution was added into the organic one by testing three different speeds: dropwise (slow), approximately 3 mL/min (medium), and one-time addition (fast). Further, 10,000, 12,000 and 16,000 rpm homogenization speed during 10 min were tested.

2.2.2. Lipid, Drug Loading, Lipid-Drug Ratio, and Organic Solvent

Chol, cocoa butter, cetyl palmitate, stearic acid, and a 1:1 mixture of Chol: cetyl palmitate were used. Once the lipid was selected, three different amounts of CUR 7.5, 10 and 15 mg as well as 12:1 and 48:1 drug-lipid ratios were tested. The organic phase composition was evaluated by testing 6 mL of the following 1:1 mixture of CH_2Cl_2: MeOH and $CHCl_3$: EtOH.

2.2.3. Surfactant Selection and Polymer Concentration

The composition of aqueous phase was kept constant during the whole optimization method using 5 mg/mL of Pluronic F-127 in acetic acid 0.1%. Tween 80 and Span 80® separately were tested as surfactants. Once selected, 2 mg/mL, and 7 mg/mL of Pluronic F 127 were evaluated.

2.2.4. Preparation, Characterization, and In Vitro Evaluation of DMC and BDM-Loaded PLHNs

A total of 250 mg of Pluronic F-127 were dissolved in 50 mL of acetic acid 0.1% and 2 g of a 1:1 mixture of Tween 80: Span80 1:1 was added. The organic phase was composed of 120 mg of Chol and 5 mg of DMC or BDM dissolved in 6 mL of MeOH: $CHCl_3$ 1:1. Then, the aqueous solution was added into the organic one at medium speed and homogenized at 16,000 rpm for 10 min to form an appropriate emulsion. The nanoparticles were collected by ultracentrifugation using a Thermo Scientific Sorvall ST 16R centrifuge (Thermo Fisher Scientific, Waltham, MA, USA) at 12,000 rpm for 40 min at 10 °C. To remove the remaining unencapsulated substrate and unreacted substances, the emulsion was washed three times with ultrapure water. The final formulation was dispersed in 5 mL of purified water containing 0.01% Tween80®; filtered through an ADVANTEC® ultrafilter unit and refrigerated up to further characterization. Blanks of PLHNs were prepared as mentioned above without adding DMC or BDM in the organic phase. In Figure 2 is presented the schematic representation of the method procedure.

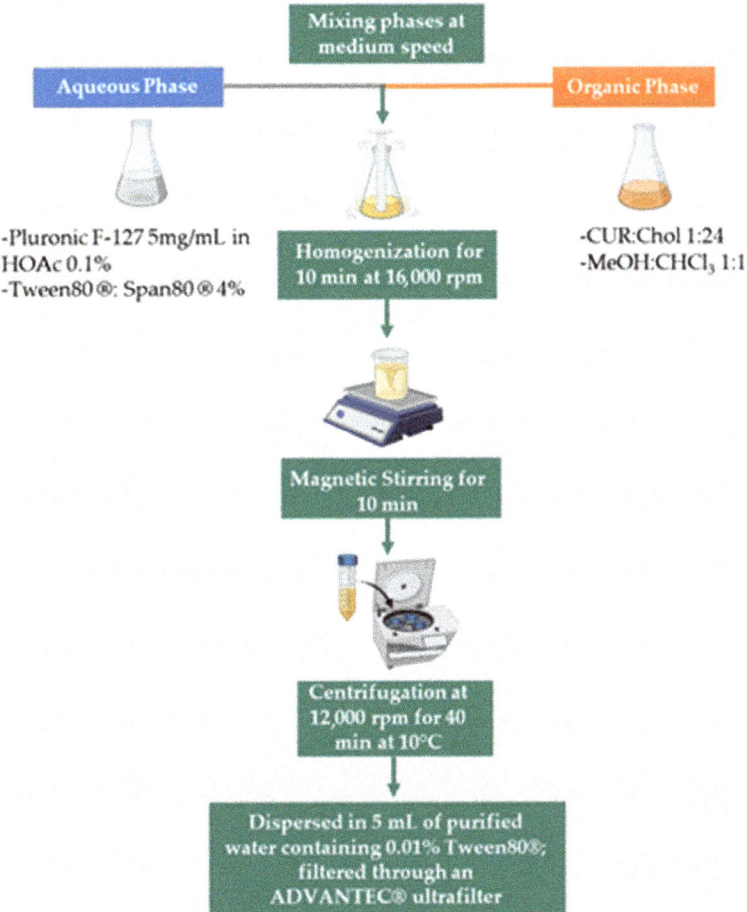

Figure 2. Schematic representation of the developed and optimized method procedure.

2.3. Characterization Techniques

2.3.1. Encapsulation Efficiency (EE)

The EE was calculated through direct and indirect methods using the Equations (1) and (2), respectively. For the direct method the amount of CUR, BDM, or DMC in the three independent formulations of nanoparticles was determined for each curcuminoid by taking 100 µL of fresh PLHNs dissolved in 900 µL of MeOH. Meanwhile, for the indirect method, the amount of free curcuminoid was determined in the supernatant collected by ultracentrifugation using a Thermo Scientific Sorvall ST 16R centrifuge (Thermo Fisher Scientific, Waltham, MA, USA) at 12,000 rpm for 40 min at 10 °C. The solutions for both the direct and indirect methods were filtered through a 0.45-µm cellulose acetate membrane placed in a Sartorius stainless steel syringe filter holder. A total of 10 µL of the samples were injected in a Dionex Ultimate 3000 UHPLC system (Thermo Fisher Scientific, Waltham, MA, USA) equipped with a variable wavelength detector, pump, variable temperature compartment column and autosampler. The chromatographic elution was carried out in a Nucleosil 100-5 C18 column (250 mm × 4.0 mm, 5 µm) at a temperature of 35 °C using 55% of MeCN and 45%

of H_3PO_4 0.1% as mobile phase at a flow rate of 1 mL/min and setting down the detection at 420 nm.

$$EE_D = \frac{\text{Drug in nanoparticle}}{\text{Total drug added}} \times 100 \quad (1)$$

$$EE_I = \frac{\text{Total drug content (mg)} - \text{free drug(mg)}}{\text{Total drug content (mg)}} \times 100 \quad (2)$$

2.3.2. Fourier Transform Infrared (FT-IR)

The FT-IR spectra of the sample were recorded on a Thermo Scientific Nicolet 6700 FT-IR spectrometer (Thermo Fisher Scientific, Waltham, MA, USA) fitted with a diamond attenuated total reflectance (ATR) accessory. The samples were placed directly into the ATR cell without further preparation and analyzed in the range of $4000-600$ cm^{-1}.

2.3.3. Dynamic Light Scattering (DLS)

The particle size (z-average) and polydispersity index (PI) were measured on the basis of the DLS technique on a Malvern Nano Zetasizer ZS90 instrument (Malvern Panalytical, Malvern, UK) using the medium refractive index of 1.33, and viscosity 0.8872 cP under 90°. The samples were diluted with deionized water to achieve the appropriate concentrations, and the measurements were performed at 25 °C.

2.3.4. High Resolution Transmission Electron Microscopy (HR-TEM)

The PLHN morphology was evaluated using a JEOL, JEM2011 HR-TEM (JEOL Ltd., Tokyo, Japan) at an acceleration voltage of 120 kV. The samples were prepared by placing 5 µL of NP suspensions and drying under a nitrogen atmosphere.

2.3.5. Differential Scanning Calorimetry (DSC)

The DSC curves of PLHNs were obtained in a TA Instruments DSC-Q200 calorimeter (TA Instruments, New Castle, DE, USA) equipped with a TA Refrigerated Cooling System 90. Approximately 2 mg of each sample were placed in aluminum pans with lids and the measurement were carried out under a dynamic nitrogen atmosphere of 50 mL/min, a heating rate of 10 °C/min and a temperature range from 40 to 250 °C.

2.3.6. In Vitro Studies

Drug Release Profile

The DMC and BDM in vitro release profile from the PLHN as well as the dissolution profile of pure DMC and BDM (used as a reference) were estimated using two different dissolution media, M1 and M2. Briefly, 1 mL of PLHN was immersed in 80 mL of phosphate buffered saline of pH 7.4, containing MeOH 20% and 2.5% of Tween 80 (M1) [37] and water (M2) maintained at 37 ± 0.5 °C and 150 rpm in a Labnet 211 DS shaking incubator (Labnet International Inc., Edison, NJ, USA). Then 4 mL of each solution were withdrawn at specific time intervals without replacing the volume. The aliquots were centrifuged at 6000 rpm for 10 min in a Thermo Scientific Sorvall ST 16R centrifuge at 37 °C. The concentration of DMC and BDM in the solutions were measured using a Shimadzu 1800 double beam UV-Vis spectrophotometer (Shimadzu Corporation, Tokyo, Japan) at a wavelength of 420 nm. The sampling was performed in triplicate.

DPPH Radical-Scavenging Activity

The DPPH evaluation was performed as previously reported [38], for both free and nanoencapsulated curcuminoids DMC and BDM. The free curcuminoid samples were evaluated in an ethanolic and aqueous solution. This last medium was also used for curcuminoid nanoparticles samples. In addition, Trolox was used as a standard [39] to assess the DPPH method applied. Briefly, a solution of 2,2-diphenyl-1-picrylhidrazyl (DPPH, 0.25 mM) was prepared using EtOH as a solvent. Next, 0.5 mL of this solution were mixed with Trolox or the respective free or nanoencapsulated curcuminoid solution at different concentrations and incubated at 25 °C in the dark for 30 min. The DPPH absorbance was measured at 517 nm using a Thermo Scientific Genesys S10 spectrophotometer (Thermo Fisher Scientific, Waltham, MA, USA). The controls were prepared for each assay using solvent instead of Trolox or samples. The percentage of the radical-scavenging activity inhibition was calculated for each concentration according to Equation (3). This percentage was plotted against the Trolox or samples concentrations to calculate IC_{50}, which is the amount required to reach 50% inhibition of DPPH radical-scavenging activity. The Trolox and samples were analyzed this way in three independent assays.

$$\text{Inhibition percentage} = \frac{(\text{Absorbance of control} - \text{Absorbance of Trolox or sample})}{(\text{Absorbance of control})} \times 100 \quad (3)$$

3. Results

3.1. Method Design and Optimization Using CUR as Model

The reported techniques to obtain PLHNs are classified into one-step and two-step approaches. The one-step synthesis is based on the nanoprecipitation and self-assembly technique [40]. According to Zhang et al. (2010), the water-miscible organic phase containing both the polymer and hydrophobic drug is added dropwise into the aqueous phase composed of the lipid and a small quantity of a water-miscible organic solvent for facilitating lipid dissolution [40]. The method proposed herein was also based on a one-step approach. However, in contrast to the existing techniques, it consisted of incorporating the aqueous phase containing the polymer into the organic one composed of the lipid and CUR, taking advantage of the components' solubilities in the respective phases. Therefore, reducing the necessity of including organic solvent in the water solution.

Designing and optimizing methods for nanoparticle preparation using traditional statistical analysis is a complex and time-consuming process that requires plenty of experiments [41]. In addition, a complete characterization of each prepared formulation is quite expensive because it requires advanced instrumentation that may not be available in every laboratory. Therefore, as an alternative, the strategy developed in this contribution was to perform a screening of variables considering first the emulsion formation which can be visually assessed [42], and a second step determining EE_I and EE_D. In this regard, high EE values are desirable for the delivery of higher amounts of drug payload. Consequently, in terms of the scalability and industrial development of these materials, high EE implies the economical usage of drugs without a decrease in their therapeutic index [41,43]. Further, EE can be influenced by the method used to carry out the encapsulation process, the partition coefficient of the target molecule in the solvents used and the size distribution of the PLHNs [43,44]. All the conditions and parameters evaluated are shown in Table 1.

3.1.1. Homogenization Technique

The structural organization of the PLHNs and the high EE depends directly on the preparation of the nanoemulsion, and the drug-polymer interactions [15]. Mixing the organic and aqueous together and applying the ultrasonication probe, neither for 30 nor 45 min, provided a visual emulsion. In contrast, using high-speed homogenization, adding aqueous solution dropwise into the organic one, and keeping the homogenization speed at 10,000 rpm for 10 min after the addition resulted in a successful emulsion with an EE of 92%. Then, instead of using the fast addition of the aqueous phase into the organic one, the medium and fast adding speeds were investigated. The optimum addition speed showed to be the medium one resulting in 98% of EE against 30% and 46% of the fast and slow speeds, respectively. Previous reports [14,30] indicate the importance of the rate of precipitation of the hydrophobic drug and the polymer, for instance, with similar or equal rates of precipitation of the two species, the homogeneous particles would be obtained while large differences between rates will force the selective precipitation of each component affecting the encapsulation of the drug [30]. Further, the tested 12,000 and 16,000 rpm homogenization speed resulted in 92% and 98% of EE respectively. Consequently, these last conditions were selected to continue the method optimization.

Table 1. Parameters and conditions tested in the development and optimization of the method.

1-Homogenization Technique [1]			N° Formulation	Emulsion Observed	EE (%)
Ultrasonic Processor One-time addition, 130 watts, 20 kHertz, 3 mm Ti probe	Sonication Time	30 min	1	No	NA
		45 min	2	No	NA
High Speed Homogenizer 10,000 rpm, 10 min	Mixing phases speed	Slow	3	Yes	46
		Medium	4	Yes	98
		Fast	5	Yes	30
2-Optimization of homogenization technique [1]				Emulsion observed	EE (%)
High speed homogenizer, mixing phases at medium speed, 10 min	Homogenization speed	12,000 rpm	6	Yes	92
		16,000 rpm	7	Yes	98
3-Lipid, drug loading, lipid-drug ratio, and organic solvent				Emulsion observed	EE (%)
Type of lipid	24:1 lipid:drug ratio	Cocoa butter	8	Yes	71
		Cetyl Palmitate	9	Yes	54
		Stearic acid	10	Yes	81
		Cholesterol: Cetyl Palmitate 1:1	11	Yes	47
Drug loading	mg of CUR	7.5	12	Yes	87
		10	13	Yes	62
		15	14	No	NA
Lipid-drug ratio	Chol:CUR	12:1	15	Yes	75
		48:1	16	Yes	70
Organic solvent	1:1 solvent mixture	CH_2Cl_2:MeOH	17	Yes	37
		CH_3Cl_3:EtOH	18	Yes	8
4-Surfactant selection and polymer concentration				Emulsion observed	EE (%)
Surfactant selection	1% surfactant concentration	Tween 80	19	No	NA
		Span 80	20	No	NA
Polymer concentration	mg/mL of Pluronic F-127	2	21	Yes	5
		7	22	Yes	16

NA: Non-applicable; [1] Starting composition: aqueous phase containing 5 mg/mL of Pluronic F-127 in acetic acid 0.1% and Tween 80: Span80 1:1 4% and organic phase containing a CUR-lipid ratio of 1:24 in a 1:1 mixture of MeOH: $CHCl_3$.

3.1.2. Lipid, Drug Loading, Lipid-Drug Ratio, and Organic Solvent

Instead of Chol, four lipids cocoa butter, cetyl palmitate, stearic acid, and a 1:1 mixture of cetyl palmitate:Chol were incorporated into the organic phase and evaluated for emulsion formation and EE. All the formulations provided emulsions and an acceptable EE; stearic acid was the best with 80%. However, the formulation containing Chol continued to exhibit the best EE.

In an attempt to achieve a high drug loading capacity into the PLHNs, increments in the amount of CUR were investigated by adding 7.5, 10, and 15 mg to the organic phase in individual experiments. Increases in drug content decreased the EE until it was not forming an emulsion when using 15 mg of CUR. Further, two additional lipid-drug ratios using a lower 12:1 and a higher 48:1 lipid quantity in relation with the CUR content resulted in 75 and 70% EE, respectively, which did not represent an improvement in comparison with the 24:1 selected as starting parameters.

Finally, two different 1:1 solvent mixtures were tested as an organic phase, CH_2Cl_2: MeOH and CH_3Cl_3: EtOH. However, the results showed a deficient EE with both solvent mixtures yielding 37% and 8% respectively.

3.1.3. Surfactant Selection and Polymer Concentration

The starting parameters in the development and optimization of this method included a 1:1 mixture of Tween 80: Span 80®. It was reported that attraction between both Tween 80 and Span 80 surfactants may affect the loading [45]. Therefore, Tween 80 and Span 80® were separately tested in the aqueous solution in order investigate the effect of using both surfactants together in the formulation. Results indicated that the use of only one of them separately did not provide an emulsion; therefore, both surfactants are needed in the formulation to successfully prepare the PLHNs. Optimizing the polymer concentration by decreasing to 2 mg/mL and increasing up to 7 mg/mL of Pluronic F-127 resulted in a considerable decline in the EE resulting in 5% for 2 mg/mL and in 16% for the 7 mg/L polymer concentration.

In sum, the parameters tested provided consequently several formulations in which acceptable EE values were observed when emulsions were successfully obtained. This can be attributed to the important role of Pluronic in the shell-core incorporation providing the forces to assemble via the attraction between alkyl groups in the polymer and aromatic groups in CUR [46]. The symmetric electron density distribution of atoms for the CUR in the Pluronic's non-polar part can lead to the assembly of electrostatic van der Waals forces. In the hydrophobic PO chain of Pluronic, there is only one $–CH_3$ per monomer while Chol can interact with the hydrophobic part of the chain [47]. This may lead to more interaction sites for CUR with the polymer, thus leading to higher loading into PLHNs.

The encapsulation was confirmed by FT-IR measurements in formulations exhibiting high EE values. FT-IR was performed to evaluate the successful incorporation of CUR loaded into PLHNs. Figure 3 shows the FT-IR spectra of the Chol, Pluronic F-127, plain PLHNs, and CUR-loaded PLHNs. The main peaks for the CUR-loaded PLHNs correspond to 3363 cm^{-1} related to the stretching vibration of hydrogen-bonded (-OH), 1649 cm^{-1} of the C=O stretching vibration, and 1270 cm^{-1} due to C-O stretching. The CUR-loaded PLHNs spectra indicated a combination of these signals from CUR as well as from Pluronic F-127 and Chol components. For instance, bands at 2897 cm^{-1} associated with C-H stretch aliphatic and at 1359 cm^{-1} corresponding to in-plane O-H bend pertain to Pluronic F-127 [48]. In addition, signals at 1449 cm^{-1}, 1043 cm^{-1} and 742 cm^{-1} correspond to CH_2 and CH_3 deformation vibrations, ring deformation, and C-H out-of-plane bending from Chol [49]. The fact that some other signals reported for CUR [50,51] are not predominant and the combination of CUR main bands with Pluronic F-127 and Chol signals in CUR-loaded PLHNs spectra confirm that CUR was successfully loaded into PLHNs.

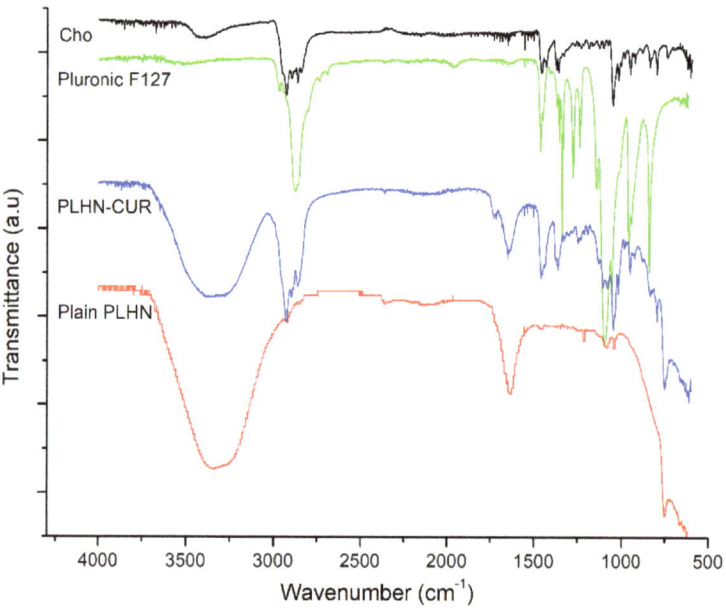

Figure 3. FT–IR spectra of PLHNs main components, CUR–loaded PLHNs and plain PLHNs.

3.2. Preparation, Characterization and In Vitro Evaluation of DMC and BDM-Loaded PLHNs

3.2.1. Encapsulation, FT-IR and EE

We then confirmed the encapsulation FT-IR measurements were performed in the nanoformulations. Comparing the FT-IR spectra of plain PLHNs, DMC, or BDM-loaded PLHNs with the ones obtained for pure DMC or BDM (Figure 4), it becomes evident that they share peaks at wavenumbers of 3318 cm^{-1} associated with stretching vibration of hydrogen-bonded (-OH), as well as at 1659 and 1707 cm^{-1} stretching vibration of conjugated carbonyl (C=O) group, 1459 cm^{-1} and 1422 cm^{-1} CH- bending, and 1370 cm^{-1} and 1347 cm^{-1} to in-plane O-H bend, respectively, for DMC and BDM. Besides, characteristic bands at 2928 cm^{-1} associated to C-H stretch aliphatic corresponding to Pluronic F-127 [48] and signals at 1048 cm^{-1} and 754 cm^{-1} corresponding to ring deformation and C-H out-of-plane bending from Chol [49] are also present in the curcuminoids loaded PLHNs spectra. Therefore, these facts are to be expected due to the combination of signals of the components and the pure drug, which is further confirm that curcuminoids DMC and BDM were successfully loaded onto PLHNs core.

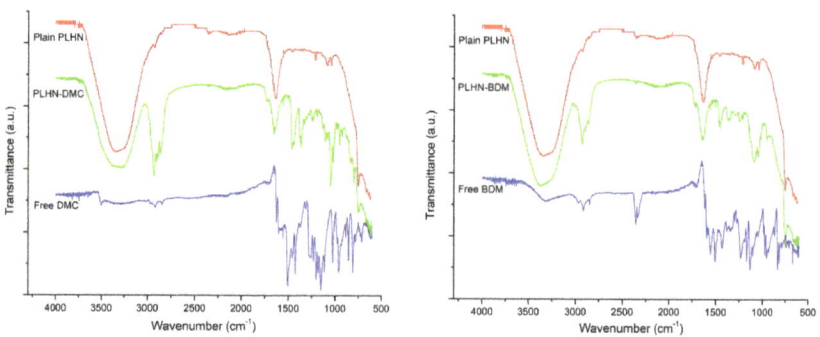

Figure 4. FT–IR spectra of DMC and BDM–loaded PLHNs, plain PLHNs and free DMC and BDM.

Concerning the EE of DMC and BDM, both DMC and BDM were efficiently loaded in PLHNs, achieving an encapsulation efficiency of 95% and 92%, respectively. Therefore, a high EE meant that the curcuminoid maximal solubility in the lipid was reached in the PLHNs and that all the molecules remained in the particles after lipid solidification [52]. The high values of EE can be attributed to phenyl groups on the curcuminoids structure loaded into HPLN. In addition, Pluronic has an important role in the shell-core incorporation, because it provides the forces to the assembly via the attraction between alkyl groups in the polymer and aromatic groups [46]. The structural organization of the PLHNs and the high EE depends directly on the preparation of nanoemulsion and the drug-polymer interactions. Previous reports [14,30] indicate the importance of the rate of precipitation of the hydro-phobic drug and the polymer, large differences between rates will force the selective precipitation of each component, consequently affecting the encapsulation of the drug [30].

3.2.2. DLS and HR-TEM Techniques

The particle size and size distribution in terms of PDI values were evaluated by DLS whereas the morphology of the prepared PLHNs was observed by HR-TEM. Figure 5 presents the HR-TEM images and histogram of the size distribution of the DMC and BDM-loaded PLHNs. For the characterization of nanosystems for drug delivery, parameters such as average size and polydispersity index (PDI) are considered one of the most important factors to evaluate the stability and the proper function of the nanoparticles due to the influence in the loading and release of the compound inside the nanoparticle [53,54]; therefore, it is important nanoparticles present high reproducibility and homogeneity [53]. In this concern, PDI defines the variation in particle size distribution within the nanoemulsion. According to Souza et al. (2014) and Valencia et al. (2021) PDI values ≤ 0.4 are considered monodisperse, which implies that there is uniformity in the sample size [53,55], and it also indicates that the formulation has a low aggregation of the sample during isolation or analysis [56].

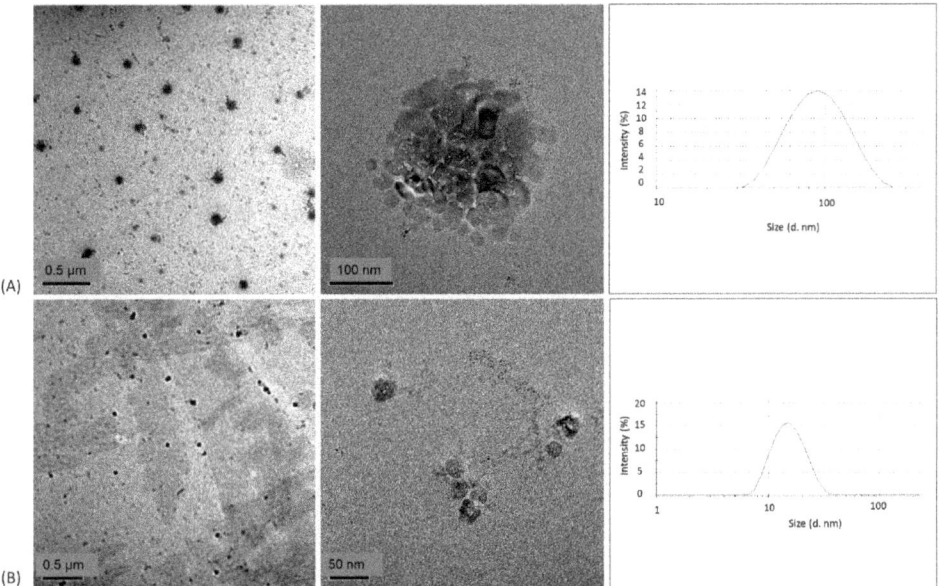

Figure 5. TEM images and size distribution histogram of (**A**) DMC and (**B**) BDM–loaded PLHNs.

The results showed that the DMC loaded nanoparticles exhibited a particle size of 75.55 ± 0.51 nm with a PDI of 0.281 ± 0.014 and the BDM nanoparticles 15.13 ± 0.014 nm with a PDI of 0.196 ± 0.032. A formulation composed of Chol, lecitin and vitamin E TPGS reports a BDM-loaded PLHNs size of 75.98 nm [57] while Dolatabi et al. [58] reported a DMC PLHNs formulation with Precirol® ATO5 and polaxamer 188 with a-size of 160.7 nm. Our findings indicate smaller particle sizes than both studies and a similar average size of (a-size) <50 nm for BDM in a PLHNs formulation with ethyl oleate and PEG-400 [59]. Due to the low size, the PLHNs can be absorbed by systemic circulation in the intestine [60] which improves the bioavailability of these small molecules.

In respect to morphology, the TEM images of DMC and BDM-loaded PLHNs present nanoparticles with a spherical appearance. These findings are in agreement with results from the literature reporting a similar shape for other curcuminoid polymeric nanoparticles [59].

3.2.3. Differential Scanning Calorimetry (DSC)

The thermal evaluation can reveal the solid state of the encapsulated drug, it provides information about the microcrystalline form and if present any polymorph change or transition change in amorphous form [61]. In addition, DSC can show any incompatibility or possible interaction between the drug and excipients, which may affect the efficacy of the encapsulated drug [62]. The DSC curves of formulation components are presented in Figure 6. DMC and BDM exhibited a unique endothermic event at 173 °C and 238 °C, respectively, in agreement with values reported in literature [63]. Further, Pluronic F-127 and Chol were observed at 58 °C and 147 °C, respectively, corresponding to their melting temperatures and in the case of Chol to its monohydrate form [64]. The DSC curves of DMC and BDM-loaded PLHNs presented similar thermal behavior. One broad endothermic event between 40 and 70 °C is coincident with the melting temperature of Pluronic F-127 and water loss. Then, a sharp endothermic one around 100 °C associated to a solvated form of Chol [65] that can be crystallized during PLHNs preparation. The endothermic event related to free curcuminoids was not observed in the nanoparticles which is an indicative that they were molecularly dispersed within the PLHNs matrix [66,67]. Moreover, there was no evidence of incompatibilities between curcuminoids and formulation constituents.

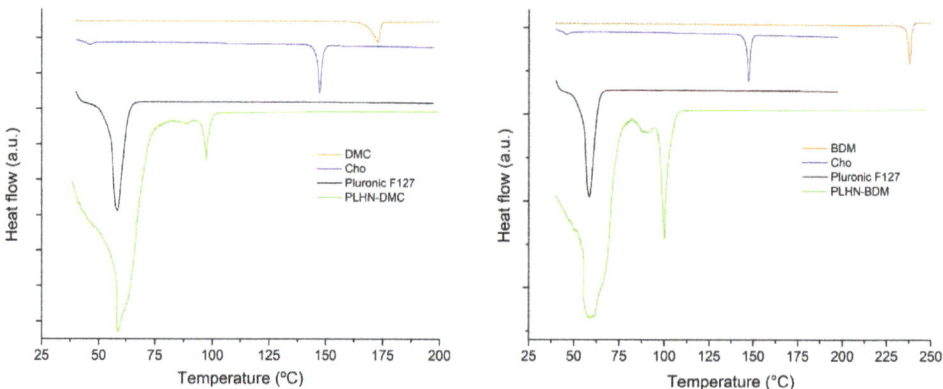

Figure 6. DSC curves of PLHNs main components and the DMC and BDM–loaded formulation.

3.2.4. In Vitro Studies

Drug Release

The in vitro release profiles of DMC and BDM from PLHNs as well as the dissolution profiles of free curcuminoids in two dissolution media are shown in Figure 7. Medium 1 (M1) composed of phosphate buffered pH 6.8 with 20% of MeOH and medium 2 (M2) was water. Several media and pH values were tested with anomalous results due to the chemical instability of curcuminoids at pHs lower than 7. Its degradation compromised the detection and quantification of curcuminoids [68,69]. In this regard, reports in the literature have used pH below 7 in the release of curcuminoids [70–72]. In addition, the absorption pH in the lumen of the intestine has been reported to be in the range of 6.8–7.4; therefore, was considered an appropriate media for curcuminoids [73]. In respect to the use of water as a medium, it was evaluated considering the main purpose of this study on preparing nanoencapsulated curcuminoids to improve their water solubility; furthermore, in light of some studies that have reported the degradation of the Pluronic backbone in water which can contribute to water solubilization and subsequent release of the drug [4–6].

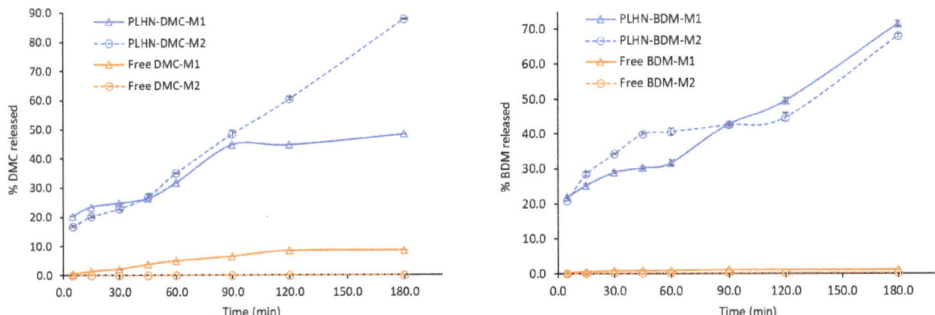

Figure 7. Release profile of DMC and BDM from the PLHNs compared with free DMC and BDM dissolution rate in two dissolution media. Error bars represent the standard deviation of DMC and BDM concentration in the triplicates.

The presence of MeOH in M1 suggested an increased dissolution rate of free curcuminoids and a higher release profile of curcuminoids from the PLHNs. However, release and dissolution profiles obtained in both media were not significantly different from BDM. Nevertheless, comparing the dissolution profile of free DMC and BDM with curcuminoids released from PLHNs in water, showed promising results. At 180 min only 0.1% of the free DMC was dissolved while its release from the hybrid nanosystem was 88% at 180 min. In turn, only 0.3% of free BDM dissolved after 180 min while 68% were released from the nanoparticle at 180 min. The difference in values in favor of DMC can be explained due to the different pka values (DMC < BDM), which can destabilize the keto-enol structure and consequently the dissociation of the enol hydrogen affecting the stability and the solubility of the curcuminoids [74]. Overall, these results suggested that this formulation would significantly improve the bioavailability of these curcuminoids.

The release of a loaded drug molecule from the shell-core largely depends on hydrophobic interactions between the inner core and drug, as previously mentioned. The increased release of DMC and BDM from the PLHNs can be attributed to the hydrophobic interaction between the curcuminoid and the bilayer, as the hydrophobic interaction becomes weak, the shell core breaks and exhibits a fast and sustained release of the molecule [75].

Antioxidant Activity Evaluation of DMC and BDMC Free and Loaded into PLHNs

The antioxidant activity of Trolox, free and PLHN-loaded DMC and BDM was studied through a DPPH analysis, as described in the Materials and Methods section. Trolox was used as a standard to assess the DPPH method, obtaining adequate results ($R^2 = 0.9956$) that allowed to determine an IC_{50} of 5.62 µg/mL. Further, results of the samples antioxidant activity are shown in Table 2.

Table 2. Antioxidant activity of free and curcuminoids DMC and BDM-loaded PLHNs.

	IC_{50} (μg/mL) [1,2,3]		
	EtOH	Water	PLHN
DMC	12.46 [a,#] ± 0.02	2143.07 [a,&] ± 0.61	12.74 [a,#] ± 0.09
BDM	17.94 [b,^] ± 0.06	1398.68 [b,*] ± 5.07	16.03 [b,^] ± 0.55

[1] IC_{50} μg/mL for each curcuminoid. [2] Values are expressed as mean ± standard deviation (S.D.). [3] Different superscript letters in the same column or different superscript signs in the same row indicate differences are significant ($p < 0.05$) using one-way analysis of variance (ANOVA) with a Tukey post hoc.

Evaluation of the antioxidant activity of the free curcuminoid dissolutions in EtOH indicated DMC ($R^2 = 0.9979$) yielded the lowest IC_{50} representing an antioxidant activity 28% higher than the antioxidant activity of BDM ($R^2 = 0.9930$). This observation is consistent with the trend previously reported for the antioxidant activity of the main curcuminoids, where DMC was found to yield higher antioxidant values than BDM [76]. A DPPH analyses of the aqueous solutions of the free curcuminoids showed a considerably higher IC_{50} for both samples; hence, much lower antioxidant activity, which is associated with their low solubility in water. Results showed an opposite trend of antioxidant activity in these aqueous samples, with BDM ($R^2 = 0.9947$) yielding the lowest IC50, indicating an antioxidant activity 35% higher than the antioxidant activity of DMC ($R^2 = 0.9932$) in water, which is in alignment with BDM relative higher aqueous solubility in respect to DMC [77].

DPPH assays of DMC and BDM-loaded PLHNs showed an important decrease in the IC_{50}, therefore a much higher 80-to 160-fold antioxidant activity than free curcuminoids in aqueous solution. As shown by the results from a one-way analysis of variance (ANOVA) (Table 1), both the DMC ($R^2 = 0.9936$) and BDM ($R^2 = 0.9960$)-loaded PLHNs presented similar antioxidant activity to the one obtained for the ethanolic solution of the free curcuminoids. Further, BDM nanoparticles showed a lower IC_{50} than free BDM ethanolic solution, with BDM-loaded PLHNs antioxidant activity improving by 11%, aligning with results using CUR that showed antioxidant activity improved by 17% [78]. In sum, our results show that the antioxidant activity of DMC and BDM was significantly enhanced by the PLHNs preparation, consistent with previous results on polymeric nanoparticle formulations of curcuminoid mixtures [79] and the individual curcuminoids DMC and BDM [59], aligning also with results obtained for other polyphenols, such as hesperetin formulations [80].

4. Conclusions

DMC and BDM-loaded polymer-lipid hybrid nanoparticles (PLHNs) composed of Pluronic F-127 and cholesterol were synthesized and characterized through a simple method designed and optimized using CUR as a model curcuminoid. The influence of the encapsulation efficiency of various processing parameters including the ultrasonic probe, high-speed homogenization, mixing phases and homogenization speed were evaluated. Organic and aqueous phases composition: lipid, drug loading, lipid-drug ratio, organic solvent, surfactant, and polymer concentration was systematically assessed. The study confirmed that Pluronic F-127 plays an important role in the shell-core incorporation providing the forces to assemble via the attraction between alkyl groups in the polymer and aromatic groups in curcuminoids. Therefore, contributing to improving the solubility, stability, and manages to successfully encapsulate the curcuminoids by the PLHNs, as well and maintain a high loading efficiency as previously reported in the literature. The method for PLHNs synthesis developed and optimized herein presents advantages in terms of preparation, appropriate nanoparticles characteristics including particle size, monodisperse size distribution, shape, and encapsulation efficiency. Further, the DMC and BDM-loaded PLHNs exhibited considerable improvements in aqueous dissolution consequently in their antioxidant activity. Therefore, this method can be extended to other curcuminoids and molecules with a similar backbone structure to improve the bioactivities associated with their limited bioavailability.

Author Contributions: Conceptualization, K.W.R. and A.M.A.-S.; methodology, K.W.R., A.M.A.-S., F.V.H., M.I.Q. and M.N.-H.; formal analysis, K.W.R., F.V.H. and M.I.Q.; investigation, K.W.R., A.M.A.-S., F.V.H. and M.N.-H.; resources, J.R.V.-B., M.N.-H. and A.M.A.-S.; data curation, K.W.R., A.M.A.-S., M.I.Q., F.V.H. and M.N.-H.; writing—original draft preparation, K.W.R., A.M.A.-S., M.I.Q. and F.V.H.; writing—review and editing, J.R.V.-B., M.N.-H. and A.M.A.-S.; supervision, A.M.A.-S., J.R.V.-B. and M.N.-H.; project administration, A.M.A.-S. and M.N.-H.; funding acquisition, J.R.V.-B., M.N.-H. and A.M.A.-S. All authors have read and agreed to the published version of the manuscript.

Funding: This research was funded by grants from the University of Costa Rica (115-B8-150 and 115-C1-501) and the National Laboratory of Nanotechnology (LANOTEC).

Institutional Review Board Statement: Not applicable.

Informed Consent Statement: Not applicable.

Data Availability Statement: Not applicable.

Acknowledgments: The authors thank the CeNAT Scholarship Program for the financial support awarded to Krissia Wilhelm and the Center for Research and Chemical and Microbiological Services from Instituto Tecnológico de Costa Rica (CEQUIATEC) for access to the Malvern Nano Zetasizer ZS90 instrument.

Conflicts of Interest: The authors declare no conflict of interest. The funders had no role in the design of the study; in the collection, analyses, or interpretation of data; in the writing of the manuscript, or in the decision to publish the results.

References

1. Rai, R.; Alwani, S.; Badea, I. Polymeric Nanoparticles in Gene Therapy: New Avenues of Design and Optimization for Delivery Applications. *Polymers* **2019**, *11*, 745. [CrossRef]
2. Vauthier, C.; Bouchemal, K. Methods for the Preparation and Manufacture of Polymeric Nanoparticles. *Pharm. Res.* **2009**, *26*, 1025–1058. [CrossRef]
3. Madkour, L.H. Nanoparticle and polymeric nanoparticle-based targeted drug delivery systems. In *Nucleic Acids as Gene Anticancer Drug Delivery Therapy*; Elsevier: Amsterdam, The Netherlands, 2019; pp. 191–240.
4. Cheow, W.S.; Hadinoto, K. Enhancing encapsulation efficiency of highly water-soluble antibiotic in poly(lactic-co-glycolic acid) nanoparticles: Modifications of standard nanoparticle preparation methods. *Colloids Surf. A Physicochem. Eng. Asp.* **2010**, *370*, 79–86. [CrossRef]
5. Hadinoto, K.; Sundaresan, A.; Cheow, W.S. Lipid–polymer hybrid nanoparticles as a new generation therapeutic delivery platform: A review. *Eur. J. Pharm. Biopharm.* **2013**, *85*, 427–443. [CrossRef]
6. Chaudhary, Z.; Ahmed, N.; .ur. Rehman, A.; Khan, G.M. Lipid polymer hybrid carrier systems for cancer targeting: A review. *Int. J. Polym. Mater. Polym. Biomater.* **2018**, *67*, 86–100. [CrossRef]
7. Jose, C.; Amra, K.; Bhavsar, C.; Momin, M.; Omri, A. Polymeric Lipid Hybrid Nanoparticles: Properties and Therapeutic Applications. *Crit. Rev. Ther. Drug Carr. Syst.* **2018**, *35*, 555–588. [CrossRef] [PubMed]
8. Cheow, W.S.; Hadinoto, K. Factors affecting drug encapsulation and stability of lipid–polymer hybrid nanoparticles. *Colloids Surf. B Biointerfaces* **2011**, *85*, 214–220. [CrossRef] [PubMed]
9. Li, Y.-Y.; Li, L.; Dong, H.-Q.; Cai, X.-J.; Ren, T.-B. Pluronic F127 nanomicelles engineered with nuclear localized functionality for targeted drug delivery. *Mater. Sci. Eng. C* **2013**, *33*, 2698–2707. [CrossRef] [PubMed]
10. Kabanov, A.V.; Alakhov, V.Y. Pluronic? Block Copolymers in Drug Delivery: From Micellar Nanocontainers to Biological Response Modifiers. *Crit. Rev. Ther. Drug Carrier Syst.* **2002**, *19*, 1–72. [CrossRef]
11. Bodratti, A.; Alexandridis, P. Formulation of Poloxamers for Drug Delivery. *J. Funct. Biomater.* **2018**, *9*, 11. [CrossRef]
12. Li, Z.; Peng, S.; Chen, X.; Zhu, Y.; Zou, L.; Liu, W.; Liu, C. Pluronics modified liposomes for curcumin encapsulation: Sustained release, stability and bioaccessibility. *Food Res. Int.* **2018**, *108*, 246–253. [CrossRef] [PubMed]
13. de Oliveira, R.S.S.; Marín Huachaca, N.S.; Lemos, M.; Santos, N.F.; Feitosa, E.; Salay, L.C. Molecular interactions between Pluronic F127 and saponin in aqueous solution. *Colloid Polym. Sci.* **2020**, *298*, 113–122. [CrossRef]
14. Sahu, A.; Kasoju, N.; Goswami, P.; Bora, U. Encapsulation of Curcumin in Pluronic Block Copolymer Micelles for Drug Delivery Applications. *J. Biomater. Appl.* **2011**, *25*, 619–639. [CrossRef] [PubMed]
15. Ganguly, R.; Kumar, S.; Kunwar, A.; Nath, S.; Sarma, H.D.; Tripathi, A.; Verma, G.; Chaudhari, D.P.; Aswal, V.K.; Melo, J.S. Structural and therapeutic properties of curcumin solubilized pluronic F127 micellar solutions and hydrogels. *J. Mol. Liq.* **2020**, *314*, 113591. [CrossRef]
16. Enumo, A.; Argenta, D.F.; Bazzo, G.C.; Caon, T.; Stulzer, H.K.; Parize, A.L. Development of curcumin-loaded chitosan/pluronic membranes for wound healing applications. *Int. J. Biol. Macromol.* **2020**, *163*, 167–179. [CrossRef] [PubMed]
17. Szymusiak, M.; Hu, X.; Leon Plata, P.A.; Ciupinski, P.; Wang, Z.J.; Liu, Y. Bioavailability of curcumin and curcumin glucuronide in the central nervous system of mice after oral delivery of nano-curcumin. *Int. J. Pharm.* **2016**, *511*, 415–423. [CrossRef]

18. Lee, Y.-S.; Cho, D.-C.; Kim, C.H.; Han, I.; Gil, E.Y.; Kim, K.-T. Effect of curcumin on the inflammatory reaction and functional recovery after spinal cord injury in a hyperglycemic rat model. *Spine J.* **2019**, *19*, 2025–2039. [CrossRef] [PubMed]
19. Siviero, A.; Gallo, E.; Maggini, V.; Gori, L.; Mugelli, A.; Firenzuoli, F.; Vannacci, A. Curcumin, a golden spice with a low bioavailability. *J. Herb. Med.* **2015**, *5*, 57–70. [CrossRef]
20. Heffernan, C.; Ukrainczyk, M.; Gamidi, R.K.; Hodnett, B.K.; Rasmuson, Å.C. Extraction and Purification of Curcuminoids from Crude Curcumin by a Combination of Crystallization and Chromatography. *Org. Process Res. Dev.* **2017**, *21*, 821–826. [CrossRef]
21. Ukrainczyk, M.; Hodnett, B.K.; Rasmuson, Å.C. Process Parameters in the Purification of Curcumin by Cooling Crystallization. *Org. Process Res. Dev.* **2016**, *20*, 1593–1602. [CrossRef]
22. Amalraj, A.; Pius, A.; Gopi, S.; Gopi, S. Biological activities of curcuminoids, other biomolecules from turmeric and their derivatives—A review. *J. Tradit. Complement. Med.* **2017**, *7*, 205–233. [CrossRef]
23. Nair, A.; Amalraj, A.; Jacob, J.; Kunnumakkara, A.B.; Gopi, S. Non-curcuminoids from turmeric and their potential in cancer therapy and anticancer drug delivery formulations. *Biomolecules* **2019**, *9*, 13. [CrossRef]
24. Lozada-García, M.C.; Enríquez, R.G.; Ramírez-Apán, T.O.; Nieto-Camacho, A.; Palacios-Espinosa, J.F.; Custodio-Galván, Z.; Soria-Arteche, O.; Pérez-Villanueva, J. Synthesis of curcuminoids and evaluation of their cytotoxic and antioxidant properties. *Molecules* **2017**, *22*, 633. [CrossRef] [PubMed]
25. Hatamipour, M.; Ramezani, M.; Tabassi, S.A.S.; Johnston, T.P.; Sahebkar, A. Demethoxycurcumin: A naturally occurring curcumin analogue for treating non-cancerous diseases. *J. Cell. Physiol.* **2019**, *234*, 19320–19330. [CrossRef] [PubMed]
26. Bhawana; Basniwal, R.K.; Buttar, H.S.; Jain, V.K.; Jain, N. Curcumin nanoparticles: Preparation, characterization, and antimicrobial study. *J. Agric. Food Chem.* **2011**, *59*, 2056–2061. [CrossRef] [PubMed]
27. Xiao, B.; Si, X.; Han, M.K.; Viennois, E.; Zhang, M.; Merlin, D. Co-delivery of camptothecin and curcumin by cationic polymeric nanoparticles for synergistic colon cancer combination chemotherapy. *J. Mater. Chem. B* **2015**, *3*, 7724–7733. [CrossRef] [PubMed]
28. Yan, Y.-D.; Kim, J.A.; Kwak, M.K.; Yoo, B.K.; Yong, C.S.; Choi, H.-G. Enhanced oral bioavailability of curcumin via a solid lipid-based self-emulsifying drug delivery system using a spray-drying technique. *Biol. Pharm. Bull.* **2011**, *34*, 1179–1186. [CrossRef]
29. Yen, F.-L.; Wu, T.-H.; Tzeng, C.-W.; Lin, L.-T.; Lin, C.-C. Curcumin Nanoparticles Improve the Physicochemical Properties of Curcumin and Effectively Enhance Its Antioxidant and Antihepatoma Activities. *J. Agric. Food Chem.* **2010**, *58*, 7376–7382. [CrossRef] [PubMed]
30. Rabanel, J.-M.; Faivre, J.; Tehrani, S.F.; Lalloz, A.; Hildgen, P.; Banquy, X. Effect of the Polymer Architecture on the Structural and Biophysical Properties of PEG–PLA Nanoparticles. *ACS Appl. Mater. Interfaces* **2015**, *7*, 10374–10385. [CrossRef]
31. Ponnusamy, L.; Natarajan, S.R.; Thangaraj, K.; Manoharan, R. Therapeutic aspects of AMPK in breast cancer: Progress, challenges, and future directions. *Biochim. Biophys. Acta-Rev. Cancer* **2020**, *1874*, 188379. [CrossRef]
32. Guo, L.Y.; Cai, X.F.; Lee, J.J.; Kang, S.C.; Shin, E.M.; Zhou, H.Y.; Jung, J.W.; Kim, Y.S. Comparison of suppressive effects of demethoxycurcumin and bisdemethoxycurcumin on expressions of inflammatory mediators In Vitro and In Vivo. *Arch. Pharm. Res.* **2008**, *31*, 490–496. [CrossRef] [PubMed]
33. Kocher, A.; Schiborr, C.; Behnam, D.; Frank, J. The oral bioavailability of curcuminoids in healthy humans is markedly enhanced by micellar solubilisation but not further improved by simultaneous ingestion of sesamin, ferulic acid, naringenin and xanthohumol. *J. Funct. Foods* **2015**, *14*, 183–191. [CrossRef]
34. Sudeep, V.H.; Gouthamchandra, K.; Chandrappa, S.; Naveen, P.; Reethi, B.; Venkatakrishna, K.; Shyamprasad, K. In vitro gastrointestinal digestion of a bisdemethoxycurcumin-rich Curcuma longa extract and its oral bioavailability in rats. *Bull. Natl. Res. Cent.* **2021**, *45*, 84. [CrossRef]
35. Rompicharla, S.V.K.; Bhatt, H.; Shah, A.; Komanduri, N.; Vijayasarathy, D.; Ghosh, B.; Biswas, S. Formulation optimization, characterization, and evaluation of in vitro cytotoxic potential of curcumin loaded solid lipid nanoparticles for improved anticancer activity. *Chem. Phys. Lipids* **2017**, *208*, 10–18. [CrossRef]
36. Udompornmongkol, P.; Chiang, B.-H. Curcumin-loaded polymeric nanoparticles for enhanced anti-colorectal cancer applications. *J. Biomater. Appl.* **2015**, *30*, 537–546. [CrossRef]
37. Godara, S.; Lather, V.; Kirthanashri, S.V.; Awasthi, R.; Pandita, D. Lipid-PLGA hybrid nanoparticles of paclitaxel: Preparation, characterization, in vitro and in vivo evaluation. *Mater. Sci. Eng. C* **2020**, *109*, 110576. [CrossRef] [PubMed]
38. Navarro, M.; Arnaez, E.; Moreira, I.; Hurtado, A.; Monge, D.; Monagas, M. Polyphenolic composition and antioxidant activity of Uncaria tomentosa commercial bark products. *Antioxidants* **2019**, *8*, 339. [CrossRef]
39. Navarro-Hoyos, M.; Arnáez-Serrano, E.; Quesada-Mora, S.; Azofeifa-Cordero, G.; Wilhelm-Romero, K.; Quirós-Fallas, M.I.; Alvarado-Corella, D.; Vargas-Huertas, F.; Sánchez-Kopper, A. Polyphenolic QTOF-ESI MS Characterization and the Antioxidant and Cytotoxic Activities of Prunus domestica Commercial Cultivars from Costa Rica. *Molecules* **2021**, *26*, 6493. [CrossRef]
40. Zhang, L.; Zhang, L. Lipid–Polymer Hybrid Nanoparticles: Synthesis, Characterization and Applications. *Nano LIFE* **2010**, *1*, 163–173. [CrossRef]
41. Malekpour, M.; Naghibzadeh, M.; Mohammad, R.; Esnaashari, S.S.; Adabi, M.; Mujokoro, B.; Khosravani, M.; Adabi, M. Effect of various parameters on encapsulation efficiency of mPEG-PLGA nanoparticles: Artificial neural network. *Biointerface Res. Appl. Chem.* **2018**, *8*, 3267–3272.
42. Aben, S.; Holtze, C.; Tadros, T.; Schurtenberger, P. Rheological Investigations on the Creaming of Depletion-Flocculated Emulsions. *Langmuir* **2012**, *28*, 7967–7975. [CrossRef] [PubMed]

43. Wischke, C.; Schwendeman, S.P. Principles of encapsulating hydrophobic drugs in PLA/PLGA microparticles. *Int. J. Pharm.* **2008**, *364*, 298–327. [CrossRef] [PubMed]
44. Peng, Z.; Li, S.; Han, X.; Al-Youbi, A.O.; Bashammakh, A.S.; El-Shahawi, M.S.; Leblanc, R.M. Determination of the composition, encapsulation efficiency and loading capacity in protein drug delivery systems using circular dichroism spectroscopy. *Anal. Chim. Acta* **2016**, *937*, 113–118. [CrossRef]
45. Sezgin, Z.; Yuksel, N.; Baykara, T. Preparation and characterization of polymeric micelles for solubilization of poorly soluble anticancer drugs. *Eur. J. Pharm. Biopharm.* **2006**, *64*, 261–268. [CrossRef]
46. Hwang, D.; Ramsey, J.D.; Kabanov, A.V. Polymeric micelles for the delivery of poorly soluble drugs: From nanoformulation to clinical approval. *Adv. Drug Deliv. Rev.* **2020**, *156*, 80–118. [CrossRef]
47. Vaidya, F.U.; Sharma, R.; Shaikh, S.; Ray, D.; Aswal, V.K.; Pathak, C. Pluronic micelles encapsulated curcumin manifests apoptotic cell death and inhibits pro-inflammatory cytokines in human breast adenocarcinoma cells. *Cancer Rep.* **2019**, *2*, e1133. [CrossRef] [PubMed]
48. Gopalan, K.; Jose, J. Development of amphotericin b Based organogels against mucocutaneous fungal infections. *Braz. J. Pharm. Sci.* **2020**, *56*. [CrossRef]
49. Gupta, U.; Singh, V.; Kumar, V.; Khajuria, Y. Spectroscopic Studies of Cholesterol: Fourier Transform Infra-Red and Vibrational Frequency Analysis. *Mater. Focus* **2014**, *3*, 211–217. [CrossRef]
50. Chen, X.; Zou, L.-Q.; Niu, J.; Liu, W.; Peng, S.-F.; Liu, C.-M. The Stability, Sustained Release and Cellular Antioxidant Activity of Curcumin Nanoliposomes. *Molecules* **2015**, *20*, 14293–14311. [CrossRef]
51. Suresh, K.; Nangia, A. Curcumin: Pharmaceutical solids as a platform to improve solubility and bioavailability. *CrystEngComm* **2018**, *20*, 3277–3296. [CrossRef]
52. Hajj Ali, H.; Michaux, F.; Bouelet Ntsama, I.S.; Durand, P.; Jasniewski, J.; Linder, M. Shea butter solid nanoparticles for curcumin encapsulation: Influence of nanoparticles size on drug loading. *Eur. J. Lipid Sci. Technol.* **2016**, *118*, 1168–1178. [CrossRef]
53. Valencia, M.S.; da Silva Júnior, M.F.; Xavier-Júnior, F.H.; Veras, B.d.O.; Albuquerque, P.B.S.d.; Borba, E.F.d.O.; Silva, T.G.d.; Xavier, V.L.; Souza, M.P.d.; Carneiro-da-Cunha, M.d.G. Characterization of curcumin-loaded lecithin-chitosan bioactive nanoparticles. *Carbohydr. Polym. Technol. Appl.* **2021**, *2*, 100119. [CrossRef]
54. Saedi, A.; Rostamizadeh, K.; Parsa, M.; Dalali, N.; Ahmadi, N. Preparation and characterization of nanostructured lipid carriers as drug delivery system: Influence of liquid lipid types on loading and cytotoxicity. *Chem. Phys. Lipids* **2018**, *216*, 65–72. [CrossRef] [PubMed]
55. Souza, M.P.; Vaz, A.F.M.; Correia, M.T.S.; Cerqueira, M.A.; Vicente, A.A.; Carneiro-da-Cunha, M.G. Quercetin-Loaded Lecithin/Chitosan Nanoparticles for Functional Food Applications. *Food Bioprocess Technol.* **2014**, *7*, 1149–1159. [CrossRef]
56. Mudalige, T.; Qu, H.; Van Haute, D.; Ansar, S.M.; Paredes, A.; Ingle, T. Characterization of Nanomaterials. In *Nanomaterials for Food Applications*; Elsevier: Amsterdam, The Netherlands, 2019; pp. 313–353.
57. Wang, Q.; Liu, J.; Liu, J.; Thant, Y.; Weng, W.; Wei, C.; Bao, R.; Adu-Frimpong, M.; Yu, Q.; Deng, W.; et al. Bisdemethoxycurcumin-conjugated vitamin E TPGS liposomes ameliorate poor bioavailability of free form and evaluation of its analgesic and hypouricemic activity in oxonate-treated rats. *J. Nanoparticle Res.* **2021**, *23*, 122. [CrossRef]
58. Dolatabadi, S.; Karimi, M.; Nasirizadeh, S.; Hatamipour, M.; Golmohammadzadeh, S.; Jaafari, M.R. Preparation, characterization and in vivo pharmacokinetic evaluation of curcuminoids-loaded solid lipid nanoparticles (SLNs) and nanostructured lipid carriers (NLCs). *J. Drug Deliv. Sci. Technol.* **2021**, *62*, 102352. [CrossRef]
59. Ahmad, N.; Umar, S.; Ashafaq, M.; Akhtar, M.; Iqbal, Z.; Samim, M.; Ahmad, F.J. A comparative study of PNIPAM nanoparticles of curcumin, demethoxycurcumin, and bisdemethoxycurcumin and their effects on oxidative stress markers in experimental stroke. *Protoplasma* **2013**, *250*, 1327–1338. [CrossRef] [PubMed]
60. Liu, J.; Wang, Q.; Omari-Siaw, E.; Adu-Frimpong, M.; Liu, J.; Xu, X.; Yu, J. Enhanced oral bioavailability of Bisdemethoxycurcumin-loaded self-microemulsifying drug delivery system: Formulation design, in vitro and in vivo evaluation. *Int. J. Pharm.* **2020**, *590*, 119887. [CrossRef]
61. Van Dooren, A.A. Design for Drug-Excipient Interaction Studies. *Drug Dev. Ind. Pharm.* **1983**, *9*, 43–55. [CrossRef]
62. Lütfi, G.; Müzeyyen, D. Preparation and characterization of polymeric and lipid nanoparticles of pilocarpine HCl for ocular application. *Pharm. Dev. Technol.* **2013**, *18*, 701–709. [CrossRef]
63. Péret-Almeida, L.; Cherubino, A.P.F.; Alves, R.J.; Dufossé, L.; Glória, M.B.A. Separation and determination of the physico-chemical characteristics of curcumin, demethoxycurcumin and bisdemethoxycurcumin. *Food Res. Int.* **2005**, *38*, 1039–1044. [CrossRef]
64. Loomis, C.R.; Shipley, G.G.; Small, D.M. The phase behavior of hydrated cholesterol. *J. Lipid Res.* **1979**, *20*, 525–535. [CrossRef]
65. Garti, N.; Karpuj, L.; Sarig, S. Correlation between crystal habit and the composition of solvated and nonsolvated cholesterol crystals. *J. Lipid Res.* **1981**, *22*, 785–791. [CrossRef]
66. Kumbhar, D.D.; Pokharkar, V.B. Physicochemical investigations on an engineered lipid–polymer hybrid nanoparticle containing a model hydrophilic active, zidovudine. *Colloids Surf. A Physicochem. Eng. Asp.* **2013**, *436*, 714–725. [CrossRef]
67. Das, S.; Chaudhury, A. Recent Advances in Lipid Nanoparticle Formulations with Solid Matrix for Oral Drug Delivery. *AAPS PharmSciTech* **2011**, *12*, 62–76. [CrossRef] [PubMed]
68. Tonnesen, H.H.; Karlsen, J. Studies on curcumin and curcuminoids. *Z. Leb. Forsch.* **1985**, *180*, 402–404. [CrossRef]
69. Wang, Y.-J.; Pan, M.-H.; Cheng, A.-L.; Lin, L.-I.; Ho, Y.-S.; Hsieh, C.-Y.; Lin, J.-K. Stability of curcumin in buffer solutions and characterization of its degradation products. *J. Pharm. Biomed. Anal.* **1997**, *15*, 1867–1876. [CrossRef]

70. Kharat, M.; Du, Z.; Zhang, G.; McClements, D.J. Physical and Chemical Stability of Curcumin in Aqueous Solutions and Emulsions: Impact of pH, Temperature, and Molecular Environment. *J. Agric. Food Chem.* **2017**, *65*, 1525–1532. [CrossRef]
71. Xie, X.; Tao, Q.; Zou, Y.; Zhang, F.; Guo, M.; Wang, Y.; Wang, H.; Zhou, Q.; Yu, S. PLGA Nanoparticles Improve the Oral Bioavailability of Curcumin in Rats: Characterizations and Mechanisms. *J. Agric. Food Chem.* **2011**, *59*, 9280–9289. [CrossRef]
72. Prajakta, D.; Ratnesh, J.; Chandan, K.; Suresh, S.; Grace, S.; Meera, V.; Vandana, P. Curcumin Loaded pH-Sensitive Nanoparticles for the Treatment of Colon Cancer. *J. Biomed. Nanotechnol.* **2009**, *5*, 445–455. [CrossRef]
73. Baek, J.-S.; Cho, C.-W. Surface modification of solid lipid nanoparticles for oral delivery of curcumin: Improvement of bioavailability through enhanced cellular uptake, and lymphatic uptake. *Eur. J. Pharm. Biopharm.* **2017**, *117*, 132–140. [CrossRef] [PubMed]
74. D'Archivio, A.A.; Maggi, M.A. Investigation by response surface methodology of the combined effect of pH and composition of water-methanol mixtures on the stability of curcuminoids. *Food Chem.* **2017**, *219*, 414–418. [CrossRef]
75. Gao, J.; Ming, J.; He, B.; Fan, Y.; Gu, Z.; Zhang, X. Preparation and characterization of novel polymeric micelles for 9-nitro-20(S)-camptothecin delivery. *Eur. J. Pharm. Sci.* **2008**, *34*, 85–93. [CrossRef] [PubMed]
76. Jayaprakasha, G.K.; Jaganmohan Rao, L.; Sakariah, K.K. Antioxidant activities of curcumin, demethoxycurcumin and bis-demethoxycurcumin. *Food Chem.* **2006**, *98*, 720–724. [CrossRef]
77. Lateh, L.; Kaewnopparat, N.; Yuenyongsawad, S.; Panichayupakaranant, P. Enhancing the water-solubility of curcuminoids-rich extract using a ternary inclusion complex system: Preparation, characterization, and anti-cancer activity. *Food Chem.* **2022**, *368*, 130827. [CrossRef]
78. Huang, X.; Huang, X.; Gong, Y.; Xiao, H.; McClements, D.J.; Hu, K. Enhancement of curcumin water dispersibility and antioxidant activity using core-shell protein-polysaccharide nanoparticles. *Food Res. Int.* **2016**, *87*, 1–9. [CrossRef]
79. Dos Santos, P.D.F.; Francisco, C.R.L.; Coqueiro, A.; Leimann, F.V.; Pinela, J.; Calhelha, R.C.; Porto Ineu, R.; Ferreira, I.C.F.R.; Bona, E.; Gonçalves, O.H. The nanoencapsulation of curcuminoids extracted from: Curcuma longa L. and an evaluation of their cytotoxic, enzymatic, antioxidant and anti-inflammatory activities. *Food Funct.* **2019**, *10*, 573–582. [CrossRef]
80. Vaz, V.M.; Jitta, S.R.; Verma, R.; Kumar, L. Hesperetin loaded proposomal gel for topical antioxidant activity. *J. Drug Deliv. Sci. Technol.* **2021**, *66*, 102873. [CrossRef]

Article

Synthesis of Water-Dispersed Sulfobetaine Methacrylate–Iron Oxide Nanoparticle-Coated Graphene Composite by Free Radical Polymerization

Suguna Perumal [1,2,*,†], Raji Atchudan [2,3,†] and Yong Rok Lee [2,*]

1. Department of Chemistry, Sejong University, Seoul 143-747, Korea
2. School of Chemical Engineering, Yeungnam University, Gyeongsan 38541, Korea
3. Department of Chemistry, Saveetha School of Engineering, Saveetha Institute of Medical and Technical Sciences, Chennai 602105, Tamil Nadu, India
* Correspondence: suguna.perumal@gmail.com (S.P.); yrlee@yu.ac.kr (Y.R.L.)
† These authors contributed equally to this work.

Abstract: Research on the synthesis of water-soluble polymers has accelerated in recent years, as they are employed in many bio-applications. Herein, the synthesis of poly[2-(methacryloyloxy)ethyl]dimethyl-(3-sulfopropyl)ammonium hydroxide (PSB) by free radical polymerization in a sonication bath is described. PSB and iron oxide nanoparticles (IONPs) were simultaneously stabilized on the graphene surface. Graphene surfaces with PSB (GPSB) and graphene surfaces with PSB and IONPs (GPSBI) were prepared. Since PSB is a water-soluble polymer, the hydrophobic nature of graphene surfaces converts to hydrophilic nature. Subsequently, the prepared graphene composites, GPSB and GPSBI, were well-dispersed in water. The preparation of GPSB and GPSBI was confirmed by X-ray diffraction, Raman spectroscopy, field emission scanning electron microscopy, transmission electron microscopy, X-ray photoelectron spectroscopy, and thermogravimetric analysis. The impacts of PSB and IONPs on the graphene surfaces were studied systematically.

Keywords: graphene composites; iron oxide nanoparticles; poly[2-(methacryloyloxy)ethyl]dimethyl-(3-sulfopropyl)ammonium hydroxide]; sonication bath; thin-layered graphene

1. Introduction

Continuous research is ongoing towards the development of diagnosis and therapeutic agents for detecting and treating cancer cells. Nanostructures and hybrid nanostructures are used for diagnosis and therapeutic applications [1–4]. These nanostructures can be tuned to accommodate the desired properties. In recent times, iron oxide nanomaterials (IONPs) attained considerable attention in various fields, including biomedical, diagnostic, and therapeutic applications [5–7]. The magnetic surface and intrinsic properties such as colloidal stability, low toxicity, and uniform size allowed researchers to use IONPs in different applications [7–9]. However, compared to IONPs alone, composites, especially those with nanoscale dimensions, have improved properties and are used in various applications [10–13]. In composites, research on graphene with IONPs is emphasized because of their excellent properties [14,15].

Graphene is a two-dimensional material composed of sp^2 carbon atoms [16,17]. Graphene is an excellent material that has superior mechanical, electrical, and thermal properties [18,19]. Because of these properties, graphene and graphene composites are used in broad applications such as supercapacitors, water treatment, and biomedical applications [20–25]. Size-controlled graphene sheets with IONPs [26] showed remarkable catalytic activity in oxygen reduction reaction (ORR) and oxygen evolution reaction (OER) [27]. Three-dimensional reduced graphene oxide surface with IONPs was reported as an effective active material for deionization electrodes [28]. Superparamagnetic IONPs with graphene oxide (GO) are used as a resonance contrast agent for magnetic resonance imaging [29]. GO with IONPs is suggested for resonance/fluorescence imaging and cancer sensing applications [29,30]. In addition, IONPs grafted on GO surfaces are used for hyperthermia applications [31]. The polymers on the graphene surface tune the properties of graphene composites [32–36]. Different types of biocompatible polymers are used to stabilize the graphene surface for cancer theranostics [37–41]. Recently, we published a study on graphene nanocarriers for treating thyroid cancer cells [42]. Doxorubicin-loaded 2-(methacryloxyloxy)ethyl phosphorylcholine and poly(ethylene glycol) monomethacrylate stabilized the graphene surface with IONPs, representing a remarkable nanocarrier [42].

In this work, [2-(methacryloyloxy)ethyl]dimethyl-(3-sulfopropyl)ammonium hydroxide (SB) monomer was polymerized on the graphene surface, and graphite (G) was exfoliated into thin-layered graphene sheets by sonication. Two composites were prepared in the absence and presence of IONPs, GPSB and GPSBI, respectively. The prepared GPSB and GPSBI composites were characterized using various studies.

2. Materials and Methods

2.1. Materials

SB (95%), G, and 4,4'-azobis(4-cyanovaleric acid) (ACVA, ≥98%) were purchased from Sigma-Aldrich, South Korea, and used as received. Deionized water (DI) was used in all experiments. Using a bath sonicator (40 kHz, 110 W, BRANSON 3800, Richmond VA, USA), in situ polymerization of SB monomer was performed on the graphene surface at 70 °C. The composites were centrifuged using VS-18000M, VISION Scientific Co., Ltd., Daejeon-Si, Korea.

2.2. Methods

PSB, GPSB, and GPSBI were characterized using various physicochemical techniques. Raman spectra for composites were obtained on the XploRA Micro-Raman spectrophotometer (Horiba) in the range between 1000 and 3000 cm^{-1}. X-ray diffraction (XRD) studies were carried out using the PANalytical X'Pert^3 MRD diffractometer with monochromatized Cu Kα radiation (λ = 1.54 Å) at 40 kV and 30 mA and were recorded in the range from 20° to 80° (2θ). Field emission scanning electron microscopy (FESEM) with energy-dispersive X-ray spectroscopy (EDS) was used to evaluate the surface morphology of the composites. Using the Hitachi S-4800 equipped with EDX at an accelerating voltage of 10 kV, FESEM and EDS measurements were carried out. Transmission electron microscopy (TEM) images were obtained from JEOL JEM with an operating accelerating voltage of 120 kV. X-ray photoelectron spectroscopy (XPS) spectra were recorded using K-Alpha (Thermo Scientific, Waltham, MA, USA), and CasaXPS software was used for the deconvolution of the high-resolution XPS spectra. Thermogravimetric analysis (TGA) measurements were carried out on SDT Q600 with nitrogen atmosphere over 0–900 °C with 10 °C/min.

2.3. Graphene-poly[2-(Methacryloyloxy)ethyl]dimethyl-(3-sulfopropyl)ammonium Hydroxide] Composite

The preparation of graphene-poly[2-(methacryloyloxy)ethyl]dimethyl-(3-sulfopropyl) ammonium hydroxide] composite (GPSB) was prepared as shown in Scheme 1. Monomer SB (500 mg, 1.78 mmol), ACVA (25.0 mg, 0.089 mmol), and 250 mg of G in 70 mL of DI water were heated at 70 °C for 6 h in a sonication bath. Then, the composite GPSB was purified by centrifugation at 5000 rpm for 15 min. Triplicate centrifugation followed by drying in a freeze dryer yielded fine powder of GPSB that was re-dispersed in DI water for further characterization.

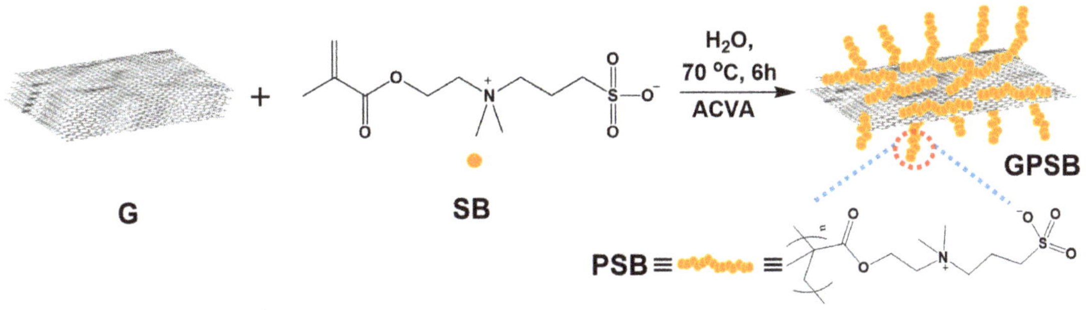

Scheme 1. Synthesis of graphene-poly[2-(methacryloyloxy)ethyl]dimethyl-(3-sulfopropyl)ammonium hydroxide] composite.

2.4. Graphene-poly[2-(Methacryloyloxy)ethyl]dimethyl-(3-sulfopropyl)ammonium Hydroxide]–Iron Oxide Nanoparticle Composite

The preparation of the graphene-poly[2-(methacryloyloxy)ethyl]dimethyl-(3-sulfopropyl) ammonium hydroxide]–iron oxide nanoparticle composite (GPSBI) was prepared as shown in Scheme 2. To prepare GPSBI, IONPs were prepared following earlier reports [43,44]. Iron acetylacetonate (3.0 g, 8.49 mmol) in 60 mL of benzyl alcohol was added to an autoclave container and heated at 180 °C for 48 h. The precipitates were purified by washing with ethanol and then centrifuged. Successively the precipitates were further washed with dichloromethane and then centrifuged. The purified IONPs were dried in a hot-air oven at 60 °C and used for the preparation of GPSBI composite. Monomer SB (500 mg, 1.78 mmol), ACVA (25.0 mg, 0.089 mmol), 250 mg of G, and 50 mg of IONPs were added to 70 mL of DI water, and the mixture was heated at 70 °C for 6 h in a sonication bath. The composite GPSBI was then purified by centrifugation at 5000 rpm for 15 min. Washing was performed three times with water and the composite was dried in a freeze dryer. The obtained fine powder of GPSBI was re-dispersed in DI water for further characterization.

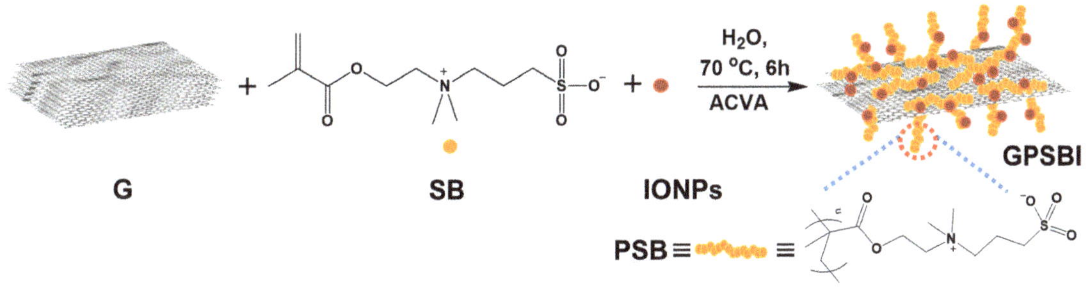

Scheme 2. Synthesis of the graphene-poly[2-(methacryloyloxy)ethyl]dimethyl-(3-sulfopropyl)ammonium hydroxide]–iron oxide nanoparticle composite.

3. Result and Discussion

In situ polymerizations of SB monomer on the graphene surface were carried out, and the graphite was exfoliated into thin layers with the smaller size of the graphene sheets. To know the molecular weight of PSB on the graphene surface, PSB was prepared by adopting the same procedure as GPSB and GPSBI by excluding graphene powder. The prepared PSB was characterized using size-exclusion chromatography (SEC) and TGA. The molecular number of PSB was measured as 24,536 g/mol using SEC (refer to supporting information). The GPSB and GPSBI composites were characterized using various physicochemical techniques. The composites were compared with G and IONPs.

X-ray diffraction (XRD) patterns have been utilized to study the defect polymer-stabilized graphene sheets in composites. The XRD patterns obtained for G, IONPs, GPSB, and GPSBI can be seen in Figure 1a. In G, the typical graphitic peaks at 2θ = 26.4° with d-spacing of 3.34 Å (Bragg law: d = n × λ/2θ; n = 1, λ = 0.154 nm) and at 2θ = ~55° correspond to (002) and (004) planes, respectively [45,46]. The diffraction peaks of IONPs observed at 30.33°, 35.77°, 43.43°, 53.81°, 57.44°, and 62.98° correspond to the planes (220), (311), (222), (400), (422), (511), and (440), respectively [47–50]. The size (Scherer formula: D = k × λ/(β × cosθ); k = 0.9, λ = 0.154 nm, β = full width half maximum) of IONPs was calculated as ~17.0 nm using the (311) plane [51]. The diffraction peak in GPSB was observed at 26.64°, corresponding to the (002) plane. The GPSBI composite shows peaks at 26.67°, 30.33°, 35.77°, 43.43°, 53.81°, 57.44°, and 62.98° that are attributed to (002-graphitic) and iron oxide peaks (220), (311), (222), (400), (422), (511), and (440), respectively. The interlayer distance of graphene sheets in GPSB and GPSBI were calculated as 3.36 and 3.30 Å; a slight increment in the interlayer distance suggests the partial exfoliation of graphite into graphene. Furthermore, compared with G, the diffraction peaks of GPSB and GPSBI show a slight shift and decrement in intensity. This might be due to defects in graphene sheets because of the presence of PSB and IONPs. However, the interlayer distance in GPSBI is higher than G but lower than the interlayer distance in GPSB; this can be explained by considering the presence of IONPs on graphene in addition to PSB.

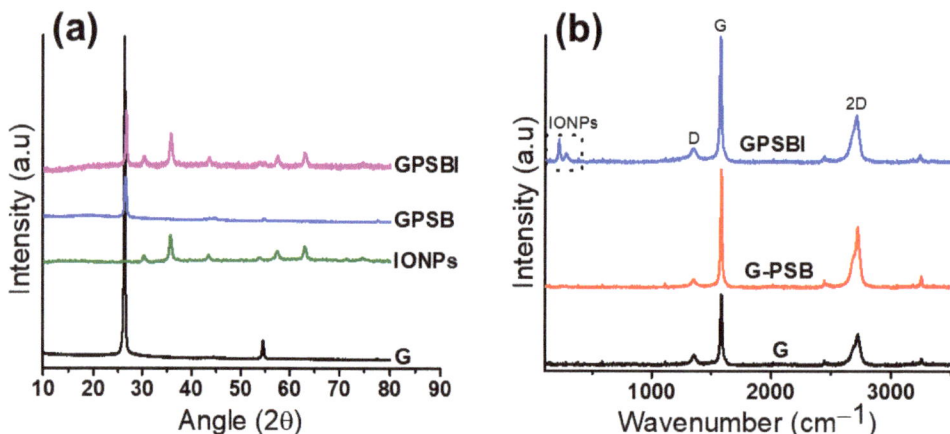

Figure 1. (**a**) XRD patterns corresponding to G, IONPs, GPSB, and GPSBI. (**b**) Raman spectra of G, GPSB, and GPSBI.

Figure 1b presents the Raman spectra of G, GPSB, and GPSBI, showing three strong peaks at ~1350, ~1580, and ~2700 cm^{-1} corresponding to D, G, and 2D bands, respectively. The D band represents the defect sites in the graphene sheets at edges and surfaces and the size of the graphitic crystals [52,53]. The G band arises from the sp^2 carbon–carbon bond from the first-order scattering of the E$_{2g}$ phonon [54]. I$_D$/I$_G$ ratios of G, GPSB, and GPSBI were calculated as 0.13, 0.15, and 0.29, respectively. The I$_D$/I$_G$ ratio represents the degree of disorder and inversely relates to the size of graphene sheets [55]. The I$_D$/I$_G$ ratio result suggests that the size of graphene is decreasing in the order of G, GPSB, and GPSBI. In addition, the disorder in the composite increases, which confirms the functionalization of the graphene surface with PSB in GPSB and with PSB and IONPs in GPSBI. The 2D band around 2700 cm^{-1} indicates the layer of graphene sheets in the composites [56,57]. Compared to G, 2D bands in GPSB and GPSBI are sharp, indicating the thin-layered graphene sheets compared to the graphene sheets in G. Additionally, GPSBI has two sharp, distinct peaks and two small, broad peaks at ~210, ~277, ~380, and ~580 cm^{-1}, attributed to the IONPs, confirmed by the Raman spectra of IONPs shown in Figure S1.

The surface morphology of the G, IONPs, GPSB, and GPSBI was examined by FESEM and TEM measurements. The FESEM images of G and corresponding elemental mapping are shown in Figure S2. The FESEM image of graphite flakes reveals graphene sheets with lateral sizes of 10 ± 4 µm. The elemental mapping measurement confirms the uniform distribution of carbon (C) elements and the fair distribution of oxygen (O) elements on the G surface. Figure S3 depicts the FESEM images of IONPs, and their elemental mapping is shown in Figure S3b. The even distribution of O and iron (Fe) elements is clear from the images (Figure S3c,d).

Figure 2 displays the FESEM images (a–c) of GPSB and their elemental mapping (d–g). GPSB possesses plate-like morphology with homogenous granular size. The size of graphene sheets is much smaller than graphene sheets in G; the lateral size of graphene sheets in GPSB was measured as 3 ± 1 µm. The elemental mapping of GPSB (Figure 2b) was further analyzed to study the existing elements. The results revealed that C (Figure 2d), nitrogen (N) (Figure 2e), O (Figure 2f), and sulfur (S) (Figure 2g) are evenly distributed on the graphene sheets. The intensity of element O is significantly improved compared to element O in G. The elements (O, N, and S) originated from PSB, indicating the successful formation of composite GPSB and functionalization of graphene sheets with PSB.

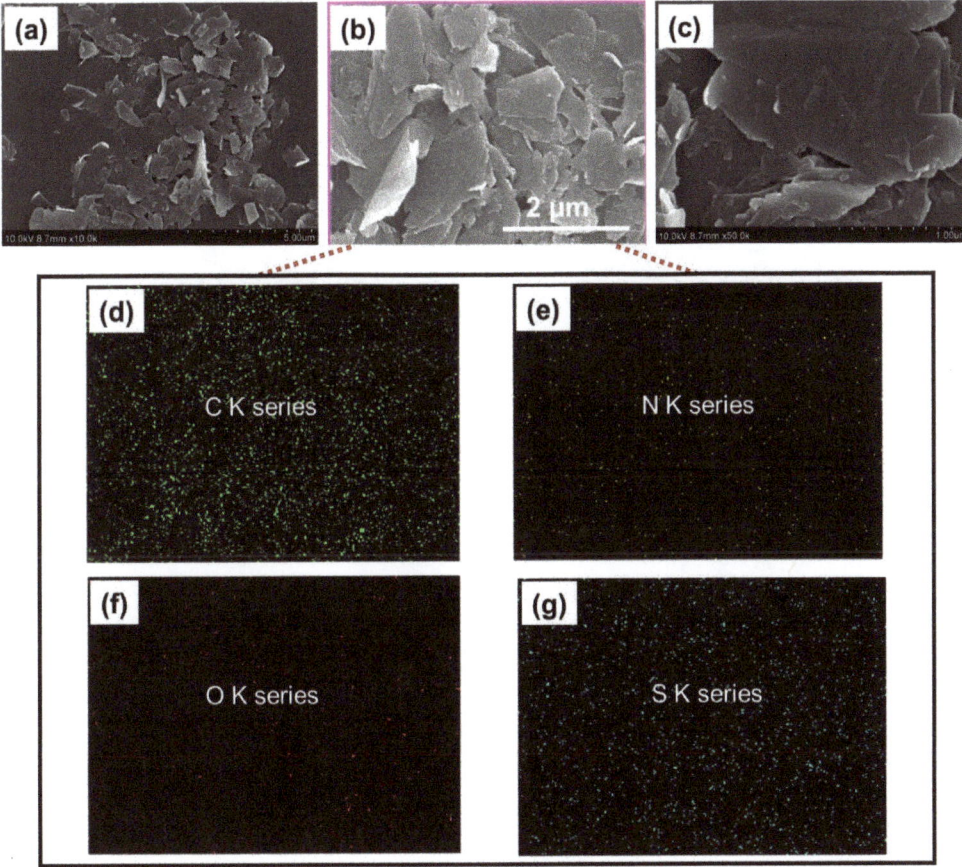

Figure 2. FESEM images of GPSB (**a–c**). Elemental mapping of GPSB C K series (**d**), N K series (**e**), O K series (**f**), and S K series (**g**).

FESEM images, along with the elemental mapping of GPSBI, are depicted in Figure 3. As can be seen in Figure 3, the prepared GPSBI exhibits the plate-like structure of graphene sheets with IONPs. Moreover, a large amount of IONPs are distributed homogeneously on the graphene sheets. The average lateral size of graphene sheets was measured as 3 ± 0.5 μm. The presence of elements C, N, O, and S from PSB and Fe and O from IONPs were confirmed from the elemental mapping images, as shown in Figure 3d–h.

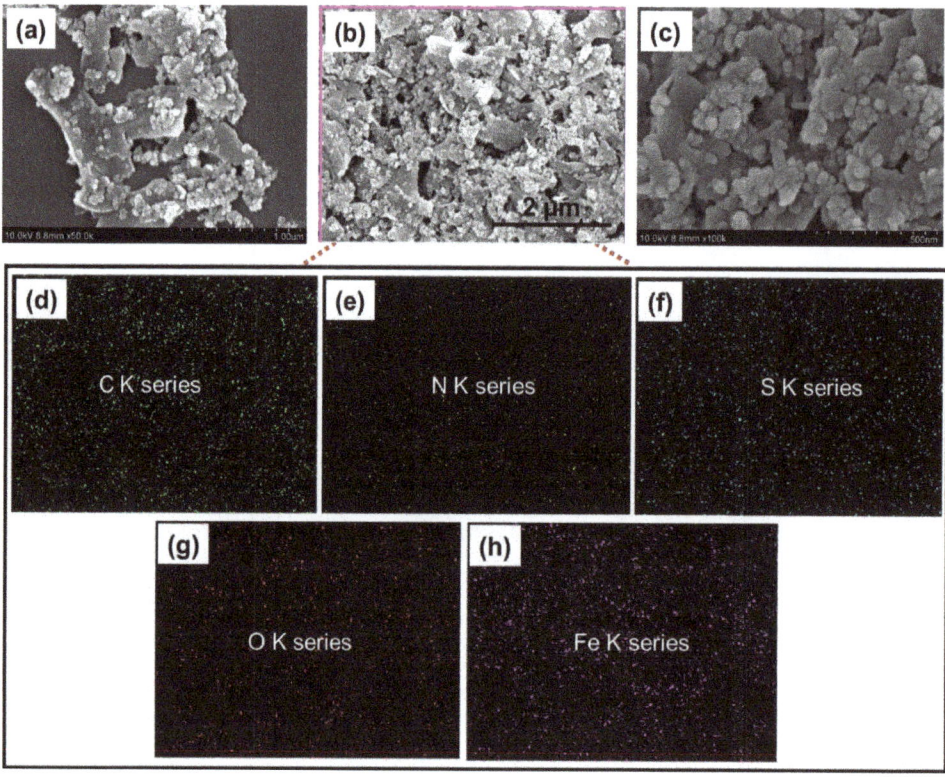

Figure 3. FESEM images of GPSBI (**a–c**). Elemental mapping of GPSBI C K series (**d**), N K series (**e**), S K series (**f**), O K series (**g**), and Fe K series (**h**).

The morphologies of G, IONPs, GPSB, and GPSBI were further investigated using TEM measurements. The TEM images of G depicted in Figure S4 reveal the plate-like, wrinkled structure of graphene sheets. The lateral size of the graphene sheets in G was measured as 9 ± 4 μm.

Figure 4 exhibits the morphology of GPSB and GPSBI composites. Figure 4a–c illustrates the TEM images of GPSB composites, displaying the layered, folded, wrinkled structure with many small graphene sheets on large graphene sheets. The lateral size of graphene sheets in GPSB was measured as 2.5 ± 0.2 μm. Figure S5 exposes the aggregated spherical shape IONPs particles with an average size of about 20 ± 8 nm. The size of IONPs measured from XRD (~17 nm) and TEM (20 ± 8 nm) shows a similar result. Aggregated IONPs (Figure S5) suggest that the surface is not well stabilized with the stabilizing molecules/benzyl alcohol used during the preparation of IONPs. However, the IONPs in GPSBI composites (Figure 4d) show a uniform distribution. In addition, Figure 4d–f reveals the thin graphene sheets compared to graphene sheets in GPSB (Figure 4a–c). This reveals that PSB and IONPs have effective interaction with graphene sheets in GPSBI. Thus, the uniform distribution of IONPs on the graphene surface is observed in the TEM image. The lateral size of graphene sheets in GPSBI was measured at 1.5 ± 0.2 µm, and IONPs were measured at 18 ± 5 nm, showing that the size of the IONPs is maintained in the stabilization of the graphene surface.

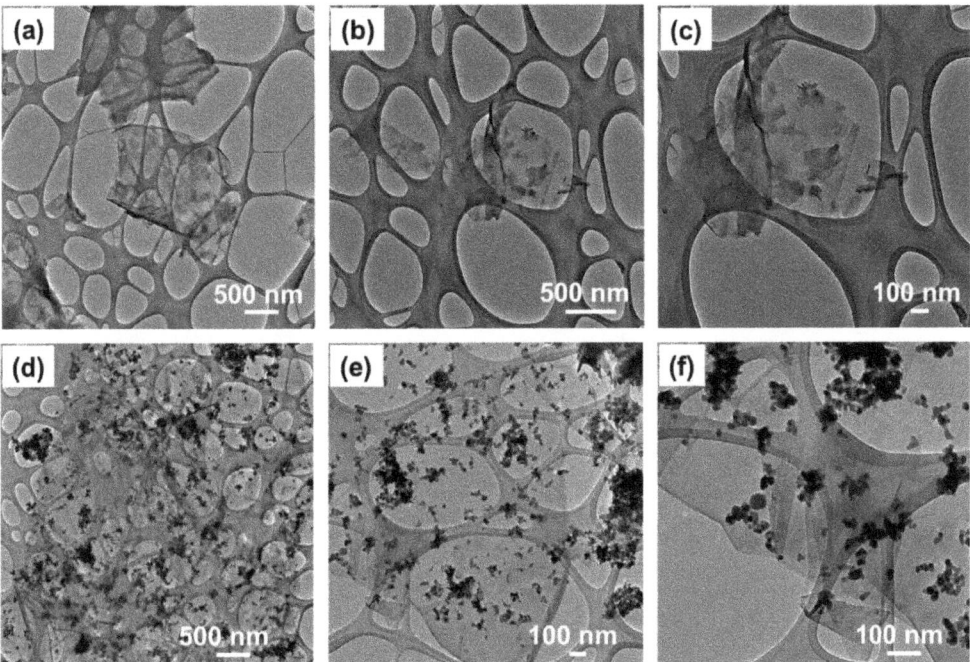

Figure 4. TEM images of GPSB (**a–c**) and GPSBI (**d–f**) with different magnifications.

The XPS data provide further information about the chemical composition and formation of GPSB and GPSBI. The survey XPS spectrum of G is depicted in Figure S6, with peaks at binding energy of ~286 and ~530 eV attributed to C 1s and O 1s, respectively (Figure S6a). The C 1s level deconvoluted into four peaks at 284.76, 285.52, 286.22, and 286.92 eV, attributed to C=C/C-C, C-O, C=O, and O=C-OH, respectively (Figure S6b). The deconvoluted XPS spectra of O 1s for G shown in Figure S6c show oxygen in different functional groups: C=O (531.80 eV), C-O (533.16 eV), and O=C-O (533.84 eV). Figure S6d shows the survey spectrum of IONPs, which contains Fe and O elements. Figure S6e displays the Fe 2p high-resolution spectrum, with two prominent peaks at ~711 and 724 eV assigned to Fe $2p_{3/2}$ and Fe $2p_{1/2}$, respectively, consistent with the reported IONPs [58,59]. Deconvolution of Fe $2p_{3/2}$ shows three peaks at 710.98, 714.54, and 719.01 eV attributed to Fe^{3+}, Fe^{2+}, and satellite peaks of Fe $2p_{3/2}$, respectively. Three peaks observed on deconvolution of Fe $2p_{1/2}$ at 724.08, 727.13, and 732.36 eV correspond to Fe^{3+}, Fe^{2+}, and satellite Fe peak of Fe $2p_{1/2}$, respectively. This result reveals that the major component of IONPs is Fe^{3+} with a minor component of Fe^{2+}.

The XPS spectra of GPSB are shown in Figure 5, and the survey spectra of GPSB are depicted in Figure 5a. The survey spectrum indicates the presence of elements such as S 2p, C 1s, N 1s, and O 1s at 170, 285, 403, and 530 eV, respectively. On deconvolution of C 1s (Figure 5b), five peaks were observed at 284.76, 285.59, 286.37, 287.13, and 289.12 eV, responsible for C=C/C-C, C-N/C-SO_3, C-O, C=O, and O=C-O/O-C-SO_3, respectively. The SO_3 arises from the sulfonate group in PSB [60,61]. The deconvoluted XPS spectra of N 1s for GPSB, shown in Figure 5c, reveal oxygen atoms in different functional groups:

−N− (399.75 eV) and −N(CH$_3$)$_3$$^+$ (402.65 eV) [60]. The O 1s level (Figure 5d) shows four peaks at 531.28, 532.40, 533.62, and 534.45 eV corresponding to C=O, C-O, O=C-OH, and H-O-H (moisture), respectively. Figure 5e shows the S 2p level with deconvolution results in four peaks at 164.06, 165.22, 167.68, and 168.89 eV attributed to oxidized S, C-SO$_3$$^{2-}$ 2p$_{3/2}$, and C-SO$_3$$^{2-}$ 2p$_{1/2}$, respectively. The presence of −SO$_3$ and −N(CH$_3$)$_3$$^+$ confirms the functionalization of the graphene surface with PSB.

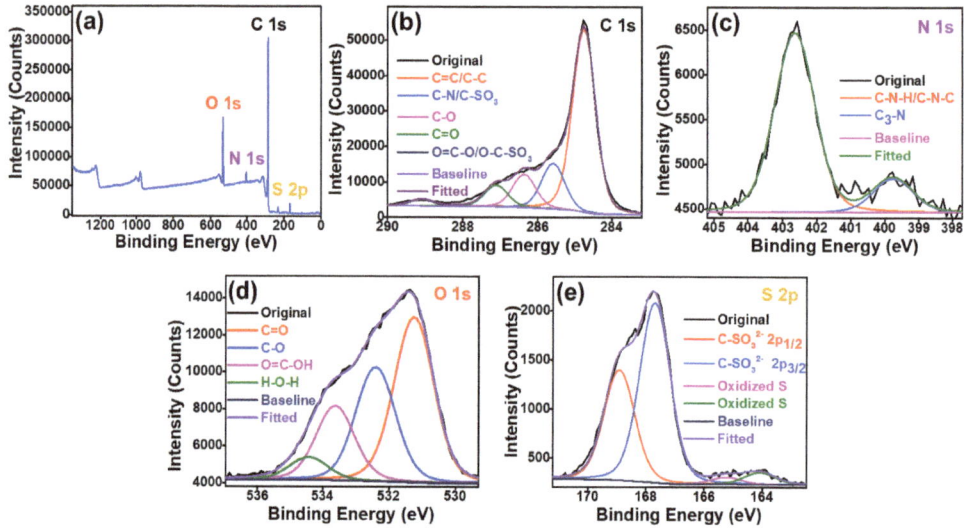

Figure 5. XPS spectra of GPSB: (**a**) survey spectra, (**b**) C 1s, (**c**) N 1s, (**d**) O 1s, and S 2p (**e**).

Figure 6 shows the XPS spectra of GPSBI. The survey spectra show distinct peaks of S 2p, C 1s, N 1s, O 1s, and Fe 2p at 170, 284, 401, 532, and 712 eV, respectively (Figure 6a). Five peaks were observed on deconvolution of the C 1s level (Figure 6b) at 284.80, 285.68, 286.51, 287.31, and 289.22 eV responsible for C=C/C-C, C-N/C-SO$_3$, C-OH/C-O-C, C=O, and O=C-O/O=C-SO$_3$, respectively. Deconvolution of the N 1s level results in two peaks at 400.14 and 402.78 eV, attributed to the presence of −N(CH$_3$)$_3$$^+$ and −N- groups in GPSBI, respectively (Figure 6c). Four peaks at 530.56, 531.65, 532.76, and 534.06 eV were observed on the deconvolution of O 1s (Figure 6d), responsible for C=O, C-O/Fe-O, O=C-OH, and H-O-H, respectively. Figure 6e depicts the high-resolution spectra of GPSBI at the S 2p level, resulting in four peaks on deconvolution. Two trace peaks of oxidized S at 164.21 and 165.36 eV and major peaks at 167.85 and 169.06 eV are responsible for C-SO$_3$$^{2-}$ 2p$_{3/2}$ and C-SO$_3$$^{2-}$ 2p$_{1/2}$, respectively. At the Fe 2p level, GPSBI showed six peaks on deconvolution, which is clear from Figure 6f. Three peaks were observed at the Fe 2p$_{3/2}$ level at 711.41, 714.46, and 719.35 eV, which are attributed to Fe^{3+}, Fe^{2+}, and Fe 2p$_{3/2}$ satellite peaks, respectively. Similarly, the Fe 2p$_{1/2}$ level showed peaks at 724.47, 727.35, and 732.81 eV, which are responsible for Fe^{3+}, Fe^{2+}, and Fe 2p$_{1/2}$ satellite peaks, respectively. This result reveals that the major composition of IONPs in GPSBI is Fe^{3+}. XPS studies reveal that the major component of IONPs is Fe$_3$O$_4$, and it has a minor component, α-Fe$_2$O$_3$. Moreover, the structure of IONPS remains the same in GPSB, and GPSBI structural changes were not observed.

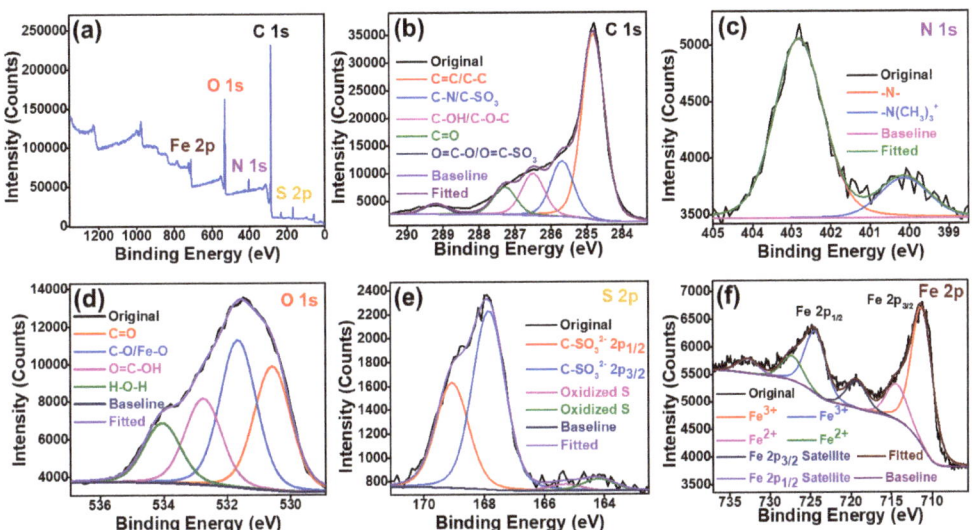

Figure 6. XPS spectra of GPSBI: (**a**) survey spectra, (**b**) C 1s, (**c**) N 1s, (**d**) O 1s, (**e**) S 2p, and (**f**) Fe 2p.

The thermal behaviors of G, IONPs, PSB, and prepared composites were studied by TGA. G, IONPs, GPSB, GPSBI, and PSB samples were subjected to thermal decomposition up to 900 °C in the inert atmosphere, as shown in Figure 7. The thermal pattern of G and IONPs clearly show residue of about 98.19% and 95.01%, respectively. This reveals the stability of G and IONPs. The 5% loss in IONPs and 2% loss in G might be responsible for the degradation of adsorbed water molecules or moisture on the surface. PSB showed three-step degradation, with weight loss of 5% below 200 °C, weight loss of 16.5% between 230 and 320 °C, and weight loss of 84.5% above 350 until 450 °C, which are attributed to adsorbed water molecules, the degradation of nitrogen atoms from quaternary ammonium salt, and sulfate groups, respectively [62–64]. The remaining polymer degraded and showed a residual weight of about 5.66% at ~900 °C. The composites GPSB and GPSBI showed three-step degradation. The 5% weight loss observed for GPSB around 100 °C was due to the degradation of adsorbed water molecules. As a second step, polymer degradation takes place with a weight loss of 23.58% between 250 and 320 °C (nitrogen groups). The sulfate groups degraded between 320 and 425 °C with a weight loss of 32.6%. Furthermore, GPSB showed residual weight loss of 60.5% observed until ~900 °C, which was attributed to the degradation of left polymer structures and decomposition of most stable oxygen functionalities [65,66]. In GPSBI composites, 5% weight loss at about 100 °C was due to the degradation of water molecules that began at about 70 °C. The PSB polymer decomposition occurred in two steps at 250–320 °C (nitrogen atoms) and 350–420 °C (sulfonate) with weight loss of ~23% and ~28%, respectively. From ~450 °C to 730 °C, weight loss might be due to the degradation of the remaining part of the polymer and degradation of IONPs left with the residual weight of about 58.8% at around 900 °C [66].

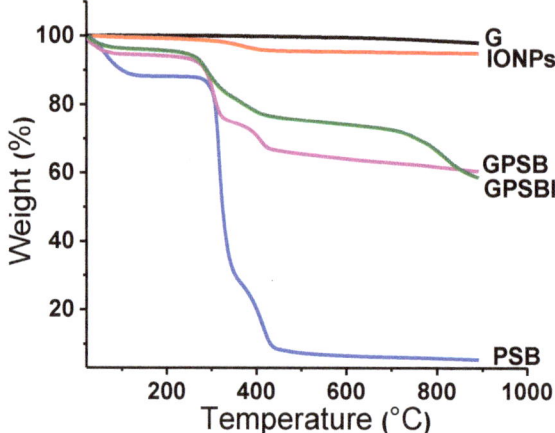

Figure 7. TGA thermograms of G, IONPs, GPSB, GPSBI, and PSB.

The overall studies revealed that PSB interacts with graphene surfaces through ester, quaternary ammonium, and oxygen from sulfonate units. Furthermore, sulfonate units from PSB interact simultaneously with IONPs and graphene surfaces. These interactions help to thoroughly disperse GPSB and GPSBI composites in water. Thus, the hydrophobic nature of the graphene surface can be converted to hydrophilic nature and can be applied to bio-applications.

4. Conclusions

We successfully prepared GPSB and GPSBI composites using a simple method. XRD and Raman studies confirm the partial exfoliation of graphite into thin-layered graphene sheets in GPSB and GPSBI composites. TEM and SEM measurements revealed the size of graphene sheets in GPSB and GPSBI as 3 ± 0.2 μm. These graphene sheets are smaller than the graphene sheets that were in G. From XRD and TEM studies, IONP size was measured as ~20 nm. The elemental mapping further showed the uniform distribution of PSB and IONPs on graphene sheets. XPS and TGA revealed the presence of PSB and IONPs in the prepared composites. TGA showed the residual weight as 60.5% and 58.8% for GPSB and GPSBI, respectively. This shows the amount of graphene that exists in the prepared composites. The combination of PSB with graphene surface showed remarkable structural and physicochemical properties with IONPs. These composites with outstanding functionalities can be potential candidates for biomedicine applications.

Supplementary Materials: The following supporting information can be downloaded at: https://www.mdpi.com/article/10.3390/polym14183885/s1. Scheme 1: synthesis of poly[2-Methacryloyloxy]ethyl] dimethyl-(3-sulfopropyl)ammonium hydroxide], Figure S1: Raman spectrum of IONPs, Figure S2: FESEM images and elemental mapping of G, Figure S3: FESEM images and elemental mapping of IONPs, Figure S4: TEM images of G, Figure S5: TEM images of INOPs, and Figure S6: XPS spectra of G and IONPs.

Author Contributions: The manuscript was written through the contributions of all authors. All authors have read and agreed to the published version of the manuscript.

Funding: This work was supported by a National Research Foundation of Korea (NRF) grant funded by the Korean government MSIT (2021R1A2B5B02002436).

Institutional Review Board Statement: Not applicable.

Informed Consent Statement: Not applicable.

Data Availability Statement: Not applicable.

Conflicts of Interest: The authors declare no competing financial interest.

References

1. Yetisgin, A.A.; Cetinel, S.; Zuvin, M.; Kosar, A.; Kutlu, O. Therapeutic Nanoparticles and Their Targeted Delivery Applications. *Molecules* **2020**, *25*, 2193. [CrossRef] [PubMed]
2. Chenthamara, D.; Subramaniam, S.; Ramakrishnan, S.G.; Krishnaswamy, S.; Essa, M.M.; Lin, F.H.; Qoronfleh, M.W. Therapeutic efficacy of nanoparticles and routes of administration. *Biomater. Res.* **2019**, *23*, 20. [CrossRef] [PubMed]
3. Mitchell, M.J.; Billingsley, M.M.; Haley, R.M.; Wechsler, M.E.; Peppas, N.A.; Langer, R. Engineering precision nanoparticles for drug delivery. *Nat. Rev. Drug Discov.* **2021**, *20*, 101–124. [CrossRef] [PubMed]
4. Tharkar, P.; Varanasi, R.; Wong, W.S.F.; Jin, C.; Chrzanowski, W. Nano-Enhanced Drug Delivery and Therapeutic Ultrasound for Cancer Treatment and Beyond. *Front. Bioeng. Biotechnol.* **2019**, *7*, 324. [CrossRef] [PubMed]
5. Arias, L.S.; Pessan, J.P.; Vieira, A.P.M.; de Lima, T.M.T.; Delbem, A.C.B.; Monteiro, D.R. Iron Oxide Nanoparticles for Biomedical Applications: A Perspective on Synthesis, Drugs, Antimicrobial Activity, and Toxicity. *Antibiotics* **2018**, *7*, 46. [CrossRef] [PubMed]
6. Hou, Z.; Liu, Y.; Xu, J.; Zhu, J. Surface engineering of magnetic iron oxide nanoparticles by polymer grafting: Synthesis progress and biomedical applications. *Nanoscale* **2020**, *12*, 14957–14975. [CrossRef] [PubMed]
7. Wu, W.; Jiang, C.Z.; Roy, V.A.L. Designed synthesis and surface engineering strategies of magnetic iron oxide nanoparticles for biomedical applications. *Nanoscale* **2016**, *8*, 19421–19474. [CrossRef] [PubMed]
8. Dave, P.N.; Chopda, L.V. Application of Iron Oxide Nanomaterials for the Removal of Heavy Metals. *J. Nanotechnol.* **2014**, *2014*, 398569. [CrossRef]
9. Sangaiya, P.; Jayaprakash, R. A Review on Iron Oxide Nanoparticles and Their Biomedical Applications. *J. Supercond. Nov. Magn.* **2018**, *31*, 3397–3413. [CrossRef]
10. Schwaminger, S.P.; Brammen, M.W.; Zunhammer, F.; Däumler, N.; Fraga-García, P.; Berensmeier, S. Iron Oxide Nanoparticles: Multiwall Carbon Nanotube Composite Materials for Batch or Chromatographic Biomolecule Separation. *Nanoscale Res. Lett.* **2021**, *16*, 30. [CrossRef]
11. Mourdikoudis, S.; Kostopoulou, A.; LaGrow, A.P. Magnetic Nanoparticle Composites: Synergistic Effects and Applications. *Adv. Sci.* **2021**, *8*, 2004951. [CrossRef] [PubMed]
12. Pillarisetti, S.; Uthaman, S.; Huh, K.M.; Koh, Y.S.; Lee, S.; Park, I.-K. Multimodal Composite Iron Oxide Nanoparticles for Biomedical Applications. *Tissue Eng. Regen. Med.* **2019**, *16*, 451–465. [CrossRef] [PubMed]
13. Zare, E.N.; Abdollahi, T.; Motahari, A. Effect of functionalization of iron oxide nanoparticles on the physical properties of poly (aniline-co-pyrrole) based nanocomposites: Experimental and theoretical studies. *Arab. J. Chem.* **2020**, *13*, 2331–2339. [CrossRef]
14. Darabdhara, G.; Borthakur, P.; Das, M.R.; Szunerits, S.; Boukherroub, R. Iron Oxide Nanoparticles-Graphene Composite Materials: Synthesis, Characterization and Applications. In *Handbook of Carbon Nano Materials*; Springer: Cham, Switzerland, 2020; pp. 265–309.
15. Rodríguez, B.A.G.; Pérez-Caro, M.; Alencar, R.S.; Filho, A.G.S.; Aguiar, J.A. Graphene nanoribbons and iron oxide nanoparticles composite as a potential candidate in DNA sensing applications. *J. Appl. Phys.* **2020**, *127*, 044901. [CrossRef]
16. Suarez-Martinez, I.; Grobert, N.; Ewels, C.P. Nomenclature of sp2 carbon nanoforms. *Carbon* **2012**, *50*, 741–747. [CrossRef]
17. Sheikhi, M.; Shahab, S.; Balali, E.; Alnajjar, R.; Kaviani, S.; Khancheuski, M.; Al Saud, S. Study of the Ribavirin drug adsorption on the surfaces of carbon nanotube and graphene nanosheet using density functional theory calculations. *Bull. Korean Chem. Soc.* **2021**, *42*, 1446–1457. [CrossRef]
18. Al Faruque, M.A.; Syduzzaman, M.; Sarkar, J.; Bilisik, K.; Naebe, M. A Review on the Production Methods and Applications of Graphene-Based Materials. *Nanomaterials* **2021**, *11*, 2414. [CrossRef] [PubMed]
19. Liao, G.; Hu, J.; Chen, Z.; Zhang, R.; Wang, G.; Kuang, T. Preparation, Properties, and Applications of Graphene-Based Hydrogels. *Front. Chem.* **2018**, *6*, 450. [CrossRef]
20. Velasco, A.; Ryu, Y.K.; Boscá, A.; Ladrón-de-Guevara, A.; Hunt, E.; Zuo, J.; Pedrós, J.; Calle, F.; Martinez, J. Recent trends in graphene supercapacitors: From large area to microsupercapacitors. *Sustain. Energy Fuels* **2021**, *5*, 1235–1254. [CrossRef]
21. Liu, C.; Yu, Z.; Neff, D.; Zhamu, A.; Jang, B.Z. Graphene-Based Supercapacitor with an Ultrahigh Energy Density. *Nano Lett.* **2010**, *10*, 4863–4868. [CrossRef]
22. Bhol, P.; Yadav, S.; Altaee, A.; Saxena, M.; Misra, P.K.; Samal, A.K. Graphene-Based Membranes for Water and Wastewater Treatment: A Review. *ACS Appl. Nano Mater.* **2021**, *4*, 3274–3293. [CrossRef]
23. Safarpour, M.; Khataee, A. Chapter 15—Graphene-Based Materials for Water Purification. In *Nanoscale Materials in Water Purification*; Thomas, S., Pasquini, D., Leu, S.-Y., Gopakumar, D.A., Eds.; Elsevier: Amsterdam, The Netherlands, 2019; pp. 383–430.
24. Shen, H.; Zhang, L.; Liu, M.; Zhang, Z. Biomedical applications of graphene. *Theranostics* **2012**, *2*, 283–294. [CrossRef] [PubMed]
25. Chung, C.; Kim, Y.-K.; Shin, D.; Ryoo, S.-R.; Hong, B.H.; Min, D.-H. Biomedical Applications of Graphene and Graphene Oxide. *Acc. Chem. Res.* **2013**, *46*, 2211–2224. [CrossRef]
26. Lyubutin, I.S.; Baskakov, A.O.; Starchikov, S.S.; Shih, K.-Y.; Lin, C.-R.; Tseng, Y.-T.; Yang, S.-S.; Han, Z.-Y.; Ogarkova, Y.L.; Nikolaichik, V.I.; et al. Synthesis and characterization of graphene modified by iron oxide nanoparticles. *Mater. Chem. Phys.* **2018**, *219*, 411–420. [CrossRef]

27. Hof, F.; Liu, M.; Valenti, G.; Picheau, E.; Paolucci, F.; Pénicaud, A. Size Control of Nanographene Supported Iron Oxide Nanoparticles Enhances Their Electrocatalytic Performance for the Oxygen Reduction and Oxygen Evolution Reactions. *J. Phys. Chem. C* **2019**, *123*, 20774–20780. [CrossRef]
28. Belaustegui, Y.; Rincón, I.; Fernández-Carretero, F.; Azpiroz, P.; García-Luís, A.; Tanaka, D.A.P. Three-dimensional reduced graphene oxide decorated with iron oxide nanoparticles as efficient active material for high performance capacitive deionization electrodes. *Chem. Eng. J. Adv.* **2021**, *6*, 100094. [CrossRef]
29. Gonzalez-Rodriguez, R.; Campbell, E.; Naumov, A. Multifunctional graphene oxide/iron oxide nanoparticles for magnetic targeted drug delivery dual magnetic resonance/fluorescence imaging and cancer sensing. *PLoS ONE* **2019**, *14*, e0217072. [CrossRef]
30. Kim, J.D.; Choi, H.C. Enzyme-free H_2O_2 Sensing at ZnO–Graphene Oxide-modified Glassy Carbon Electrode. *Bull. Korean Chem. Soc.* **2021**, *42*, 25–28. [CrossRef]
31. Umar, A.A.; Patah, M.F.A.; Abnisa, F.; Daud, W.M.A.W. Preparation of magnetized iron oxide grafted on graphene oxide for hyperthermia application. *Rev. Chem. Eng.* **2020**, *38*, 569–601. [CrossRef]
32. Pinto, A.M.; Magalhães, F.D. Graphene-Polymer Composites. *Polymers* **2021**, *13*, 685. [CrossRef]
33. Ganguly, S.; Kanovsky, N.; Das, P.; Gedanken, A.; Margel, S. Photopolymerized Thin Coating of Polypyrrole/Graphene Nanofiber/Iron Oxide onto Nonpolar Plastic for Flexible Electromagnetic Radiation Shielding, Strain Sensing, and Non-Contact Heating Applications. *Adv. Mater. Interfaces* **2021**, *8*, 2101255. [CrossRef]
34. Ibrahim, A.; Klopocinska, A.; Horvat, K.; Abdel Hamid, Z. Graphene-Based Nanocomposites: Synthesis, Mechanical Properties, and Characterizations. *Polymers* **2021**, *13*, 2869. [CrossRef] [PubMed]
35. Ramachandra Kurup Sasikala, A.; Thomas, R.G.; Unnithan, A.R.; Saravanakumar, B.; Jeong, Y.Y.; Park, C.H.; Kim, C.S. Multifunctional Nanocarpets for Cancer Theranostics: Remotely Controlled Graphene Nanoheaters for Thermo-Chemosensitisation and Magnetic Resonance Imaging. *Sci. Rep.* **2016**, *6*, 20543. [CrossRef] [PubMed]
36. Liu, J.; Yang, W.; Tao, L.; Li, D.; Boyer, C.A.; Davis, T.P. Thermosensitive graphene nanocomposites formed using pyrene terminal polymers made by RAFT polymerization. *J. Polym. Sci. Part A* **2010**, *48*, 425–433. [CrossRef]
37. Sivakumar, M.P.; Islami, M.; Zarrabi, A.; Khosravi, A.; Peimanfard, S. Polymer-Graphene Nanoassemblies and their Applications in Cancer Theranostics. *Anti-Cancer Agents Med. Chem.* **2020**, *20*, 1340–1351. [CrossRef]
38. Liu, L.; Ma, Q.; Cao, J.; Gao, Y.; Han, S.; Liang, Y.; Zhang, T.; Song, Y.; Sun, Y. Recent progress of graphene oxide-based multifunctional nanomaterials for cancer treatment. *Cancer Nanotechnol.* **2021**, *12*, 18. [CrossRef]
39. Dash, B.S.; Jose, G.; Lu, Y.-J.; Chen, J.-P. Functionalized Reduced Graphene Oxide as a Versatile Tool for Cancer Therapy. *Int. J. Mol. Sci.* **2021**, *22*, 2989. [CrossRef]
40. Gurunathan, S.; Jeyaraj, M.; Kang, M.-H.; Kim, J.-H. Graphene Oxide–Platinum Nanoparticle Nanocomposites: A Suitable Biocompatible Therapeutic Agent for Prostate Cancer. *Polymers* **2019**, *11*, 733. [CrossRef]
41. Sattari, S.; Adeli, M.; Beyranvand, S. Functionalized Graphene Platforms for Anticancer Drug Delivery. *Int. J. Nanomed.* **2021**, *16*, 5955–5980. [CrossRef]
42. Perumal, S.; Gangadaran, P.; Bae, Y.W.; Ahn, B.-C.; Cheong, I.W. Noncovalent Functionalized Graphene Nanocarriers from Graphite for Treating Thyroid Cancer Cells. *ACS Biomater. Sci. Eng.* **2021**, *7*, 2317–2328. [CrossRef]
43. Atchudan, R.; Perumal, S.; Jebakumar Immanuel Edison, T.N.; Lee, Y.R. Facile synthesis of monodisperse hollow carbon nanospheres using sucrose by carbonization route. *Mater. Lett.* **2016**, *166*, 145–149. [CrossRef]
44. Pinna, N.; Grancharov, S.; Beato, P.; Bonville, P.; Antonietti, M.; Niederberger, M. Magnetite Nanocrystals: Nonaqueous Synthesis, Characterization, and Solubility. *Chem. Mater.* **2005**, *17*, 3044–3049. [CrossRef]
45. Zhang, K.; Zhang, Y.; Wang, S. Enhancing thermoelectric properties of organic composites through hierarchical nanostructures. *Sci. Rep.* **2013**, *3*, 3448. [CrossRef]
46. Çakmak, G.; Öztürk, T. Continuous synthesis of graphite with tunable interlayer distance. *Diam. Relat. Mater.* **2019**, *96*, 134–139. [CrossRef]
47. Girod, M.; Vogel, S.; Szczerba, W.; Thünemann, A.F. How temperature determines formation of maghemite nanoparticles. *J. Magn. Magn. Mater.* **2015**, *380*, 163–167. [CrossRef]
48. Schwaminger, S.P.; Syhr, C.; Berensmeier, S. Controlled Synthesis of Magnetic Iron Oxide Nanoparticles: Magnetite or Maghemite? *Crystals* **2020**, *10*, 214. [CrossRef]
49. Karami, H. Synthesis and Characterization of Iron Oxide Nanoparticles by Solid State Chemical Reaction Method. *J. Clust. Sci.* **2010**, *21*, 11–20. [CrossRef]
50. Ajinkya, N.; Yu, X.; Kaithal, P.; Luo, H.; Somani, P.; Ramakrishna, S. Magnetic Iron Oxide Nanoparticle (IONP) Synthesis to Applications: Present and Future. *Materials* **2020**, *13*, 4644. [CrossRef]
51. He, K.; Chen, N.; Wang, C.; Wei, L.; Chen, J. Method for Determining Crystal Grain Size by X-Ray Diffraction. *Cryst. Res. Technol.* **2018**, *53*, 1700157. [CrossRef]
52. Kim, S.-G.; Park, O.-K.; Lee, J.H.; Ku, B.-C. Layer-by-layer assembled graphene oxide films and barrier properties of thermally reduced graphene oxide membranes. *Carbon Lett.* **2013**, *14*, 247–250. [CrossRef]
53. Bîru, E.I. *Graphene Nanocomposites Studied by Raman Spectroscopy*; IntechOpen: London, UK, 2018.
54. Mohan, V.B.; Bhattacharyya, M.S.; Liu, D.; Jayaraman, K. Improvements in Electronic Structure and Properties of Graphene Derivatives. *Adv. Mater. Lett.* **2016**, *7*, 421–429. [CrossRef]

55. Strankowski, M.; WBodarczyk, D.; Piszczyk, A.; Strankowska, J. Polyurethane Nanocomposites Containing Reduced Graphene Oxide, FTIR, Raman, and XRD Studies. *Spectroscopy* **2016**, *2016*, 1–6. [CrossRef]
56. Ferrari, A.C. Raman spectroscopy of graphene and graphite: Disorder, electron–phonon coupling, doping and nonadiabatic effects. *Solid State Commun.* **2007**, *143*, 47–57. [CrossRef]
57. Frank, O.; Mohr, M.; Maultzsch, J.; Thomsen, C.; Riaz, I.; Jalil, R.; Novoselov, K.S.; Tsoukleri, G.; Parthenios, J.; Papagelis, K.; et al. Raman 2D-Band Splitting in Graphene: Theory and Experiment. *ACS Nano* **2011**, *5*, 2231–2239. [CrossRef] [PubMed]
58. Arunima, R.; Madhulika, S.; Niroj, K.S. Assessing magnetic and inductive thermal properties of various surfactants functionalised Fe_3O_4 nanoparticles for hyperthermia. *Sci. Rep.* **2020**, *10*, 15045. [CrossRef]
59. Li, S.; Xia, X.; Vogt, B.D. Microwave-Enabled Size Control of Iron Oxide Nanoparticles on Reduced Graphene Oxide. *Langmuir* **2021**, *37*, 11131–11141. [CrossRef]
60. Chien, H.-W.; Lin, H.-Y.; Tsai, C.-Y.; Chen, T.-Y.; Chen, W.-N. Superhydrophilic Coating with Antibacterial and Oil-Repellent Properties via NaIO(4)-Triggered Polydopamine/Sulfobetaine Methacrylate Polymerization. *Polymers* **2020**, *12*, 2008. [CrossRef]
61. Yang, W.J.; Neoh, K.-G.; Kang, E.-T.; Lay-Ming Teo, S.; Rittschof, D. Stainless steel surfaces with thiol-terminated hyperbranched polymers for functionalization via thiol-based chemistry. *Polym. Chem.* **2013**, *4*, 3105–3115. [CrossRef]
62. Hildebrand, V.; Laschewsky, A.; Päch, M.; Müller-Buschbaum, P.; Papadakis, C.M. Effect of the zwitterion structure on the thermo-responsive behaviour of poly(*sulfobetaine methacrylates*). *Polym. Chem.* **2017**, *8*, 310–322. [CrossRef]
63. Galin, M.; Marchal, E.; Mathis, A.; Meurer, B.; Soto, Y.M.M.; Galin, J.C. Poly(sulphopropylbetaines): 3. Bulk properties. *Polymer* **1987**, *28*, 1937–1944. [CrossRef]
64. Schönemann, E.; Laschewsky, A.; Wischerhoff, E.; Koc, J.; Rosenhahn, A. Surface Modification by Polyzwitterions of the Sulfabetaine-Type, and Their Resistance to Biofouling. *Polymers* **2019**, *11*, 1014. [CrossRef] [PubMed]
65. Shen, J.; Hu, Y.; Shi, M.; Lu, X.; Qin, C.; Li, C.; Ye, M. Fast and Facile Preparation of Graphene Oxide and Reduced Graphene Oxide Nanoplatelets. *Chem. Mater.* **2009**, *21*, 3514–3520. [CrossRef]
66. Ibrahim, G.P.S.; Isloor, A.M.; Ismail, A.F.; Farnood, R. One-step synthesis of zwitterionic graphene oxide nanohybrid: Application to polysulfone tight ultrafiltration hollow fiber membrane. *Sci. Rep.* **2020**, *10*, 6880. [CrossRef] [PubMed]

Article

Semiconducting Soft Submicron Particles from the Microwave-Driven Polymerization of Diaminomaleonitrile

Marta Ruiz-Bermejo [1,*], Pilar García-Armada [2], Pilar Valles [3] and José L. de la Fuente [3]

1 Departamento de Evolución Molecular, Centro de Astrobiología (INTA-CSIC), Ctra. Torrejón-Ajalvir, km 4, Torrejón de Ardoz, 28850 Madrid, Spain
2 Department of Industrial Chemical Engineering, Escuela Técnica Superior de Ingenieros Industriales, Universidad Politécnica de Madrid, José Gutiérrez Abascal, 2, 28006 Madrid, Spain
3 Instituto Nacional de Técnica Aeroespacial "Esteban Terradas" (INTA), Ctra. Torrejón-Ajalvir, km 4, Torrejón de Ardoz, 28850 Madrid, Spain
* Correspondence: ruizbm@cab.inta-csic.es; Tel.: +34-915206458

Abstract: The polymers based on diaminomaleonitrile (DAMN polymers) are a special group within an extensive set of complex substances, namely HCN polymers (DAMN is the formal tetramer of the HCN), which currently present a growing interest in materials science. Recently, the thermal polymerizability of DAMN has been reported, both in an aqueous medium and in bulk, offering the potential for the development of capacitors and biosensors, respectively. In the present work, the polymerization of this plausible prebiotic molecule has been hydrothermally explored using microwave radiation (MWR) via the heating of aqueous DAMN suspensions at 170–190 °C. In this way, polymeric submicron particles derived from DAMN were obtained for the first time. The structural, thermal decomposition, and electrochemical properties were also deeply evaluated. The redox behavior was characterized from DMSO solutions of these highly conjugated macromolecular systems and their potential as semiconductors was described. As a result, new semiconducting polymeric submicron particles were synthetized using a very fast, easy, highly robust, and green-solvent process. These results show a new example of the great potential of the polymerization assisted by MWR associated with the HCN-derived polymers, which has a dual interest both in chemical evolution and as functional materials.

Keywords: HCN polymers; DAMN; electrochemical properties; soft submicron particles; microwave-driven polymerization

Citation: Ruiz-Bermejo, M.; García-Armada, P.; Valles, P.; de la Fuente, J.L. Semiconducting Soft Submicron Particles from the Microwave-Driven Polymerization of Diaminomaleonitrile. *Polymers* 2022, 14, 3460. https://doi.org/10.3390/polym14173460

Academic Editor: Suguna Perumal

Received: 29 July 2022
Accepted: 15 August 2022
Published: 24 August 2022

Publisher's Note: MDPI stays neutral with regard to jurisdictional claims in published maps and institutional affiliations.

Copyright: © 2022 by the authors. Licensee MDPI, Basel, Switzerland. This article is an open access article distributed under the terms and conditions of the Creative Commons Attribution (CC BY) license (https://creativecommons.org/licenses/by/4.0/).

1. Introduction

Diaminomaleonitrile, DAMN, is a π-conjugated compound with electron donor (-NH$_2$) and electron acceptor (-CN) substituents, and a high electron affinity, which features a strong intramolecular charge transfer interaction. Its high nitrogen content has meant that it has often been used as a versatile precursor for heterocycles, for example, imidazoles, triazoles, porphyrazines, pyrimidines, pyrazines, diazepines, and purines. DAMN can be used as synthetic precursor of fluorescent dyes, jet-printing links, hair dyes, and biologically active compounds such as insecticides and anticancer agents [1]. In addition, DAMN has potential in the area of chemosensors for the sensing of ionic and neutral species because of its ability to act as a building block for well-defined molecular architectures and scaffolds for preorganized arrays of functionality, as it has been reviewed in the literature [2]. Moreover, there is currently a growing interest in the development of macromolecular systems based on DAMN. In this sense, DAMN is the formal tetramer of hydrogen cyanide, HCN, and the potential applications of the DAMN polymers should be understood in the light of the up-to-date attention of the HCN polymers in the field of materials science. The DAMN polymers are a particular group within the more general set of the HCN-derived polymers.

HCN-derived polymers, or simply HCN polymers, can be currently considered as a new promising type of multifunctional soft materials [3–5]. This heterogeneous family of complex substances present interesting properties that consequently encourage the development of emergent photocatalyzers [6], semiconductors, nanowires, ferroelectric materials [7], capacitors [8], coatings with potential biomedical applications [9–11], protective films against corrosion [12,13] and bacteria [14], antimicrobial media for passive filters [15], and biosensors [16] based on the HCN oligomerization/polymerization chemistry.

HCN polymers have been obtained under multiple conditions, i.e., using net liquid HCN together with a basic catalyzer such as ammonia or trimethylamine [17,18]; by the irradiation of HCN gas or aqueous solutions [19–21]; from alkaline water solutions of pure HCN or their soluble salts such as NaCN or KCN (see e.g., [22–28]); through the wet polymerization of its trimer, aminomalononitrile (AMN), at room temperature [29]; by DAMN, using thermal activation, both in bulk [16,30–32] and in an aqueous solution [8]; and even by the heating of its hydrolysis product, formamide [33]. Moreover, the generation of HCN polymers assisted by the microwave radiation (MWR) of formamide [34] and cyanide [35] has been recently described.

The advantages of using MW dielectric heating for performing polymerization reactions are well known (e.g., the remarkable reduction in polymerization time, improved monomer conversion, and clearer reactions than the ones conducted under a conventional thermal system (CTS)). In addition, the production of HCN polymers at the nanoscale was only achieved when this innovative synthetic methodology was applied to aqueous NH_4CN polymerizations varying the reaction times and the temperatures between 130 and 205 °C [36]. In the present work, the production of soft submicron particles from the MW-driven polymerization of DAMN is described, whereas conventional thermal polymerization conditions were found to be ineffective in this regard [8]. On the other hand, it has been suggested that the integration of polymeric HCN-derived nanoparticles as fillers in different matrixes could lead to the design of novel composite materials [8].

Beyond these morphological aspects, the redox behavior of the several HCN polymers is remarkable. The NH_4CN polymers can be considered as insolating materials; on the contrary, the DAMN polymers synthetized under hydrothermal conditions present a capacitive performance [8]. Even DAMN polymers produced in bulk (in the absence of a solvent), in both solid and molten state, have shown to be efficient materials for the modification of electrodes and the development of biosensors [16]. Thus, it seems interesting to explore the DAMN polymers for the design of electronic devices and the potential generation of new composites with appropriate electrochemical properties.

While considering the potential of the assisted MW syntheses for the production of polymeric HCN-derived nanoparticles and the singular electrochemical characteristics of the DAMN polymers previously reported; herein, a new generation of submicron-sized highly conjugated polymeric materials are described together with a detailed study of their structural, thermal, and redox behaviors. In concrete, DAMN polymers were synthetized at 170 and 190 °C under the presence of air or an inert atmosphere of nitrogen. Therefore, the results reported in the present work not only can be of interest for the design of coming multifunctional materials, but also in the field of the origins of life, since they could enrich the understanding of prebiotic chemistry [37].

2. Materials and Methods

2.1. Synthesis of the DAMN Polymers

All the DAMN polymers synthetized in this work were synthetized from aqueous suspensions of 135 mg of commercial DAMN (98%, purchased from Sigma Aldrich (St. Louis, MO, USA) and used as received) in 5 mL of distilled water, under air or N_2, using 20 mL capacity vials fitted with Teflon/silicone septa. The reaction times, temperatures, and the work pressures for each experiment are indicated in the Table 1. A Biotage Initiator + microwave reactor purchased by Biotage (Sweden) was used in the present study. The power supplied by the reactor is in the range from 0 to 400 W from a magnetron at 2.45 GHz, with a power of about 70–95 W for the temperatures chosen in this work. No additional efforts were made regarding the heating ramp to obtain the desired work temperature, as was the case for the NH$_4$CN polymerizations [35]. No pressure peaks were observed for the release of gaseous species such as ammonia. The final dark suspensions were vacuum filtered using glass fiber filters (Merck Millipore Ltd. (Burlington, MA, USA) and washed with water. For details of the MW reactor and of the filtration system, please see Figure 1 in [36]. In addition, Scheme 1 shows a graphical diagram for the synthesis of the DAMN polymers described herein. The gel fractions were collected and dried under reduced pressure. Polymeric conversions, α, were calculated as α (%) = [(final weight of insoluble dark polymer/initial weight of DAMN)·100].

Table 1. Experimental conditions for the production of DAMN polymers assisted by MWR.

Polymer	T (°C)	t (min)	Air	P (bar) [a]	α (%)	C/H	C/N	N/O [b]
1	170	16	-	12	36 ± 3	1.00 ± 0.17	1.28 ± 0.02	2.47 ± 0.18
2	190	3.2	-	17	33 ± 3	1.02 ± 0.09	1.25 ± 0.02	2.83 ± 0.11
3	170	16	+	12	38 ± 2	1.08 ± 0.16	1.28 ± 0.01	2.66 ± 0.20
4	190	3.2	+	17	33 ± 2	1.02 ± 0.02	1.26 ± 0.01	2.70 ± 0.04

[a] Pressures reached using the hydrothermal conditions discussed herein. At least three independent experiments were carried out to calculate α values. In the same way, at least three independent samples were measured to determinate the molar relationships of C/H, C/N, and N/O. [b] The O percentage was calculated by difference from the results of the CHN elemental analysis.

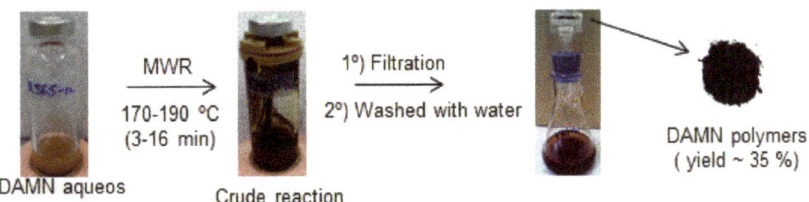

Scheme 1. General synthetic procedure for the MW-driven polymerization of DAMN. In particular, the imagines show the synthesis of polymer **4** (Table 1).

2.2. Characterization of the DAMN Polymers

2.2.1. Elemental Analysis

Approximately 5 mg of polymer sample was examined for determination of the mass fractions of carbon, hydrogen, and nitrogen by using a PerkinElmer elemental analyzer, model CNHS-2400. The oxygen content of the samples was calculated by means of the difference.

2.2.2. FTIR Spectroscopy

Diffuse-reflectance spectra were acquired in the 4000–400 cm^{-1} spectral region with an FTIR spectrometer (Nicolet, model NEXUS 670) configured with a DRIFT reflectance accessory (Harrick, model Praying Mantis DRP) mounted inside the instrument compartment. The spectra were obtained in CsI pellets, and the spectral resolution was 2 cm^{-1}.

2.2.3. C Solid-State Cross Polarization/Magic Angle Spinning Nuclear Magnetic Resonance (CP/MAS) NMR

^{13}C CP MAS NMR spectra were obtained with a Bruker Advance 400 spectrometer and a standard cross-polarization pulse sequence. Samples were spun at 10 kHz, and the spectrometer frequency was set to 100.62 MHz. A contact time of 1 ms and a period between successive accumulations of 5 s were used. The number of scans was 5000, and the chemical shift values were referenced to TMS.

2.2.4. Powder XRD

Powder XRD was performed with a Bruker D8 Eco Advance with $Cu_{K\alpha}$ radiation (λ = 1.542 Å) and a Lynxeye XE-T linear detector. The X-ray generator was set to an acceleration voltage of 40 kV and a filament emission of 25 mA. Samples were scanned between 5 (2θ) and 50° (2θ) with a step size of 0.05° and count time of 1 s in Bragg–Brentano geometry.

2.2.5. UV-Vis Spectroscopy

UV-Vis spectra were obtained using an Agilent 8453 spectrophotometer. All spectra were recorded in DMSO [38].

2.2.6. Thermal Analysis

Thermogravimetry (TG), differential scanning calorimetry (DSC), and differential thermal analysis (DTA) were performed with a simultaneous thermal analyzer model SDTQ-600/Thermo Star from TA Instruments. Non-isothermal experiments were carried out under dynamic conditions from room temperature to 1000 °C at a heating rate of 10 °C min^{-1} under argon atmosphere. The average sample weight was approximately 20 mg, and the argon flow rate was 100 mL min^{-1}. In the present case, only the TGA data were discussed. A coupled TG-mass spectrometer (TG-MS) system equipped with an electron-impact quadrupole mass-selective detector (model Thermostar QMS200 M3) was employed to analyze the main species that evolved during the dynamic thermal decomposition of all of the samples.

2.2.7. Scanning Electron Microscopy (SEM)

The surface morphologies of the polymers were determined by a ThermoScientific Apreo C-LV field emission electron microscope (FE-SEM) equipped with an Aztec Oxford energy-dispersive X-ray microanalysis system (EDX). The samples were coated with 4 nm of chromium using a sputtering Leica EM ACE 600. The images were obtained at 10 kV.

2.2.8. Particle Size Analysis

The values of the Z-average and polydispersity index (PdI) were registered using a Zetasizer Nano instrument (Malvern Instruments Ltd., Almelo, the Netherlands), employing ethanol as a solvent to scatter the samples. Approximately 0.2 mg of the polymeric sample was suspended in 2 mL of ethanol and sonicated for 15 min, and the resulting solution was then measured.

2.2.9. Surface Area

Brunauer–Emmett–Teller (BET) surface areas were evaluated by nitrogen adsorption–desorption isotherms obtained at −196 °C using a Micromeritics ASAP 2010 device. Before each measurement, the samples were degassed at 250 °C for at least 3 h.

2.2.10. Density

Density for DAMN polymers' data were obtained in a helium pycnometer AccuPyc 1340 of Micromeritics.

2.2.11. Gel-Permeation Chromatography (GPC)

The apparent molecular weights of the DAMN polymers were estimated by gel-permeation chromatography (GPC) using poly(methyl methacrylate) as standard (Polymer Laboratories, Laboratories, Ltd. (USA) ranging from 2.4×10^6 to 9.7×10^2 g/mol) from suspensions of 8 mg/mL of DAMN polymers in dimethylformamide (DMF).

2.3. Electrochemical Measurements

An Ecochemie BV Autolab PGSTAT 12 with a conventional three-electrode cell at 20–21 °C was used in the electrochemical studies. The electrodes were a Pt disc (3 mm diameter) as working electrode, a Pt wire as auxiliary, and an Ag/AgCl/KCl 3M (E = 0.205 vs. ENH) as reference electrode. All electrochemical measurements were in dimethyl sulfoxide (DMSO) with NaClO$_4$ 0.1 M as supporting electrolyte. All the polymer DMSO solutions were stable and showed no changes even during or after the electrochemical experiments.

3. Results and Discussion

3.1. Synthetic Conditions and Chemical Composition of the DAMN Polymeric Submicron Particles

Considering the technical specifications of the MW reactor's manufacture, the reaction times used here in each case are roundly equivalent to a reaction time of 144 h at 80 °C using a CTS. It is known that in the case of cyanide polymers, if equivalent reaction times are used, the main structural characteristics are preserved when the hydrothermal polymerization reactions are carried out using a CTS or under MWR, and only the morphology (shape and size) of the polymeric particles is modified [36]. Observe that the properties of the HCN polymers are greatly affected in the synthetic conditions [3]. Therefore, it is mandatory to check if the MWR could only alter the morphology of the DAMN polymeric particles (as in the case of the cyanide polymers) or if it would also have an influence on the structural characteristics of the macromolecular system. For that, a previously well-studied analogous DAMN polymer synthetized using CTS was used as a reference polymer (RP) for comparative purposes. Note that no submicron particles were observed for this RP synthetized under these particular conditions (0.25 M water suspensions of DAMN heating at 80 °C for 144 h under atmospheric pressure) [8].

Table 1 shows the experimental conditions used for the production of soft submicron particles based on the MW-driven DAMN polymerization, and two working temperatures, 170 and 190 °C, were chosen since the temperature could also have a significant influence on the morphological properties of the DAMN polymers [36]. On the other hand, since the final properties of the materials are directly dependent on their structure, it is also mandatory to have as much control as possible on the experimental conditions of polymerization, including the working atmosphere. It has been shown that the atmosphere seems to have a notable influence on the crystallographic properties of HCN polymers, increasing the crystallinity when the reactions are carried out using an inert atmosphere [35]. Thus, experiments under nitrogen (Table 1, polymers **1** and **2**), or under atmospheric air (Table 1, polymers **3** and **4**) were carried out. In addition, the experiments conducted under anoxic conditions could simulate prebiotic hydrothermal conditions, because DAMN can be considered as an important prebiotic precursor [24].

The conversion (α) obtained for all the reactions were all similar and about 35% was found (Table 1), independently of the pressure atmospheric work (air or nitrogen). Therefore, the hydrothermal polymerizations of DAMN do not seem to be influenced by the presence of air, as was the case of the MW-driven polymerizations of cyanide. In that case, the yield of the reactions was substantially diminished under an inert nitrogen atmosphere [36]. However, the lower conversions reached for polymers **1–4** compared to the RP (~75%) seems to indicate that the secondary oxidation processes from DAMN to diiminosuccinonitrile, and the hydrolysis reactions from DAMN to the formation of formamide, glycine, and aminomalonic acid, could be notably increased by the high temperatures and pressures used in the present study, in agreement with a previous proposal made by Ferris and Edelson [39] (Scheme 2). Moreover, it is important to mention that we also observed experimentally via gas chromatography-mass spectrometry (GC-MS) the formation of urea, oxalic acid, glycine, and formamide in the DAMN polymerization processes under hydrothermal conditions using a CTS (data not shown). Thus, the minor available amount of DAMN likely leads to a diminution of the precipitated DAMN polymers.

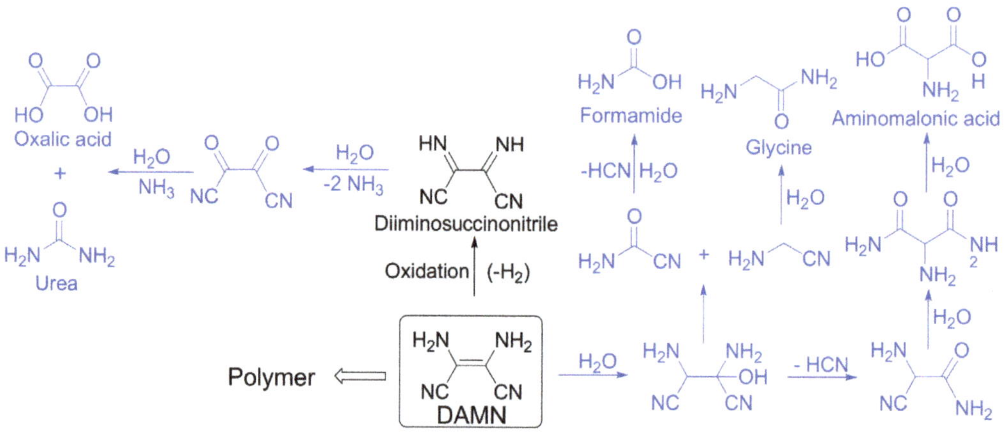

Scheme 2. Products of hydrolysis and oxidation of DAMN proposed by Ferris and Edelson [39], and according to our recent results.

On the other hand, the results of the elemental analysis indicate very similar chemical compositions for the four polymers (Table 1). Moreover, the N/O ratios are comparable to the RP, while the C/N values are slightly greater; however, the C/H data are somewhat lower please see [8]. The roughly empirical formulas for polymers **1–4** are $C_3H_3N_2O$, $C_{13}H_{13}N_{10}O_4$, $C_{15}H_{14}N_{12}O_5$, $C_{12}H_{12}N_{10}O_4$, and respectively; and $C_5H_5N_4O_2$ for the RP.

3.2. Structural Characterization

The structural characteristics of polymers **1–4** inferred by FTIR, ^{13}C NMR, and XRD corroborated the high similarity between them and also with the RP [8] (Figure 1). The IR spectra of polymers **1–4** show four regions centered at 3343, 2202, 1653, and 640 cm^{-1}, which have been largely described elsewhere (see e.g., [38,40,41]). In brief, these bands can be related to amine and hydroxyl groups, nitrile, and imine bonds (Figure 1a). The nitrile bands are more intense than that observed in the equivalent RP sample. This observation can be quantitatively calculated by the EOR values (extension of the reaction, [42]) (Table 2). However, no other significant differences were observed between the FTIR spectra from polymers **1–4** and from the RP. The values of the EOC (extension of the conjugation [35]) were comparable for all the polymers (Table 2). Thus, at first glance, this quantitative spectroscopic analysis seems to indicate that the DAMN polymers synthetized using MWR or a CTS present similar macrostructures, only with a slight increase in the EOR value. These differences can also be observed by ^{13}C NMR and XRD.

Table 2. Summary of some quantitative parameters from the FTIR spectra, such as EOR and EOC, taking into account the intensity of the band centered around 3340, 2200, and 1650 cm^{-1}. EOR = $[I_{1650}/(I_{1650} + I_{2200})] \times 100$ and EOC = $[I_{1650}/(I_{1650} + I_{3340})] \times 100$.

	EOR (%)		EOC (%)	
Conditions	170 °C	190 °C	170 °C	190 °C
N$_2$/MWR	76 ± 2	74 ± 2	47 ± 1	48 ± 1
Air/MWR	77 ± 3	73 ± 1	48 ± 2	48 ± 1
	EOR (%)		EOC (%)	
Air/CTS	86 ± 1		45 ± 1	

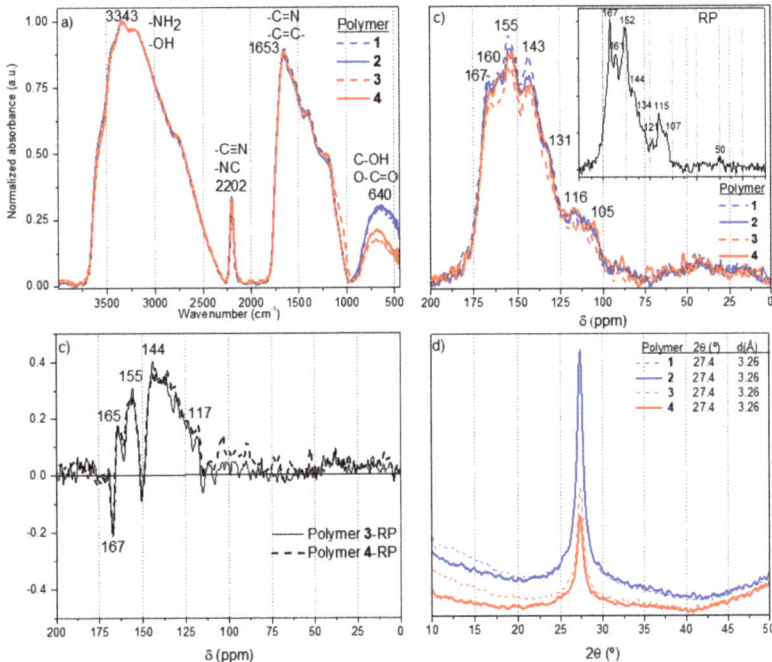

Figure 1. (**a**) Representative FTIR spectra of polymers **1–4**. (**b**) Solid state ^{13}C NMR spectra of polymers **1–4**. In the insert plot, the ^{13}C NMR spectrum of the RP is shown [8]; (**c**) Subtractions of the ^{13}C NMR spectra; (**d**) Representative XRD patters of polymers **1–4**.

The solid-state CP MAS ^{13}C NMR spectra for polymers **1–4** are very similar between themselves and no notable differences were found, showing broad and unresolved resonances at 167, 160, 155, 143, 131, 116, and 105 ppm (Figure 1b). These same resonances were also observed in the ^{13}C NMR spectrum of the analogous RP (insert plot in Figure 1b). These signals can be assigned to carbon in nitrogenous functional groups, such as imines and nitriles, or being part of imidazoles or pyrazines rings; additionally, they can also be assigned to the carbon of the oxygenated groups such as amides and on heterocycles such as oxazoles [8]. The subtractions of these spectra of polymers **3** and **4** to the ^{13}C NMR spectrum of the RP are indicated in Figure 1c. The macrostructures of the polymers **3** and **4** present additional peaks at 165, 155, 144, and 117 ppm with respect to the RP. The first one can be attributed to C=O from esters of the amides, the second resonance to –C=N- from azomethine groups and/or from N-heterocycles, the broad signal with a maximum at 144 ppm can be assigned to C=C in heteroaromatic systems, and the last one at 117 ppm with nitrile groups (in good agreement with the FTIR spectra). Thus, it seems that the DAMN polymers from the MWR syntheses present a greater portion of N-heterocycles and aromatic carbons than the RP. It was proposed that the DAMN polymers synthetized under conventional hydrothermal conditions are complex mixtures of extended conjugated systems, lineal or cyclic, based on their NMR spectra [8], as it is shown in Scheme 3. The comparison between the FTIR data and the subtractions of the ^{13}C NMR spectra appears to indicate that the MWR improves the formation of the N-heterocyclic fractions and the polyamide portion in the structures of the DAMN polymers. The XRD patters of polymers **1–4** show only one peak at 2θ = 27.4°, which can be related to the (002) diffraction of the graphitic layers [37], with interlayer distances of 3.26 Å in all the cases (Figure 1d). However, they are considerably different to those found for the RP [8], since the weak peak related to the (100) reflections is missing when the polymers are synthesized by MWR.

This result may be related to a different in-plane structural packing motif. In this way, polymers **1–4** could present amorphous, disordered, in-plane and out-plane structures with only a nanoscopic range order by the stacking of the main structural motif due to the lack of peaks at lower 2θ [43,44]. On the other hand, when the polymerization of DAMN is carried out under a nitrogen atmosphere, the intensity of the peaks is increased while their width is decreased. This fact probably points out a higher order in the stacking of the two-dimensional (2-D) macrostructures for polymers **1** and **2** (Figure 1d), in agreement with previous results [35] as commented above.

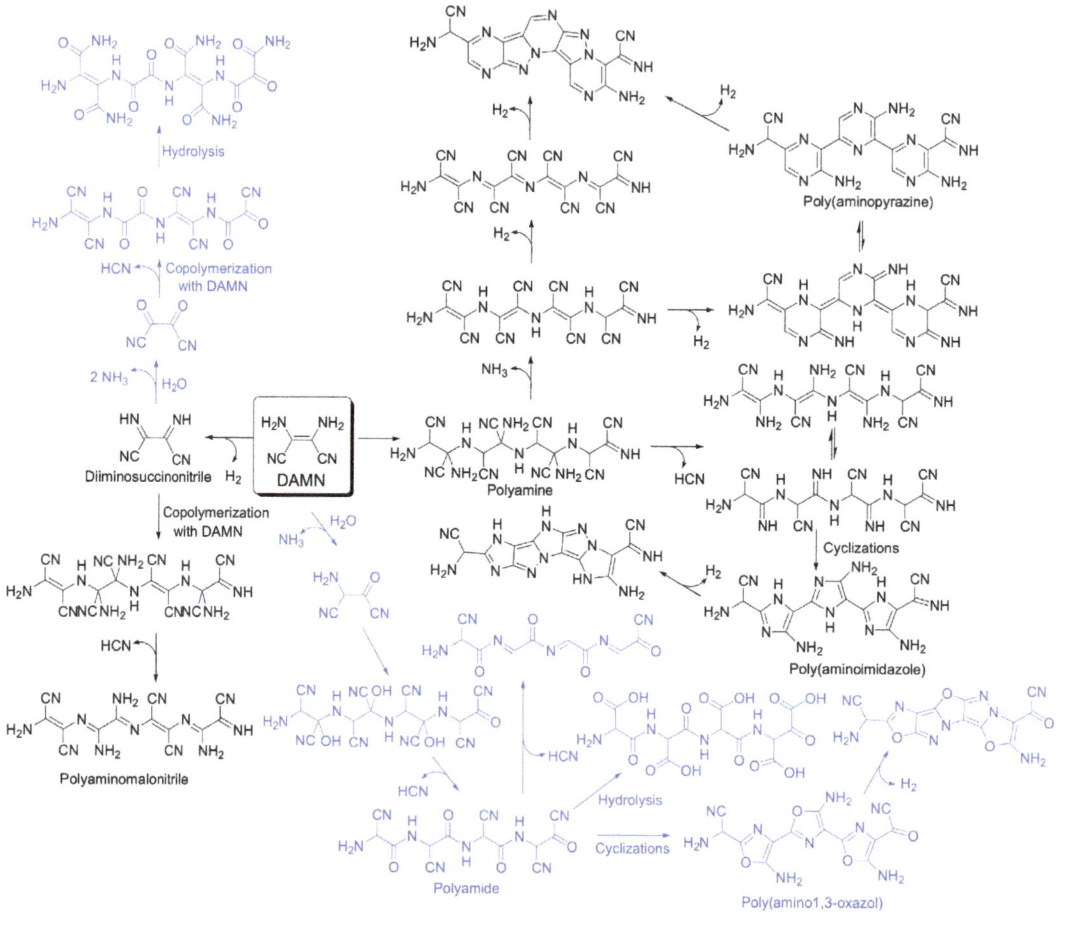

Scheme 3. Structures proposed for the DAMN polymers synthetized under hydrothermal conditions, based on their FTIR and ^{13}C NMR spectra (for details, please see [8]).

Finally, the UV-Vis spectra of polymers **1–4** show a main absorption at around 430 nm, which may be related to a π–π * transition along the polymeric backbone (Figure 2) [45]. Moreover, the samples synthetized at 170 °C (polymers **1** and **3**) exhibit an additional weak band centered at ~310 nm, which is likely due to aromatic fragments that have a naphthalene-like frame [46].

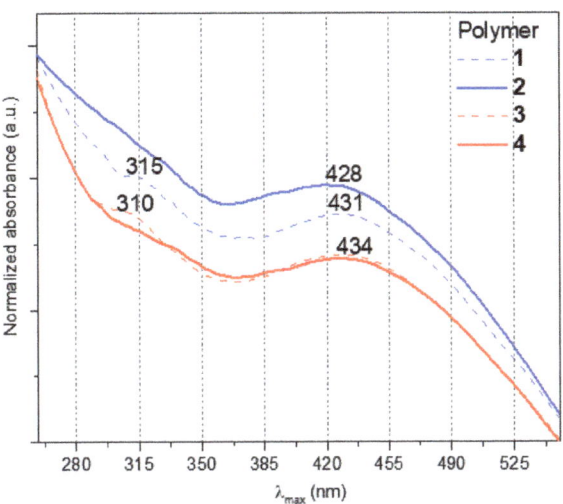

Figure 2. UV-Vis spectra of polymers **1–4** registered using DMSO as solvent.

3.3. Thermal Stability

The derivative thermogravimetry (DTG) curves can be considered as excellent fingerprints for HCN-derived polymers for samples showing similar spectroscopic data, but synthetized under different experimental conditions; thus, the derived fragments from the thermogravimetry-mass spectrometry (TG-MS) analysis may be very useful for the structural elucidation of these types of complex systems [47]. In Figure 3, the DTG-grams and the TG-MS curves of polymers **1–4** are shown. For all the samples, three thermal degradation steps are observed (drying, <225 °C; main decomposition, 225–450 °C; carbonization, >450 °C) (Figure 3a,b), displaying temperatures of the DTG maxima around 80–90, 240, 420, 680, and 830 °C, and leaving char residues of ~25% in weight, in agreement with previous results from the RP [8], with the exception of a DTG peak around 330 °C, which is not found in the polymers synthesized herein. These results denote a thermo-structural resemblance, but no identical structures for the DAMN polymers synthetized under hydrothermal conditions using MWR against a CTS, according to the spectroscopic data discussed above. The lack of the thermolabile groups at ~330 °C could be related to the greater proportion of structures based on N-heterocycles and could therefore be more thermally stable than the corresponding lineal blocks or segments on the polymeric backbone found in the DAMN polymers synthesized using MWR, as commented above.

Similar conclusions can be inferred from the TG-MS traces. Polymers **1–4** show the same fragmentation pattern (Figure 3c–f). The fragments at m/z = 16, 17, 18, 26, 27, 28, 43, 44, 45, and 52 can be assigned to $^+NH_2$, $^+OH/^+NH_3$, ^+H_2O, ^+CN, ^+HCN, $^+N_2$, $^+H_3CN_2$, formamidine $^+H(C=NH)NH_2/^+CO_2$, formamide $^+HCONH_2$, and $^+$-CH=CH-CN, respectively. All these signals were also identified previously in other HCN polymers [16,47], but the profiles of the curves are different depending on the experimental synthetic conditions used for the HCN polymers' production. This once again reveals the structural similarity in the macromolecular backbone of all the HCN polymers with the main carbonization process associated with the decyanation of the structure (m/z = 26 and 27, ^+CN and ^+HCN, respectively) and related to a structure based mainly on a conjugated –C=N– system. In addition, considering the oxygenated fragments (m/z = 44 and 45, $^+CO_2$ and formamide, respectively) and the shape of the corresponding curves suggests that there are at least two classes of oxygenated functional groups. On the one hand, the weaker bonds are present in the pendant groups (for example, -COOH or -CONH$_2$ from the hydrolysis of –CN groups) or from the generation of lineal polyamides and/or polyacids (<450 °C); on the other hand, the strongest bonds are present in the macrostructure formed, for instance, by oxazole rings (>450 °C) (Scheme 3). However, in any case, the presence of oxygenated functional groups only can be related to the role of the water in the pathways of the formation of the macrostructures, being independent of the occurrence of the air in the reaction environment. In addition, all these fragments can be correlated with the macrostructures proposed in Scheme 3 [47].

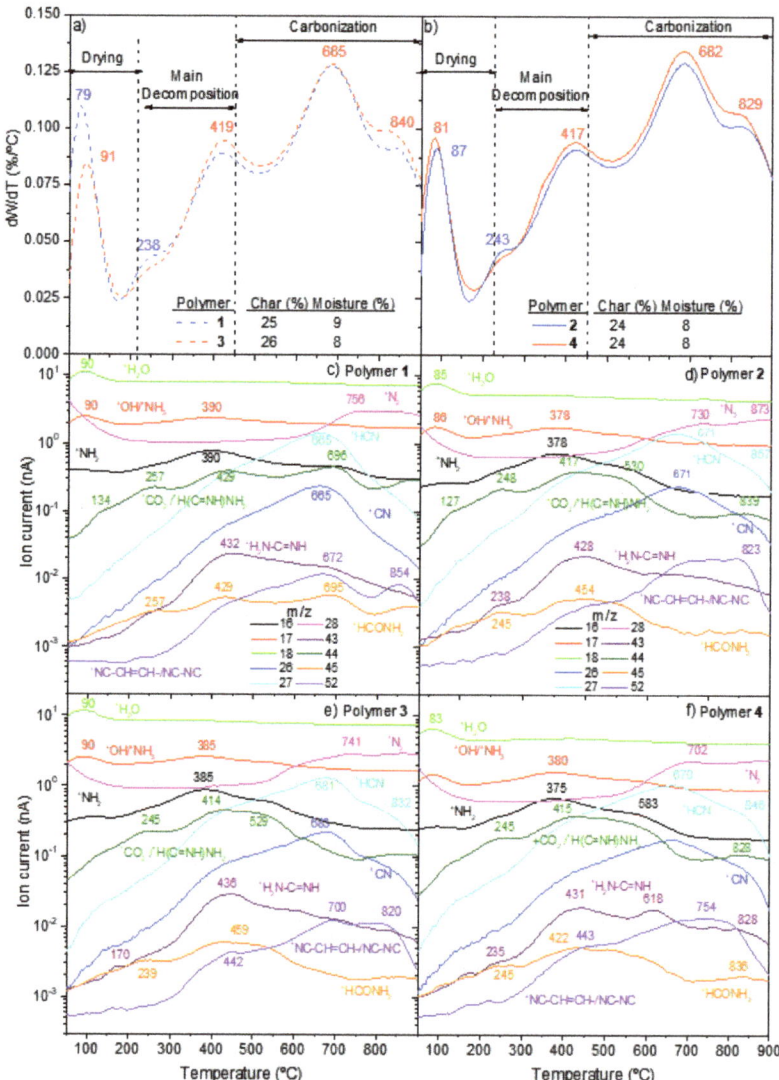

Figure 3. (a) DTG curves of polymers **1** and **3** and (b) of polymers **2** and **4**; (c–f) ion intensity curves for the main fragments from polymers **1**, **2**, **3**, and **4**, respectively, heated at 10 °C/min and under an Ar atmosphere.

3.4. Morphological and Textural Properties

As it was indicated above, the MWR entails two important modifications both in the shape and the size of the HCN-derived polymeric particles [35,36]. Thus, the RP presents particles with no well-defined morphology at the micrometric scale. However, polymers **1–4** are much smaller and present uniform shapes, especially the polymers **1** and **3** (Figure 4). In polymers **2** and **4**, a few long, isolated nanofibers were observed together with particles with a long rice-like shape similar to those previously described for the MW-driven polymerization of cyanide [35]. However, for the first time, it was found that the DAMN polymeric submicron particles from polymers **1** and **3** exhibit only one pure morphology. Previously, only mixtures of different morphologies and sizes for insoluble HCN-derived polymeric particles were described, from microspheres to laminated particles and with no well-defined shape microparticles [8,35,36].

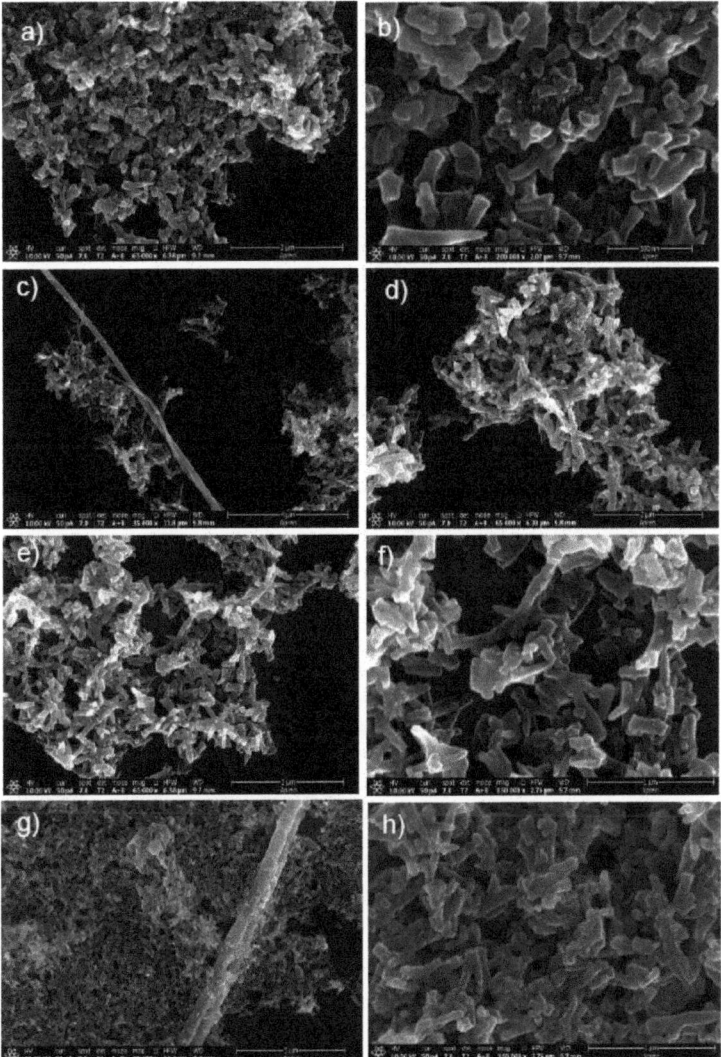

Figure 4. SEM images for representative polymers: (a,b) **1**; (c,d) **2**; (e,f) **3**; (g,h) **4**.

In addition, the size distributions of polymers **1–4** present a Z-average of ~230 nm (Figure 5, Table 3), with a narrower distribution for the samples prepared at 190 °C (polymers **2** and **4**), despite the presence of single isolated nanofibers. On the other hand, it was statistically proven that the molecular sizes for the cyanide polymers were inversely related on the O content present on the macrostructure when the reactions were conducted by MWR [36]. However, as it can be seen in the insert plot in Figure 5, for the DAMN polymers, there seems to be a contrary behavior, i.e., a higher O percentage leads to larger particles. Again, it is demonstrated that the properties of the HCN polymers are sensitive to the experimental conditions chosen for their production, in this case, the reagent used for their syntheses. Moreover, the size distributions seem to depend on a greater grade of the polymerization temperature than the reaction atmosphere (air or N_2), which is in good agreement with the independence of the O proportion due to the presence or lack of air during the polymerization processes.

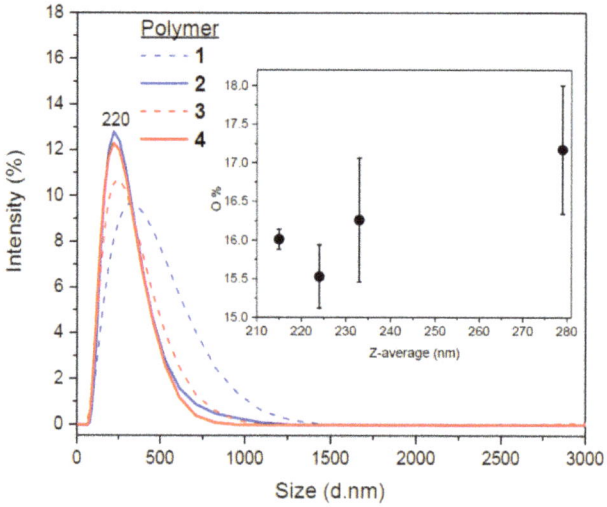

Figure 5. Size distributions of the particles from polymers **1–4** measured by DLS. In the insert plot, the relationship between the Z-average and the oxygen content in the corresponding samples is represented.

Table 3. Morphological and textural characteristics and other physical properties of the DAMN polymers **1–4**. [a] Polydispersity index calculated from the results shown in Figure 5. [b] LP = Low pressure; [c] HP = High pressure; [d] M_n = number-average molecular weight by GPC analysis.

		Polymer			
		1	2	3	4
Z-average (nm)		279	224	233	215
PdI [a]		0.24	0.16	0.24	0.18
BET	LP [b]	33.2	27.8	29.8	32.4
Surface area (m²/g)	HP [c]	34.6	29.2	31.3	33.5
Density (g/cm³)		2.04	2.42	2.29	2.09
M_n [d]		5300	7000	5900	6000

On the other hand, other textural and physical properties were measured (Table 3). The BET surface area of polymers **1–4** have values around 30 m²/g. Moreover, the density values of polymers **1–4** are roughly 2.2 g/cm³, being higher than the only value reported in the bibliography for a HCN polymer, 1.62 g/cm³, which was obtained during the manufacturing process of the HCN in the DuPont gas-manufacturing system [41]. Finally, the number-average molecular weights (M_n) obtained by the GPC analysis for all the samples were determined. These apparent molecular weights must be taken with care because the solubility of the DAMN polymers is limited in DMF, and it is difficult to have a similar standard due to the high complexity of these systems, as it was shown in Scheme 3. In any case, a high polymerization temperature could increase the M_n values, at least in those samples prepared under an N_2 atmosphere.

As a summary of all these multi-technical characterizations for polymers **1–4**, it can be said that the DAMN polymerization assisted by MWR leads to the generation of complex mixtures of lineal and cyclic structures based mainly on C=N bonds (Scheme 3) with an apparent greater proportion of N-heterocycles and amide groups with respect to the analogous polymer synthetized using a CTS. At least a fraction of these heterogeneous polymeric systems are 2-D macrostructures that are able to form stackings. This layer arrangement is influenced by the use of an inert atmosphere during the polymerization processes. Moreover, the MWR has allowed for the preparation, for the first time, of uniform DAMN polymeric submicron particles with surface areas around 30 m^2/g and a density of ~2.2 g/cm^3.

The close resemblance of the chemical and structural characteristics of polymers **1–4** is also reflected in their electrochemical properties, showing a similar redox behavior, which will be discussed in the next section. Noteworthily, the decrease in the particle size of the DAMN polymers described herein has enabled the measurement of their electrochemical properties in DMSO solutions against the voltammetry measurements made using pressing powders of the RP [8].

3.5. Electrochemistry

In order to complete the characterization of the new DAMN polymers, their electrochemical properties were studied. All the polymers showed similar cyclic voltammograms (CVs) with two stable redox systems, a quasi-reversible one at around −0.7 V, and an irreversible one at 0.4 V approximately. Figure 6 shows the CV of polymer **1** as an example.

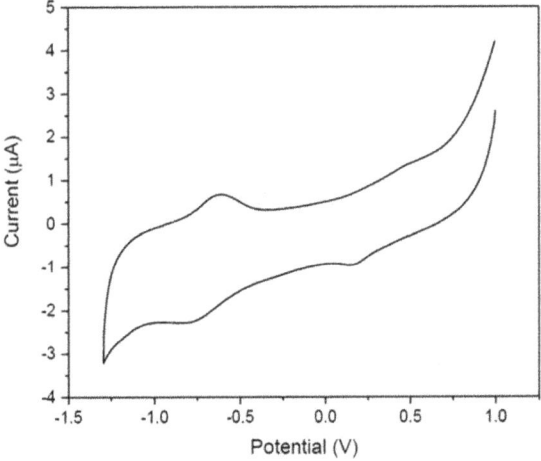

Figure 6. Cyclic voltammogram of polymer **1** in 0.1 M NaClO$_4$/DMSO supporting electrolyte at scan rate 100 mV s^{-1}.

In order to determine the electrochemical properties of the four polymers, the reversible redox system has been chosen for a detailed study. Figure 7 shows the cyclic voltammograms of about 3 × 10^{-5} M solutions of polymers **1–4** in NaClO$_4$/DMSO. As it can be seen, all the polymers present similar CVs, with a quasi-reversible electrochemical system at very close potentials, corresponding to the same electroactive functional groups, which is in agreement with their very similar macrostructures. The polymers' corresponding Tafel plots (overpotential vs. log i) [48] are shown in Figure 8. The slopes of the Tafel plots allowed us to calculate the electron transfer coefficient, α, the number of electrons exchanged, n, and the exchange current density, j_0. The results shown in Table 4 indicate that, in all cases, the corresponding redox systems exchange one electron, and regarding the α values, we can conclude that their symmetry is similar. That is to say, this electrochemical system is the same in the four DAMN polymers.

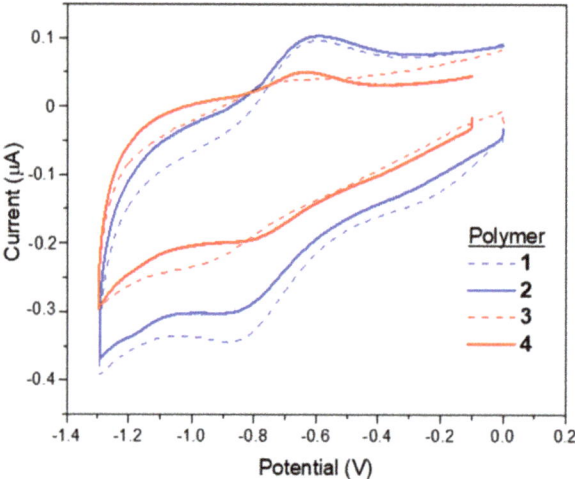

Figure 7. Cyclic voltammograms of redox system of polymers **1–4** in about 0.1 M NaClO$_4$/DMSO supporting electrolyte at scan rate 100 mV s^{-1}.

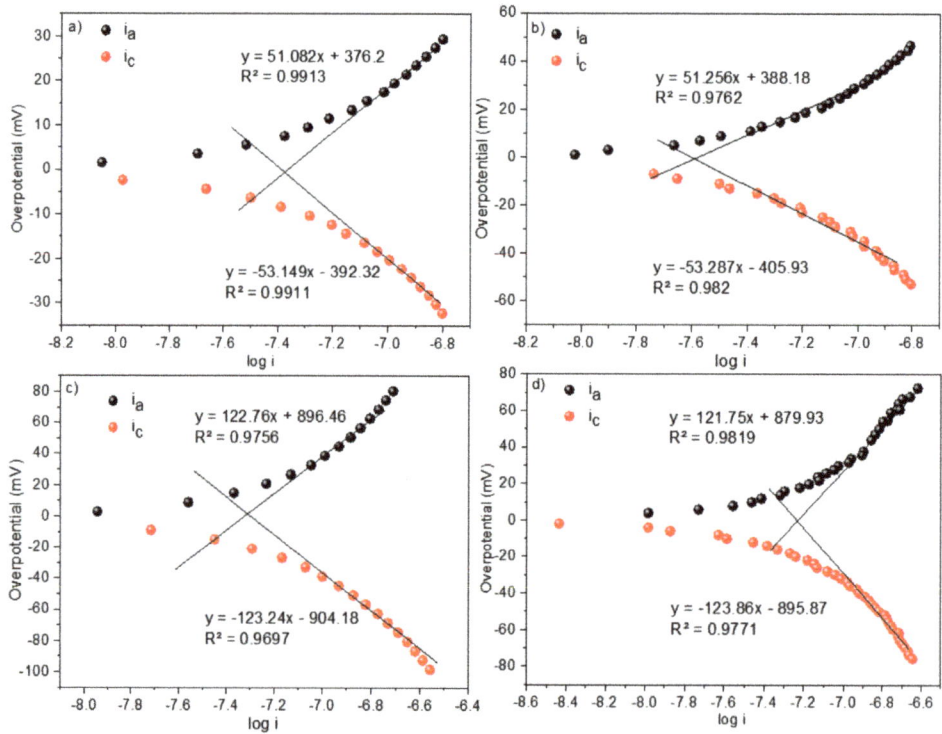

Figure 8. Tafel plots for the redox systems of the polymers: (**a**) **1**; (**b**) **2**; (**c**) **3**; (**d**) **4**.

Table 4. Results of the electrochemical characterization of the DAMN polymers.

Polymer	$E^{0\prime}$ (V)	n	α	D (cm^2 s^{-1})	k_f (cm s^{-1})	k^0 (cm s^{-1})	j_0 (A cm^{-2})
1	−0.77	0.97	0.50	1.52×10^{-4}	0.27	3.58×10^{-4}	1.35×10^{-7}
2	−0.77	0.97	0.49	1.05×10^{-4}	0.29	3.19×10^{-4}	8.44×10^{-7}
3	−0.77	0.96	0.50	1.47×10^{-4}	0.29	2.91×10^{-4}	6.84×10^{-7}
4	−0.77	0.96	0.50	1.43×10^{-4}	0.26	3.19×10^{-4}	8.41×10^{-7}

In addition, the exchange current density, j_0, equal to j_c or j_a at equilibrium ($E = E^0$) when the net current is zero, and indicative of the reversibility of the redox systems, allowed us to estimate the standard rate constant of the electrochemical reaction, k_0, from the $j_0 = n F k_0 C$ equation. The obtained values are very close, confirming the similar electrochemical behavior of the four solved polymers.

On the other hand, the peak current for a reversible system (at 25 °C) follows the Randles–Sevcik equation [49]:

$$i_p = (2.69 \times 10^5)\, n^{3/2}\, A\, C\, D^{1/2}\, v^{1/2}$$

where n is the exchanged number of electrons, A is the electrode area, C is the concentration (in mol/cm^3), D is the diffusion coefficient, and v is the potential scan rate. That is to say, the current must be directly proportional to the concentration and increases with the square root of the scan rate. Thus, the dependence on the scan rate is indicative of an electrode reaction controlled by mass transport. Figure 9 shows the obtained plots of peak current density, normalized with the corresponding concentrations, vs. the square root of the scan rate. All polymers show a linear relationship with the scan rate and the slopes allow us to calculate the diffusion coefficients. The slopes of the lineal fits in Figure 9 suggest that the diffusion coefficients are different and related to the different molecular weights of the polymers. These results also indicate the electrochemical stability of the polymers.

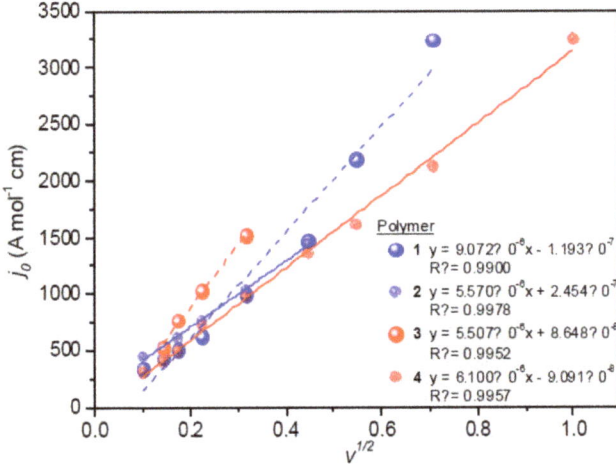

Figure 9. Variation of normalized cathodic current densities with the square root of scan rate.

In addition, the study of the diffusion with rotating electrode by the Koutecky–Levich treatment [48] allows us to determine the influence of the kinetics in the global process and the value of the rate constant of the forward reaction (reduction) at the measurement potential, and to calculate the diffusion coefficient. The limiting current for the voltametric wave is expressed by the Koutecky–Levich equation:

$$\frac{1}{i_l} = \frac{1}{nFAk_fC} + \frac{1}{0.62nFAD^{2/3}\nu^{-1/6}C\omega^{1/2}}$$

where n is the number of electrons, F the Faraday constant, A the electrode area, ν the DMSO kinematic viscosity (0.0214 cm$^2\cdot$s^{-1}), D the diffusion coefficient, ω the angular velocity, k_f the heterogeneous second-order rate constant, and C the bulk polymer concentration (in mol cm^{-3}). The slopes of the plots $1/i_l$ vs. $1/\omega^{1/2}$ allow us to calculate the diffusion coefficients and the intercepts give us the rate constants. Figure 10 shows the Koutecky–Levich plots obtained with the four polymers at $E = -0.8$ V. The obtained diffusion coefficients are close to those obtained by the Randles–Sevcik equation, and the mean values are collected in Table 4. The values are in concordance with the polymers' size (see data of the Table 3). On the other hand, the obtained k_f values are very close and notably higher than the standard value k_0, indicating a very fast electrochemical reaction at $E = -0.8$ V.

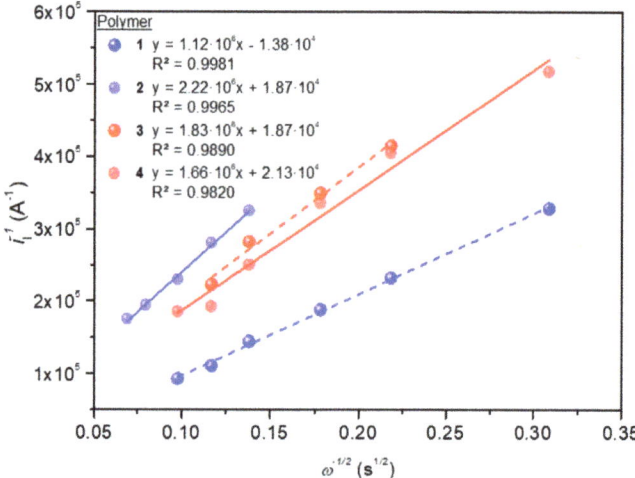

Figure 10. Koutecky-Levich plots obtained for the four polymers at -0.8 V.

Finally, in order to complete the electrochemical study and consider the shape of the CVs (Figure 6), we have considered it to be of interest to study the possibility of applying these materials as semiconductors. The formal potentials of the two systems showed in the CVs can be related to the p- and n-doping status of a semiconductor material and can be used to calculate the HOMO (highest occupied molecular orbital), the LUMO (lowest unoccupied molecular orbital), the band gap (the gap between the highest occupied molecular orbital (HOMO)), and the lowest unoccupied molecular orbital (LUMO)) [50,51]. For these calculations, the values of the absolute potential for the standard hydrogen electrode (SHE) and the potential difference between the Ag/AgCl and SHE of 4.43 eV and 0.22 V were used. The formal potentials are usually taken as the average of the anodic and cathodic peak potentials for both p- and n-doping systems. As expected, all the polymers showed the same values for the formal potentials. Table 5 collects the results obtained, which demonstrates that the synthesized polymers could be applied as good semiconductors.

Table 5. Electrochemical data determined from the peak potential values.

E^0_{ox} (V)	E^0_{red} (V)	HOMO (eV)	LUMO (eV)	E_g (eV)
0.34	−0.77	4.99	3.88	1.11

In this work, the electrochemical characterization of the new polymers has been carried out. The preparation of the modified electrodes and their application for the development of electrochemical devices is beyond our aim at this moment. In view of the results obtained in the electrochemical characterization shown in this work, it is possible that we will address the studies of modified electrodes in further work.

3.6. New Insights in Prebiotic Chemistry and Materials Science

Classically, HCN polymers have been considered preferential prebiotic precursors of biomonomers [52] and the DAMN as a key intermediate in the oligomerization/polymerization of HCN [3,24,39]. Beyond the interest in the generation of organics with biological applications, it was shown herein that DAMN under the simulated conditions of plausible prebiotic-subaerial-hydrothermal environments (high temperatures, relatively high pressure, and a lack of oxygen) lead to the generation of semiconducting submicron particles, which could notably enrich the chemical space and enable the discovery of new protobiological reaction networks due to their potential as catalyzers and photocatalyzers. These results are especially interesting when considering the hypothesis about the role of hydrothermal scenarios as good niches for the origins of life or at least for the advancement of molecular complexity [53,54]. On the other hand, the robustness of the possible prebiotic syntheses of HCN polymers offers additional competitive environmental and economic advantages for the development of a new class of multifunctional polymeric materials using low-cost, easy-to-produce, and green-solvent processes. Moreover, the DAMN polymerization is not affected by the presence of air, thereby improving the handle of this one-pot process. On the contrary, it was also shown that the DAMN is highly sensitive to hydrolysis and oxidation reactions (Scheme 1). This fact leads to the decrease in the available amount of DAMN susceptible to polymerization. A further research work is in progress to solve this weakness of the hydrothermal polymerization of DAMN, using solvents other than water. The preliminary results indicate that it is possible to obtain conversions for DAMN polymers around ~90% with a narrower particle size distribution than reported herein. In any case, this research showed the high capacity of the MWR for reducing the reaction time of polymerization to a few minutes, driving the formation of DAMN soft submicron particles as semiconductors as a first step in the development of fillers in new composite materials and others important applications such as medicine [55]. In addition, the results discussed in the present work complete and extend the knowledge about related π-conjugated systems based on diaminoanthraquinones [56], oligothiophenes with dicyanomethylene groups [57], and C=N based systems with donor-acceptor intramolecular charge transfer [58] with optoelectronic, amphoteric redox and semiconductive properties, respectively.

4. Conclusions

Bearing in mind that the possibility of tuning the morphology of HCN polymers synthesized through MWR has been recently discovered, in the present study, DAMN polymerization was explored with this technique. To achieve this, DAMN aqueous suspensions were subjected to MWR at 170 and 190 °C and short reaction times, 3 and 16 min, respectively. Although moderate yields of insoluble black polymers were found, their textural and morphological characteristics at the submicron scale allowed us to approach the study, for the first time, of their redox properties in solution. Thus, the cyclic voltammetry of these highly conjugated polymeric submicron particulate systems in DMSO solution provided relevant information concerning their electrochemical behavior. In addition, the diffusion coefficients and different kinetic parameters were estimated and confirmed the high similarity of the four samples produced in this work, agreeing with the conclusions reached after the complete structural and thermal analyses were carried out. Moreover, the new submicron particles synthetized from the air tolerant and very fast MW-driven polymerization of DAMN can be considered as semiconductors based on the values of the band gaps found. This study presents novel results that on the one hand increase the prebiotic chemical space to be explored, and on the other hand show significant improvement in the production of the HCN polymers. As a general conclusion, DAMN is a π-conjugated compound with a great potential for developing novel soft materials in modern polymers science.

Author Contributions: M.R.-B., conceptualization, methodology, formal analysis, investigation, writing—original draft preparation, writing—review and editing; P.G.-A., investigation, formal analysis and writing—original draft preparation; P.V., formal analysis and resources; J.L.d.l.F., writing—review, editing, and supervision. All authors have read and agreed to the published version of the manuscript.

Funding: This research was funded by the projects PID2019-104205GB-C21, PID2019-107442RBC32 from the Spanish Ministerio de Ciencia e Innovación, and by the Spanish State Research Agency, project MDM-2017-0737 Centro de Astrobiología (CSIC-INTA), Unidad de Excelencia María de Maeztu.

Institutional Review Board Statement: Not applicable.

Informed Consent Statement: Not applicable.

Data Availability Statement: Not applicable.

Acknowledgments: The authors used the research facilities of the Centro de Astrobiología (CAB), CSIC-INTA, and were supported by the Instituto Nacional de Técnica Aeroespacial "Esteban Terradas" (INTA), by the projects PID2019-104205GB-C21 and PID2019-107442RB-C32 from the Spanish Ministry of Science and Innovation, and by the Spanish State Research Agency (AEI) project MDM-2017-0737 CAB (CSIC-INTA), Unidad de Excelencia María de Maeztu. Additionally, the authors are grateful to Mª Teresa Fernández for performing the FTIR and XRD measurements, to Alexandra Muñoz Bonilla for the support in the GPC analysis and to the Services of Thermal and Chemical Analysis and Nuclear Magnetic Resonance of the Instituto de Ciencias de Materiales de Madrid (ICMM, CSIC). M.R.-B. also thanks Sabino Veintemillas for his useful helpful in the DLS measurements.

Conflicts of Interest: The authors declare no conflict of interest. The funders had no role in the design of the study; in the collection, analyses, or interpretation of data; in the writing of the manuscript; or in the decision to publish the results.

References

1. Szady-Chełmieniecka, A.; Kołodziej, B.; Morawiak, M.; Kamieński, B.; Schilf, W. Spectroscopic studies of the intramolecular hydrogen bonding in o-hydroxy Schiff bases, derived from diaminomaleonitrile, and their deprotonation reaction products. *Spectrochim. Acta Part A Mol. Biomol. Spectrosc.* **2018**, *189*, 330–341. [CrossRef] [PubMed]
2. Aruna; Rani, B.; Swami, S.; Agarwala, A.; Behera, D.; Shrivastava, R. Recent progress in development of 2,3-diaminomaleonitrile (DAMN) based chemosensors for sensing of ionic and reactive oxygen species. *RSC Adv.* **2019**, *9*, 30599–30614. [CrossRef] [PubMed]
3. Ruiz-Bermejo, M.; de la Fuente, J.L.; Pérez-Fernández, C.; Mateo-Martí, E. A Comprehensive Review of HCN-Derived Polymers. *Processes* **2021**, *9*, 597. [CrossRef]
4. d'Ischia, M.; Manini, P.; Moracci, M.; Saladino, R.; Ball, V.; Thissen, H.; Evans, R.A.; Puzzarini, C.; Barone, V. Astrochemistry and Astrobiology: Materials Science in Wonderland? *Int. J. Mol. Sci.* **2019**, *20*, 4079. [CrossRef]
5. Thissen, H.; Evans, R.A.; Ball, V. Films and Materials Derived from Aminomalononitrile. *Processes* **2021**, *9*, 82. [CrossRef]
6. Zhou, X.; Fang, Y.; Su, Y.; Ge, C.; Jin, B.; Li, Z.; Wu, S. Preparation and Characterization of Poly-Hydrogen Cyanide Nanofibers with High Visible Light Photocatalytic Activity. *Catal. Commun.* **2014**, *46*, 197–200. [CrossRef]
7. Rahm, M.; Lunine, J.I.; Usher, D.A.; Shalloway, D. Polymorphism and Electronic Structure of Polyimine and Its Potential Significance for Prebiotic Chemistry on Titan. *Proc. Natl. Acad. Sci. USA* **2016**, *113*, 8121–8126. [CrossRef]
8. Ruiz-Bermejo, M.; de la Fuente, J.L.; Carretero-González, J.; García-Fernández, L.; Aguilar, M.R. A Comparative Study on HCN Polymers Synthesized by Polymerization of NH$_4$CN or Diaminomaleonitrile in Aqueous Media: New Perspectives for Prebiotic Chemistry and Materials Science. *Chem. A Eur. J.* **2019**, *25*, 11437–11455. [CrossRef]
9. Thissen, H.; Koegler, A.; Salwiczek, M.; Easton, C.D.; Qu, Y.; Lithgow, T.; Evans, R.A. Prebiotic-Chemistry Inspired Polymer Coatings for Biomedical and Material Science Applications. *NPG Asia Mater.* **2015**, *7*, e225. [CrossRef]

10. Menzies, D.J.; Ang, A.; Thissen, H.; Evans, R.A. Adhesive Prebiotic Chemistry Inspired Coatings for Bone Contacting Applications. *ACS Biomater. Sci. Eng.* **2017**, *3*, 793–806. [CrossRef]
11. Cheng, S.-Y.; Chiang, Y.-L.; Chang, Y.-H.; Thissen, H.; Tsai, S.-W. An Aqueous-Based Process to Bioactivate Poly(ε-Caprolactone)/Mesoporous Bioglass Composite Surfaces by Prebiotic Chemistry-Inspired Polymer Coatings for Biomedical Applications. *Colloids Surf. B Biointerfaces* **2021**, *205*, 111913. [CrossRef]
12. Ball, V. Antioxidant Activity of Films Inspired by Prebiotic Chemistry. *Mater. Lett.* **2021**, *285*, 129050. [CrossRef]
13. Pérez-Fernández, C.; Ruiz-Bermejo, M.; Gálvez-Martínez, S.; Mateo-Martí, E. An XPS Study of HCN-Derived Films on Pyrite Surfaces: A Prebiotic Chemistry Standpoint towards the Development of Protective Coatings. *RSC Adv.* **2021**, *11*, 20109–20117. [CrossRef]
14. Liao, T.-Y.; Easton, C.D.; Thissen, H.; Tsai, W.-B. Aminomalononitrile-Assisted Multifunctional Antibacterial Coatings. *ACS Biomater. Sci. Eng.* **2020**, *6*, 3349–3360. [CrossRef]
15. Jung, J.; Menzies, D.J.; Thissen, H.; Easton, C.D.; Evans, R.A.; Henry, R.; Deletic, A.; McCarthy, D.T. New Prebiotic Chemistry Inspired Filter Media for Stormwater/Greywater Disinfection. *J. Hazard. Mater.* **2019**, *378*, 120749. [CrossRef]
16. Ruiz-Bermejo, M.; García-Armada, P.; Mateo-Martí, E.; de la Fuente, J.L. HCN-derived polymers from thermally induced polymerization of diaminomaleonitrile: A non-enzymatic peroxide sensor based on prebiotic chemistry. *Eur. Polym. J.* **2022**, *162*, 110897. [CrossRef]
17. Matthews, C.N.; Moser, R.E. Peptide Synthesis from Hydrogen Cyanide and Water. *Nature* **1967**, *215*, 1230–1234. [CrossRef]
18. Budil, D.E.; Roebber, J.L.; Liebman, S.A.; Matthews, C.N. Multifrequency Electron Spin Resonance Detection of Solid-State Organic Free Radicals in HCN Polymer and a Titan Tholin. *Astrobiology* **2003**, *3*, 323–329. [CrossRef]
19. Vujošević, S.I.; Negrón-Mendoza, A.; Draganić, Z.D. Radiation-Induced Polymerization in Dilute Aqueous Solutions of Cyanides. *Orig. Life Evol. Biosph.* **1990**, *20*, 49–54. [CrossRef]
20. Negrón-Mendoza, A.; Draganić, Z.D.; Navarro-González, R.; Draganić, I.G. Aldehydes, Ketones, and Carboxylic Acids Formed Radiolytically in Aqueous Solutions of Cyanides and Simple Nitriles. *Radiat. Res.* **1983**, *95*, 248–261. [CrossRef]
21. Mizutani, H.; Mikuni, H.; Takahasi, M.; Noda, H. Study on the Photochemical Reaction of HCN and Its Polymer Products Relating to Primary Chemical Evolution. *Orig. Life Evol. Biosph.* **1975**, *6*, 513–525. [CrossRef]
22. Villafañe-Barajas, S.A.; Ruiz-Bermejo, M.; Rayo Pizarroso, P.; Galvez-Martinez, S.; Mateo-Martí, E.; Colín-García, M. A serpentinite lizardite-HCN interaction leading the increasing of molecular complexity in an alkaline hydrothermal scenario: Implications for the origin of life studies. *Life* **2021**, *11*, 661. [CrossRef]
23. Ferris, J.P.; Joshi, P.C.; Edelson, E.H.; Lawless, J.G. HCN: A Plausible Source of Purines, Pyrimidines and Amino Acids on the Primitive Earth. *J. Mol. Evol.* **1978**, *11*, 293–311. [CrossRef]
24. Ferris, J.P.; Hagan Jr, W.J. HCN and Chemical Evolution: The Possible Role of Cyano Compounds in Prebiotic Synthesis. *Tetrahedron* **1984**, *40*, 1093–1120. [CrossRef]
25. Schwartz, A.W.; Voet, A.B.; Van der Veen, M. Recent Progress in the Prebiotic Chemistry of HCN. *Orig. Life* **1984**, *14*, 91–98. [CrossRef]
26. Eastman, M.P.; Helfrich, F.S.E.; Umantsev, A.; Porter, T.L.; Weber, R. Exploring the Structure of a Hydrogen Cyanide Polymer by Electron Spin Resonance and Scanning Force Microscopy. *Scanning* **2003**, *25*, 19–24. [CrossRef]
27. Fernández, A.; Ruiz-Bermejo, M.; de la Fuente, J.L. Modelling the Kinetics and Structural Property Evolution of a Versatile Reaction: Aqueous HCN Polymerization. *Phys. Chem. Chem. Phys.* **2018**, *20*, 17353–17366. [CrossRef] [PubMed]
28. Mas, I.; de la Fuente, J.L.; Ruiz-Bermejo, M. Temperature Effect on Aqueous NH$_4$CN Polymerization: Relationship between Kinetic Behaviour and Structural Properties. *Eur. Polym. J.* **2020**, *132*, 109719. [CrossRef]
29. Toh, R.J.; Evans, R.; Thissen, H.; Voelcker, N.H.; d'Ischia, M.; Ball, V. Deposition of Aminomalononitrile-Based Films: Kinetics, Chemistry, and Morphology. *Langmuir* **2019**, *35*, 9896–9903. [CrossRef] [PubMed]
30. Mas, I.; Hortelano, C.; Ruiz-Bermejo, M.; de la Fuente, J.L. Highly Efficient Melt Polymerization of Diaminomaleonitrile. *Eur. Polym. J.* **2021**, *143*, 110185. [CrossRef]
31. Hortelano, C.; Ruiz-Bermejo, M.; de la Fuente, J.L. Solid-state polymerization of diaminomaleonitrile: Toward a new generation of conjugated functional materials. *Polymer* **2021**, *223*, 123696. [CrossRef]
32. Mamajanov, I.; Herzfeld, J. HCN Polymers Characterized by SSNMR: Solid State Reaction of Crystalline Tetramer (Diaminomaleonitrile). *J. Chem. Phys.* **2009**, *130*, 134504. [CrossRef]
33. Cataldo, F.; Lilla, E.; Ursini, O.; Angelini, G. TGA–FT-IR Study of Pyrolysis of Poly(Hydrogen Cyanide) Synthesized from Thermal Decomposition of Formamide. Implications in Cometary Emissions. *J. Anal. Appl. Pyrolysis* **2010**, *87*, 34–44. [CrossRef]
34. Enchev, V.; Angelov, I.; Dincheva, N.; Stoyanova, N.; Slavova, S.; Rangelov, M.; Markova, N. Chemical Evolution: From Formamide to Nucleobases and Amino Acids without the Presence of Catalyst. *J. Biomol. Struct. Dyn.* **2020**, *39*, 5563–5578. [CrossRef]
35. Hortal, L.; Pérez-Fernández, C.; de la Fuente, J.L.; Valles, P.; Mateo-Martí, E.; Ruiz-Bermejo, M. A Dual Perspective on the Microwave-Assisted Synthesis of HCN Polymers towards the Chemical Evolution and Design of Functional Materials. *Sci. Rep.* **2020**, *10*, 22350. [CrossRef]
36. Pérez-Fernández, C.; Valles, P.; González-Toril, E.; Mateo-Martí, E.; de la Fuente, J.L.; Ruiz-Bermejo, M. Tuning the morphology in the nanoscale of NH$_4$CN polymers synthesized by microwave radiation: A comparative study. *Polymers* **2022**, *14*, 57. [CrossRef]
37. Yang, C.; Wang, B.; Zhang, L.; Yin, L.; Wang, X. Synthesis of Layered Carbonitrides from Biotic Molecules for Photoredox Transformations. *Angew. Chem. Int. Ed.* **2017**, *56*, 6627–6631. [CrossRef] [PubMed]

38. Ruiz-Bermejo, M.; de la Fuente, J.L.; Rogero, C.; Menor-Salván, C.; Osuna-Esteban, S.; Martín-Gago, J.A. New Insights into the Characterization of 'Insoluble Black HCN Polymers. *Chem. Biodiver.* **2012**, *9*, 25–40. [CrossRef]
39. Ferris, J.P.; Edelson, E.H. Chemical Evolution. 31. Mechanism of the Condensation of Cyanide to Hydrogen Cyanide Oligomers. *J. Org. Chem.* **1978**, *43*, 3989–3995. [CrossRef]
40. De la Fuente, J.L.; Ruiz-Bermejo, M.; Nna-Mvondo, D.; Minard, R. Further progress into the thermal characterization of HCN polymers. *Polym. Degrad. Stab.* **2014**, *110*, 241–251. [CrossRef]
41. Khare, B.N.; Sagan, C.; Thompson, W.R.; Arakawa, E.T.; Meisse, C.; Tuminello, P.S. Optical Properties of Poly-HCN and Their Astronomical Applications. *Can. J. Chem.* **1994**, *72*, 678–694. [CrossRef] [PubMed]
42. Dalton, S.; Heatley, F.; Budd, P.M. Thermal Stabilization of Polyacrylonitrile Fibres. *Polymer* **1999**, *40*, 5531–5543. [CrossRef]
43. Miller, T.S.; Jorge, A.B.; Suter, T.M.; Sella, A.; Corà, F.; McMillan, P.F. Carbon nitrides: Synthesis and characterization of a new class of functional materials. *Phys. Chem. Chem. Phys.* **2017**, *19*, 15613–15638. [CrossRef]
44. Ciria-Ramos, I.; Navascués, N.; Diaw, F.; Furgeaud, C.; Arenal, R.; Ansón-Casaos, A.; Juarez-Perez, E.J. Formamidinium halide salts as precursors of carbon nitrides. *Carbon* **2022**, *196*, 1035–1046. [CrossRef]
45. Yamamoto, T.; Lee, B.-L. New Soluble, Coplanar Poly(naphthalene-2,6-diyl)-Type π-Conjugated Polymer, Poly(pyrimido[5,4-d]pyrimidine-2,6-diyl), with Nitrogen Atoms at All of the o-Positions. Synthesis, Solid Structure, Optical Properties, Self-Assembling Phenomena, and Redox Behavior. *Macromolecules* **2002**, *35*, 2993–2999. [CrossRef]
46. Wait, S.C.; Wesley, J.W. Azanaphthalenes: Part I. Huckel orbital calculations. *J. Mol. Spectrosc.* **1966**, *19*, 25–33. [CrossRef]
47. Ruiz-Bermejo, M.; de la Fuente, J.L.; Marín-Yaseli, M.R. The Influence of Reaction Conditions in Aqueous HCN Polymerization on the Polymer Thermal Degradation Properties. *J. Anal. Appl. Pyrolysis* **2017**, *124*, 103–112. [CrossRef]
48. Bard, A.J.; Faulkner, L.R. Electrochemical Methods. In *Fundamentals and Applications*, 2nd ed.; John Wiley & Sons, Inc.: New York, NY, USA, 2001; ISBN 0-471-04372-9.
49. Wang, J. *Analytical Electrochemistry*, 3rd ed.; Wiley & Sons, Inc.: Hoboken, NJ, USA, 2006; p. 32. ISBN 13 978-0-471-67879-3.
50. Admassie, S.; Inganäs, O.; Mammo, W.; Perzon, E.; Andersson, M.R. Electrochemical and optical studies of the band gaps of alternating polyfluorene copolymers. *Synth. Met.* **2006**, *156*, 614–623. [CrossRef]
51. Johansson, T.; Mammo, W.; Svensson, M.; Andersson, M.R.; Inganäs, O. Electrochemical bandgaps of substituted polythiophenes. *J. Mater. Chem.* **2003**, *13*, 1316–1323. [CrossRef]
52. Ruiz-Bermejo, M.; Zorzano, M.-P.; Osuna-Esteban, S. Simple organics and biomonomers identified in HCN polymers: An overview. *Life* **2013**, *3*, 421–448. [CrossRef]
53. Rimmer, P.B.; Shorttle, O. Origin of Life's Building Blocks in Carbon- and Nitrogen-Rich Surface Hydrothermal Vents. *Life* **2019**, *9*, 12. [CrossRef] [PubMed]
54. Deamer, D. Where did Life Begin? Testing Ideas in Prebiotic Analogue Conditions. *Life* **2021**, *11*, 134. [CrossRef] [PubMed]
55. Li, J.; Rao, J.; Pu, K. Recent Progress on Semiconducting Polymer Nanoparticles for Molecular Imaging and Cancer Phototherapy. *Biomaterials* **2018**, *155*, 217–235. [CrossRef] [PubMed]
56. Masilamani, G.; Batchu, H.; Amsallem, D.; Bedi, A. Novel Series of Diaminoanthraquinone-Based π-Extendable Building Blocks with Tunable Optoelectronic Properties. *ACS Omega* **2022**, *7*, 25874–25880. [CrossRef]
57. Takahashi, T.; Matsuoka, K.; Takimiya, K.; Otsubo, T.; Aso, Y. Extensive Quinoidal Oligothiophenes with Dicyanomethylene Groups at Terminal Positions as Highly Amphoteric Redox Molecules. *J. Am. Chem. Soc.* **2005**, *127*, 8928–8929. [CrossRef]
58. Bedi, A.; Senanayak, S.P.; Das, S.; Narayan, K.S.; Zade, S.S. Cyclopenta[c]thiophene oligomers based solution processable D–A copolymers and their application as FET materials. *Polym. Chem.* **2012**, *3*, 1453–1460. [CrossRef]

Article

Evaluation of Antidiabetic Activity of Biogenic Silver Nanoparticles Using *Thymus serpyllum* on Streptozotocin-Induced Diabetic BALB/c Mice

Maryam Wahab [1,2], Attya Bhatti [1,*] and Peter John [1]

1. Department of Healthcare Biotechnology, Atta-ur-Rahman School of Applied Biosciences (ASAB), National University of Sciences and Technology (NUST), Islamabad 44000, Pakistan; maryam.wahab@jacks.sdstate.edu (M.W.); pjohn@asab.nust.edu.pk (P.J.)
2. Department of Dairy and Food Science, South Dakota State University, Brookings, SD 57007, USA
* Correspondence: attyabhatti@asab.nust.edu.pk; Tel.: +92-51-886-6128

Citation: Wahab, M.; Bhatti, A.; John, P. Evaluation of Antidiabetic Activity of Biogenic Silver Nanoparticles Using *Thymus serpyllum* on Streptozotocin-Induced Diabetic BALB/c Mice. *Polymers* 2022, 14, 3138. https://doi.org/10.3390/polym14153138

Academic Editor: Suguna Perumal

Received: 5 July 2022
Accepted: 29 July 2022
Published: 1 August 2022

Publisher's Note: MDPI stays neutral with regard to jurisdictional claims in published maps and institutional affiliations.

Copyright: © 2022 by the authors. Licensee MDPI, Basel, Switzerland. This article is an open access article distributed under the terms and conditions of the Creative Commons Attribution (CC BY) license (https://creativecommons.org/licenses/by/4.0/).

Abstract: Type 2 Diabetes Mellitus is one of the most common metabolic disorders, and is characterized by abnormal blood sugar level due to impaired insulin secretion or impaired insulin action—or both. Metformin is the most commonly used drug for the treatment of Type 2 Diabetes Mellitus, but due to its slow mode of action and various side effects it shows poor and slow therapeutic response in patients. Currently, scientists are trying to tackle these limitations by developing nanomedicine. This research reports novel synthesis of silver nanoparticles using aqueous extract of *Thymus serpyllum* and aims to elucidate its therapeutic potential as an antidiabetic agent on streptozotocin induced diabetic BALB/c mice. *Thymus serpyllum* mediated silver nanoparticles were characterized through UV, SEM, XRD, and FTIR. The alpha amylase inhibition and antioxidant activity were checked through α amylase and DPPH radical scavenging assay, respectively. To check the effect of silver nanoparticles on blood glucose levels FBG, IPGTT, ITT tests were employed on STZ induced BALB/c mice. To assess the morphological changes in the anatomy of liver, pancreas, and kidney of BALB/c mice due to silver nanoparticles, histological analysis was done through H&E staining system. Finally, AMPK and IRS1 genes expression analysis was carried out via real time PCR. Silver nanoparticles were found to be spherical in shape with an average size of 42 nm. They showed an IC50 of 8 µg/mL and 10 µg/mL for α amylase and DPPH assay, respectively. Our study suggests that silver nanoparticles—specifically 10 mg/kg—cause a significant increase in the expression of AMPK and IRS1, which ultimately increase the glucose uptake in cells. *Thymus serpyllum* mediated silver nanoparticles possess strong antioxidant and antidiabetic potential and can further be explored as an effective and cheaper alternative option for treatment of Type 2 Diabetes Mellitus.

Keywords: Type 2 Diabetes Mellitus; silver nanoparticles; IRS1 and AMPK; BALB/c mice; *Thymus serpyllum*; nanomedicine

1. Introduction

Type 2 Diabetes Mellitus, also known as Insulin Independent Diabetes Mellitus (IDDM), is a multifactorial, chronic disorder responsible for high co-morbidity rates across the globe [1]. T2DM involves failure of pancreatic β cells, thus causing insulinopenia and insulin resistance in liver, skeletal muscles, and adipose tissues, ultimately leading towards their metabolic derangement and failure [2]. Blood glucose levels are majorly regulated by pancreas, which releases the enzymes in accordance with the signals. Approximately 422 million people have diabetes now, accounting for 1.6 million causalities in the year 2016, as per the World Health Organization (WHO), hence making diabetes the 7th driving reason of morbidity and mortality worldwide. The number of diabetic patients worldwide is expected to rise to 640 million by 2040 [3]. Multiple factors and mechanisms have been found to play a part in causing Diabetes Mellitus, yet the exact causing factors are uncertain [4]. Being a disorder of multiple aetiology, the genetic susceptibility superimposed by the environmental factors is undoubtedly involved in causing Type 2 Diabetes Mellitus [5]. Symptoms of diabetes include excessive secretion of urine, known as polyuria; thirst, known as polydipsia; increased hunger; fatigue; sores that do not heal; blurred vision; numbness in hands or feet; weight loss; and tiredness [6]. The first line of treatment for T2DM is diet control, weight management, and physical activity. High caloric food intake and built up of excess adipose tissue induces insulin resistance and leads to decreased glucose uptake by cells and reduced glycogen synthesis [7]. Currently available drugs for treatment include various classes of drugs such as sulfonylureas, thiazolidinedione, α-Glucosidase Inhibitors, Repaglinide, and insulin therapies, but themost commonly used drug is metformin, which unfortunately possess various side effects and shows slow therapeutic response [8].

The frontiers of research on diabetes are focused to update the best method of diabetes analysis, monitoring, and cure. Defects in various molecular signaling pathways are shown to be associated with pathogenesis of type 2 diabetes [9]. Insulin signaling, Adipocytokine signaling, and glycation haxosamine signaling are some of the major pathways involved in regulating blood glucose homeostasis [10]. Genes contributing in these pathways involve GLUT2, GLUT4, IRS, IRS1, PI3K, AKT, TNFα, mTOR, protein kinases and many more. Activation of upstream gene phosphorylates is another target and hence leads to highly regulated signaling in the body to control glucose homeostasis. Any impairment in this signaling can lead to pathogenesis of type 2 diabetes via insulin resistance or β cell dysfunction.

In normal conditions GLUT4 translocate from cytoplasm to cell membrane. This happens due to the binding of insulin peptide to the insulin receptor that initiates the signaling cascade based on phosphorylation. GLUT4 helps in the uptake of glucose molecules into the cells and prevents the usage of stored fats for energy [11]. On the other hand in diabetes, insulin receptors remain inactive in the cells and there is no translocation of GLUT4 from cytoplasm to the membrane. Therefore, there is no absorption of glucose molecules by the cells, which hence leads to the development of chronic hyperglycaemia. Combination of insulin resistance and inhibition of insulin secretion results in T2DM as influenced by genetic determinants, dietary pattern, lifestyle, level of physical activity, and aging. Nutrient overload and imbalance, which is caused by excessive intake of sugars, fats, and oils can develop hyperlipidaemia and hyperglycaemia [12]. There is glycation and lipid peroxidation due to the persistent exposure of carbohydrates and fats. These factors results in insulin resistance and are the key contributors to cause T2DM [13].

AMPK i.e., AMP activated protein kinase, is the key sensor and regulator of the energy status of cells in all eukaryotes. AMPK is regulated by the AMP:ATP ratio, so when this ratio increases, this causes activation of AMPK, so it is activated during energy depletion and regulates many different processes in the cell, and thus it is known as the master regulator of energy metabolism. There are several factors activating AMPK, like AMP in starvation, hunger, and exercise, which are nucleotide dependent activations, while the two main upstream serine/threonine kinases LKB1 and CAMKKβ cause the nucleotide independent activation of AMPK by phosphorylating it at Thr172. AMPK deactivates all energy consuming pathways and activates the energy producing pathways.

The downstream effects of AMPK involves the negative regulation of mTOR (Mammalian Target of Rapamycin) signaling by phosphorylating it, which in fact is an activator of protein synthesis so AMPK blocks the protein synthesis. Another function of AMPK is activation of ULK1, which further activates the macro autophagy, i.e., degradation and recycling of cellular contents by vacuoles and lysosomes. The other downstream function of AMPK is the activation of fatty acid catabolism by activating ATGL (Adipose Triglyceride Lipase), which is the first enzyme responsible for the release of fatty acids from triglycerides. In another mechanism, AMPK inhibits the Acetyl CoA Carboxylase (ACC), which is absolutely necessary for fatty acid synthesis. AMPK also stops the cholesterol synthesis by blocking HMG-CoA reductase. The most important function of AMPK, which makes it the molecule of interest in diabetes, is its activation of glucose uptake by GLUT 4 transporters through the process involving TBC1D1, so that an insulin sensitive cell may be able to uptake and utilize the glucose [14]. Several studies on animal models of metabolic diseases such as diabetes have shown decreased AMPK activity in muscle, liver, and adipose tissues [15]. Metformin, which is the standard drug for diabetes, indirectly activates the AMPK by inhibition of mitochondrial function [16]. Although Metformin is the drug of choice for many T2DM patients, it can cause abdominal discomfort, diarrhea, anorexia, flatulence, and also decreases vitamin B12 intestinal absorption [17].

Silver nanoparticles can be a potential source of insulin sensitization as they increase the cytosolic calcium ions concentration and activates the AMPK by phosphorylating it via CAMKKβ pathway in SH-SY5Y cells and also in rats [18]. AMPK activation enhances the sensitivity towards the insulin and it could mediate the insulin by increasing its action [19]. Insulin binds to its receptor and activates the phosphorylation cascade from IRS1, which induces the transport of glucose into the cells [20]. Studies have shown that animal models lacking IRS1 developed hyperglycaemia or Type 2 Diabetes Mellitus, hence increasing the protein levels of IRS1 will ultimately reduce the hyperglycaemia complications [21]. Silver nanoparticles lead towards the reduction in blood glucose levels by increasing the IRS1 and GLUT2 expression levels. Besides, silver nanoparticles elevate the expression levels of insulin and its secretion [22].

T2DM is also caused due to an imbalance among the antioxidants produced by the body's natural mechanism and the cellular reactive oxygen species produced, thus declaring diabetes as an oxidative stress-based disorder. Due to the excessive production of ROS, apoptosis and maturation of β cells increases while the synthesis and secretion of insulin decreases [23]. Antioxidants are used to cure the oxidative stress, and interest is diverting to the natural antioxidants rather than synthetic antioxidants [24]. Silver nanoparticles are a rich source of antioxidants and they are readily available for action into the tissues, as they can easily penetrate deep down into the tissues [25]. It has been proved that free radicals, especially oxygen-based, are effectively scavenged by silver nanoparticles [26]. The synthesis of metallic nanoparticles from precursor salts occur via the oxidation reduction reactions. The reducing materials, which are present in the plant extracts, shifts the electrons to the ions of the metal precursor, thus producing the nanoparticles [27]. Silver nanoparticles synthesis from plants is more beneficial as compared to microbes and algae, especially because they do not require the tedious stages of growing the cultures on media, hence they are less biohazardous and can be easily improved [28]. Plants possess an industry of compounds like phenols, flavonoids, terpenoids, and a lot more, which act as reducing, stabilizing, as well as capping agents for the nanoparticles and enhance their biomedical properties [29]. The current focus of anti-hyperglycaemic drugs is turning towards the inhibition of intestinal enzymes such as alpha amylase and alpha-glucosidase, which would in turn decrease the elevation in the post prandial blood glucose level [29]. Biogenic silver nanoparticles thus serve as potent inhibitors of such digestive enzymes [30]. The synthesis and efficiency of nanoparticles depends on the amount of reducing compounds like phenols, flavonoids and terpenes, etc., in the plant extracts. One of these plants is *Thymus serpyllum*, commonly known as Breckland thyme, which belongs to the family Lamiaceae and it possess several important compounds like minerals, phenols, flavonoids, and many other reducing agents [31]. Traditionally, its leaves and flowers in dried form are used as tea and infusions against fever, bronchitis, cold, and cough [32]. Several studies have shown that it depicts anti-rheumatic, ant-inflammatory, and hypoglycaemic activities [31,33].

In the present study, *Thymus serpyllum* will be analyzed for its antidiabetic efficacy both in vitro and in vivo on streptozotocin induced diabetic mice models. Current research is focused on biogenic synthesis of silver nanoparticles from *Thymus serpyllum* and their efficacy as antidiabetic agents in streptozotocin-induced diabetic mice models.

2. Materials and Methods

2.1. Plant Selection and Storage

Thymus serpyllum natural plant was collected from Rakaposhi Base Camp, Gilgit Baltistan for silver nanoparticles synthesis. The aerial parts of plant were dried and then ground into a fine powder by an automated electric grinder and stored at room temperature in a sterile sealed container.

2.2. Preparation of Thymus serpyllum Extract

Plant extract was prepared by modification of the protocol of Sun et al. [34]. A total of 10 g of plant powder was soaked in 100 mL deionized water in an Erlenmeyer flask at room temperature for a few minutes and then heated at 60 °C for 15 min on a hotplate. The extract was made to cool for half an hour and then the supernatant was collected and twice filtered through 0.45 μm pore size filter paper using vacuum filtration assembly. The filtrate obtained was stored at 4 °C as a stock solution to be used within 1 week.

2.3. Synthesis of Thymus serpyllum Mediated Silver Nanoparticles

The stock solution of plant, acting as a reducing and capping agent for the precursor, was diluted to 15% (v/v). A total of 850 μL of silver nitrate (10 mM) was added drop by drop per second into the 14.25 mL solution taken from 15% diluted extract of *Thymus serpyllum* under magnetic stirring at 300 rpm and 25 °C temperature. This working solution was then kept in the dark in a rotary orbital shaker for 5 h at 700 rpm and 25 °C temperature, followed by monitoring after every 1 h using UV Spectrophotometer for SPR band.

2.4. Purification of Nanoparticles

The reaction mixture was centrifuged at 15,000 rpm for 30 min at 4 °C in a refrigerated centrifuge. The supernatant was discarded and the pellet was re-suspended thrice in deionized water and centrifuged with the above conditions to obtain a thicker pellet. The pellet was then air dried, collected, and stored in Eppendorf tubes. This dried mass of particles was weighed and used for further activities.

2.5. Characterization of Silver Nanoparticles

The biosynthesized nanoparticles were characterized using various analytical techniques. UV Visible spectra of AgNP reaction mixture and plant extract was recorded using LABOMED, Inc., Los Angeles, CA, USA, Model UVD-2950 spectrophotometer. The functional groups were analyzed by using Perkin-Elmer spectrum 100 FTIR instrument of Waltham, MA, USA over the wavelength range from 450 to 4000 cm^{-1}. The composition and molecular structure of silver nanoparticles crystals were studied using an X-Ray Diffractometer (D8 ADVANCE BRUKER, AXS, Munich, Germany) using Cu K alpha as the radiation source over the scanning range of Bragg angle set as 20–80 theta. AgNP were sputter coated with gold to be conductive for Scanning Electron Microscopy (SEM model No 51-ADD0007 TESCAN VEGA3, sensor 51-1385-046 Kohoutovice, Brno, Czech Republic) in order to know the size and shape of nanoparticles. The elemental analysis was carried out using an EDX instrument (Oxford X-act, Tubney Woods Abingdon, Oxfordshire, UK).

2.6. Determination of Antioxidant Activity (DPPH Assay)

To evaluate the radical scavenging property of AgNP, DPPH (2, 2 Di Phenyl, 1 Picryl Hydrazyl) radical scavenging assay was carried out with some modifications in the protocol followed by Sanganna et al. [35]. The protocol for this assay started by making a fresh solution of DPPH (0.1 mM) in methanol. Different concentrations of both ascorbic acid and AgNPs such as 10 μg/mL, 20 μg/mL, 40 μg/mL, 60 μg/mL, 80 μg/mL, and 100 μg/mL were prepared. A total of 200 μL was taken from each of various concentrations of ascorbic acid and AgNPs and mixed thoroughly with 800 μL of DPPH solution in separate tubes to make a final volume of 1 mL. It was then followed by an incubation period of 30 min under dark condition at room temperature. After 30 min of incubation, the absorbance was measured at 517 nm with the help of UV/Vis spectrophotometer. The formula used for percent scavenging/inhibition was: Percentage Scavenging/Inhibition = [(control sample absorbance value − test sample absorbance value)/ control absorbance value] × 100. This procedure was repeated thrice and the values obtained were used to plot a graph and relative comparison of the free radicals scavenging ability of different concentrations of silver nanoparticles with respect to ascorbic acid was done. Because of the high antioxidant activity, ascorbic acid was used as a reference.

2.7. Anti-Diabetic Study

2.7.1. Alpha Amylase Inhibitory Assay

The effect of silver nanoparticles on alpha amylase activity was determined by making various concentrations of AgNPs and acarbose in separate tubes as 10, 20, 40, 60, 80, and 100 µg/mL in 0.02 M PBS with adjusted pH of 6.9, to which 500 µL of α-amylase was added and the reaction mixture was allowed to incubate for 10 min at 37 °C. Afterwards, 500 µL of starch (1% solution) was poured into each of the tubes containing the reaction mixture of varying concentrations and incubated for another 10 min, as per protocol followed by Haritha et al. [36]. Then, 1 mL of DNS was added in all tubes in order to terminate the reaction, followed by placing the tubes in a water bath at 60 °C for 15 min. The buffered mixture was cooled, followed by the addition of 10 mL distilled water in each tube. Then the spectrophotometer was set at 540 nm to take absorption readings of each concentration and make comparisons with the standard amylase inhibitor, i.e., acarbose.

2.7.2. Experimental Animals Acclimatization & Selection

The study was conducted on 4 weeks old male BALB/c mice (n = 50), purchased from the National Institute of Health and bred and housed in animal house of Atta-ur-Rahman School of Applied Biosciences (ASAB), National University of Sciences and Technology (NUST). These mice were kept in cages (5 mice per cage) at a constant temperature (25 ± 2 °C) and natural light-dark cycle (12–12 h) and were given distilled water *ad libitum* and fed a basic chow diet. The approval for all the protocols carried out during research was obtained from internal review board (IRB) of ASAB, (NUST). All the tests and experiments performed were according to the guidelines provided by the Institute of Laboratory Animal Research, Division on Earth and Life Sciences, National Institute of Health, Washington, DC, USA (Guide for the Care and Use of Laboratory Animals: Eighth Edition, 2011).

2.7.3. Streptozotocin Induced T2DM Mice Model Construction and Treatment Design

The T2DM mice model was constructed by combination of High Fat Diet (HFD) and low doses of Streptozotocin [37]. The normal control mice (n = 10) were fed with a basic diet (crude fiber 4%, crude fat 9%, and crude protein 30%) while the other mice models (n = 40) for the purpose of diabetes induction, after weaning when they were 3 weeks of age, were switched to a high fat diet (basic mice feed 59%, sugar 20%, animal fat 18%, and egg yolk 3%) and were given intraperitoneal STZ injections (100 mg/kg) in 0.1 M citrate buffer (pH 4.5) at the 6th (Figure 1) and 9th week of age after overnight fasting. The control animals however received citrate buffer injections only. After 2 days of 2nd STZ injection i.e., on the 44th day, the fasting blood glucose levels of all mice were checked using On-Call EZ II Blood Glucose Monitoring System (Blood ACON International, San Diego, CA, USA). Mice with blood glucose levels greater than 126 mg/dL were considered as diabetic [38]. Afterwards, mice were divided into 5 groups and each group had total of 10 animals of almost 10 weeks of age, which were further taken through the experiment and were categorized as follows:

Group 1: Mice served as the normal control and received normal diet and water throughout the experimentation.

Group 2: 10 mice were assigned as the negative control group for diabetes. They were left untreated throughout the experiment and were used as comparison for the rest of treatment groups.

Group 3: 10 mice served as the positive control group, which received the standard drug Metformin (100 mg/kg) orally in feed for 28 consecutive days.

Group 4: Allocated as the low dose silver nanoparticles treatment group. These mice were given silver nanoparticles 5 mg/kg orally through feed for a period of 28 days.

Group 5: The mice of this group were given silver nanoparticles 10 mg/kg in a normal diet for 28 days. Body weight and fasting blood glucose level were measured day 7, 14, 21, and 28.

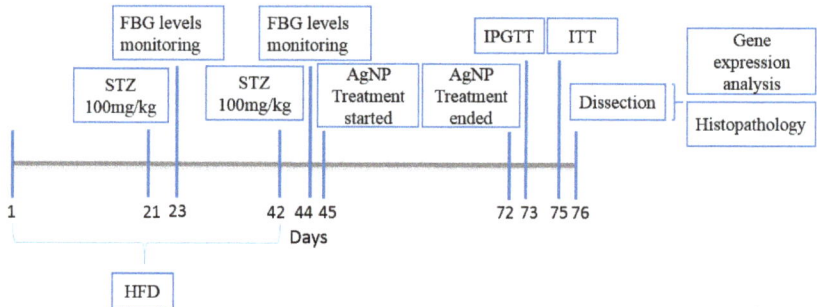

Figure 1. Timeline depicting the diabetic animal model construction and treatment period.

2.7.4. Glucose Level Estimation Tests

Fasting Blood Glucose Test

The glucose in the blood of normal and that of streptozotocin induced mouse models i.e., all groups, was monitored on a weekly basis. For the fasting blood glucose test, the animals were fasted for 8 h. The blood was taken from the tail of the mice and measured using the ON Call glucometer.

Intraperitoneal Glucose Tolerance Test (IGTT)

On the 29th day of treatment, mice from each group were subjected to the Intraperitoneal Glucose Tolerance Test (IPGTT). Mice were kept fasting for 8–10 h. A total of 2.5 g/kg of glucose was given to the animals. Blood samples of about 200 µL from the vein of each mouse were collected at different time intervals, i.e., 0, 30, 60, 90, and 120 min, in order to measure the level of glucose. The glucose-time curve was calculated using the values of Areas under the Curve (AUC).

Insulin Tolerance Test (ITT)

On the 31st day of model development, insulin tolerance test was done. For this test, mice were fasted for 4 h and then injected with 0.5 U/kg of human insulin subcutaneously. For the determination of blood glucose concentration, the blood samples were collected after 0, 30, 60, and 120 min of insulin injection and the values of Area under the Curve (AUC) were calculated.

2.7.5. Expression Analysis of AMPK and IRS1 Gene by Quantitative Real-Time Polymerase Chain Reaction (RT-PCR)

In order to evaluate the effect of AgNPs on the expression of IRS1 and AMPK gene, total RNA was isolated from liver tissue in Trizol reagent and cDNA was synthesized using first strand cDNA synthesis kit (Thermofisher Scientific, Waltham, MA, USA). Quantitative detection of these specific genes were carried out in Real time PCR (Applied Biosystems, Waltham, MA, USA). The primer pairs that were used are listed in Table 1. All samples for RT-PCR were assayed in triplicate, and data was normalized to the relative levels of GAPDH as a housekeeping gene in the same experiment.

Table 1. Primer sequences used for the expression analysis of AMPK, IRS1, and GAPDH.

Gene	Primer Sequence 5′ to 3′	Product Size
GAPDH	F-ACCCAGAAGACTGTGGATGG R-CACATTGGGGGTAGGAACAC	175 bp
IRS1	F-ACATCACAGCAGAATGAAGACC R-CCGGTGTCACAGTGCTTTCT	232 bp
AMPK	F-GTCGACGTAGCTCCAAGACC R-ATCGTTTTCCAGTCCCTGTG	250 bp

2.8. Histological Examination of Kidney, Liver, and Pancreas

Histological analysis was done to check the effects of biosynthesized silver nanoparticles on the anatomy of sample tissues. Right after anaesthesia, liver, pancreas and kidney tissues of mice were isolated to avoid decomposition and post-mortem autolysis. The tissues were preserved in 10% formalin solution. Sample tissues were cut into thin sections of about 4 microns using microtome and followed by dehydration, and H&E (Haematoxylin and Eosin) staining system was used for staining of the sections. The slides were then examined under light microscope.

2.9. Statistical Analysis

Statistical analysis of the expression of the AMPK, IRS1, and GAPDH genes in different diabetic mice and those treated with nanoparticles in comparison to normal control group (healthy mice) was done using One Way ANOVA and Post Hoc Bonferroni test on Graph Pad Prism version 5.01, GraphPad, San Diego, CA, USA. Statistical significance was set $p < 0.05$ and expressed as mean ± standard deviation.

3. Results and Discussion

3.1. Characterization of Biosynthesized AgNPs

A change in color from transparent to light brown and then dark brown was observed gradually within 4 h of reaction, which preliminary confirms the formation of biogenic silver nanoparticles (Figure 2A). After air drying the nanoparticles solution in a petri plate for 24 h, the nanoparticles were obtained in the form of solid brownish black powder (Figure 2B). UV Visible absorption spectroscopy of silver nanoparticles showed the Surface Plasmon Resonance peak at 425 nm (Figure 3).

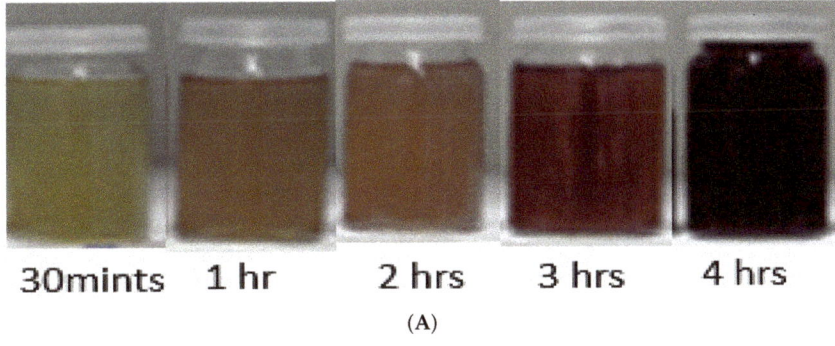

(A)

Figure 2. *Cont.*

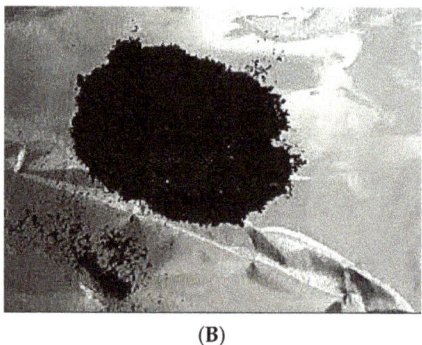

(B)

Figure 2. Visual confirmation of biosynthesized nanoparticles. (**A**) The color change represents the preliminary confirmation of silver nanoparticles. (**B**) Purified AgNPs obtained after drying.

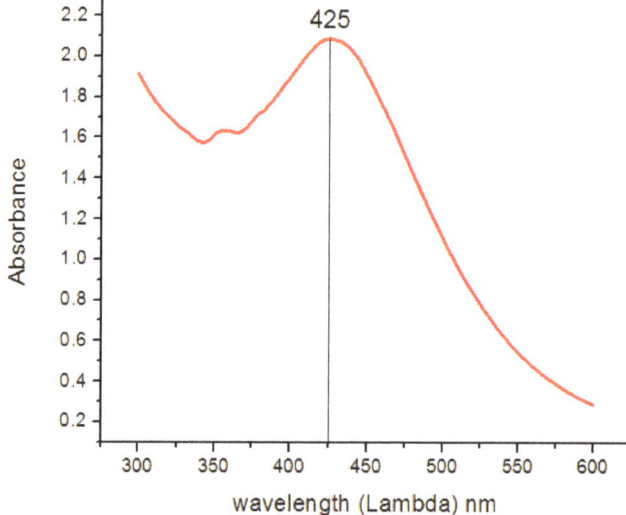

Figure 3. UV-Vis spectroscopy of reaction mixture indicating typical absorbance peak of AgNPs around 425 nm.

The crystalline nature of AgNPs was known through the XRD pattern, i.e., seven intensive diffraction peaks obtained at 2θ were 28°, 32°, 39°, 46°, 55°, 57° and 68°, which are indexed at (111), (200), (220), (311), (222), (400), and (331) facets of silver, respectively (Figure 4). The sharp peaks indicate the crystallinity and purity of silver nanoparticles and are in agreement with the database of the Joint committee on Powder Diffraction Standards (JCPDS Card Number 00-04-0783). The rest of the intense peaks at 2θ angles are due to the involvement of AgNO3 used for the synthesis of silver nanoparticles [39]. The FTIR spectra depicted the similar functional groups among the extract and silver nanoparticles, which enhances the stability and reduces the cytotoxicity of nanoparticles. The infrared spectral bands of *T. serpyllum* plant extract at 2923 cm^{-1}, 1607 cm^{-1}, 1515 cm^{-1}, and 1179 cm^{-1} (black) were shifted to 2924 cm^{-1}, 1620 cm^{-1}, 1510 cm^{-1}, and 1175 cm^{-1} in the spectra of biogenic silver nanoparticles (red), indicating the involvement of C-H, N-H, N-O and C-O groups in the biosynthesis of silver nanoparticles (Figure 5). The SEM analysis revealed smaller, spherical, and monodispersed AgNPs and their average particle size measured from the SEM image turned out to be 42 nm (Figure 6). The identification and quantification of elemental composition of AgNPs was done by EDX analysis at a magnification of 90× with 512 × 384 pixel and probe current of 20 Kv. EDX analysis of biogenic silver nanoparticles via SEM machine revealed the presence of silver in the colloidal solution of *Thymus serpyllum* mediated silver nanoparticles. Silver exists in mass of 9.04% (Figure 7). The other elements present were carbon, oxygen, and chlorine (Table 2), which attributes the presence of the organic source, i.e., plant extract.

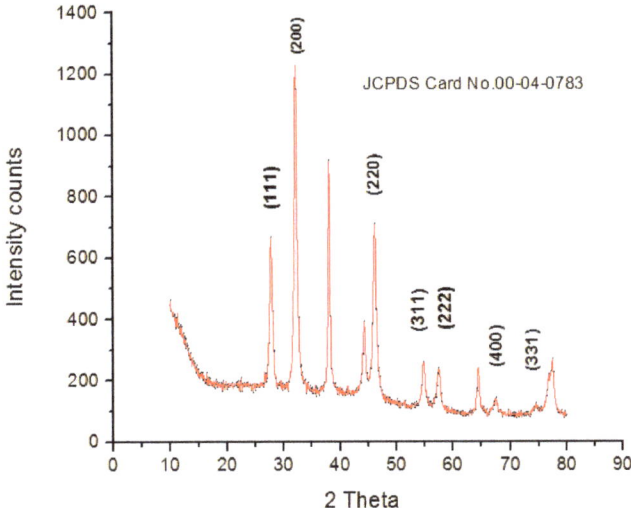

Figure 4. X-ray diffraction pattern of biogenic silver nanoparticles.

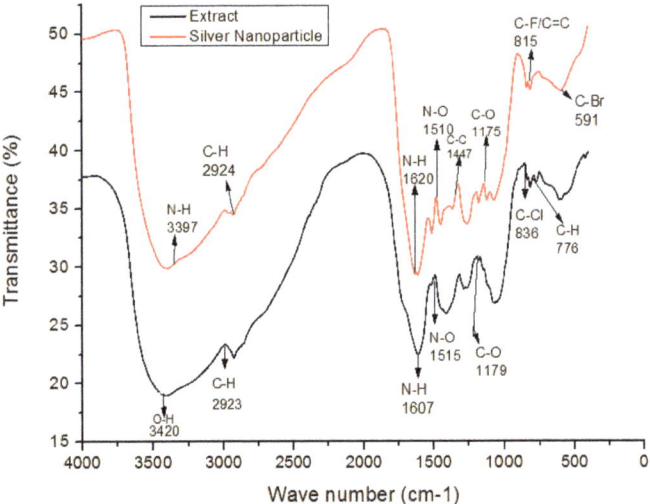

Figure 5. FTIR depicting functional groups of silver nanoparticles vs. thyme extract. FTIR peaks of AgNPs reveals few similar functional groups to that of thyme extract as 2924 cm^{-1} (C-H stretching), 1620 cm^{-1} (N-H bending), 1510 cm^{-1} (N-O stretching), and 1175 cm^{-1} (C-O bending), which potentially act as surface capping and stabilizing groups.

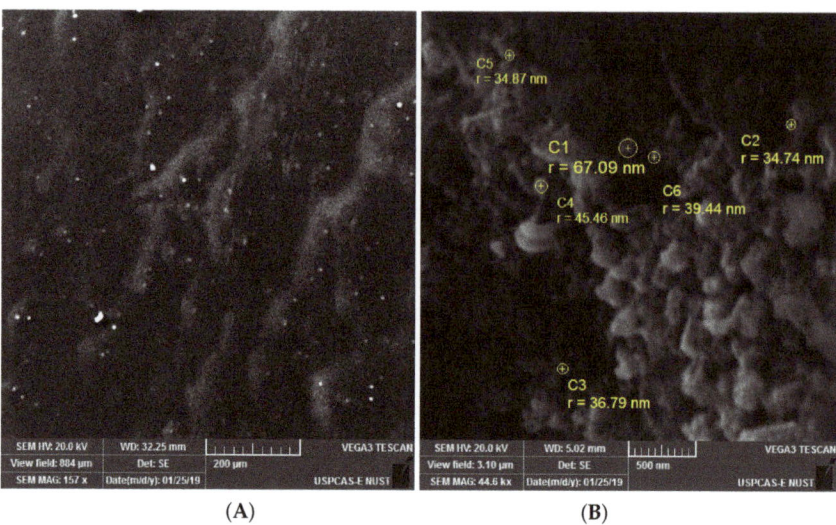

(A) (B)

Figure 6. (**A**) SEM image of biogenic AgNPs at 157 kx and 200 µm indicating the formation of spherical nanoparticles. (**B**) SEM image of biogenic AgNPs at 44.6 kx and 500 nm, indicating various sizes of nanoparticles.

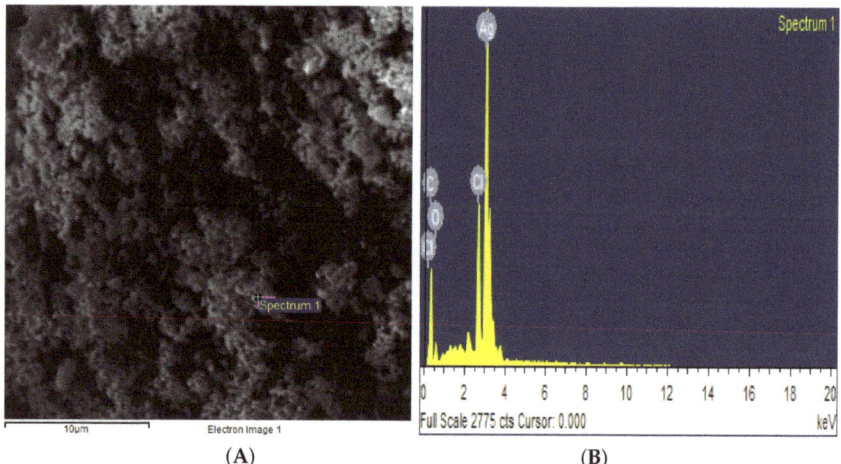

(A) (B)

Figure 7. Energy Dispersive X-ray Spectroscopy. (**A**) SEM region of the sample from where the EDS graph was taken. (**B**) The EDS graph, with energy on X-axis and counts on Y-axis, shows the elemental composition of AgNPs in colloidal solution and confirms the presence of AgNPs at the highest spectrum peak of 3 KeV.

Table 2. Atomic percent and weight of the elements in AgNPs.

Element	Weight%	Atomic%
Silver	43.82	9.04
Oxygen	10.63	14.78
Chlorine	6.66	4.18
Carbon	38.89	72.02

3.2. Antioxidant and Alpha Amylase Inhibitory Activity

DPPH assay was employed to evaluate the antioxidant ability of biosynthesised silver nanoparticles. The assay revealed that the scavenging ability of these nanoparticles increases as the concentration of nanoparticles increases (Figure 8). The standard ascorbic acid showed a maximum free radicals scavenging ability of 80% at a concentration of 100 µg/mL while the silver nanoparticles at the same concentration scavenged 78% of the free radicals. The IC50 of silver nanoparticles was found to be 8 µg/mL while that of standard ascorbic acid showed IC50 of 5.5 µg/mL. This suggests that the biologically synthesized AgNPs can be used as antioxidants.

The enzyme, alpha amylase, catalyzes the hydrolysis of α 1-4 glycosidic linkages in polysaccharides, so its inhibition possesses a significant role in controlling hyperglycemia. The assay was performed to assess the enzyme inhibition potential of the synthesized silver nanoparticles. Acarbose served as a control while the sample without AgNPs and containing enzyme was used as standard. The inhibition potential was expressed as percent inhibition. The synthesized silver nanoparticles inhibited the alpha amylase in a dose dependent manner comparable to acarbose (Figure 9), which showed a maximum enzyme inhibition of 90% at a concentration of 80 µg/mL with an IC50 value at 7.5 µg/mL while AgNPs inhibited the enzyme up to 83% at a concentration of 80 µg/mL with an IC50 value of 10 µg/mL.

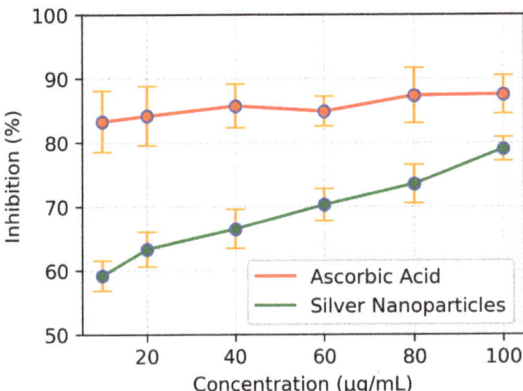

Figure 8. Comparisons of absorbance values of different concentrations of ascorbic acid and AgNP in the DPPH assay.

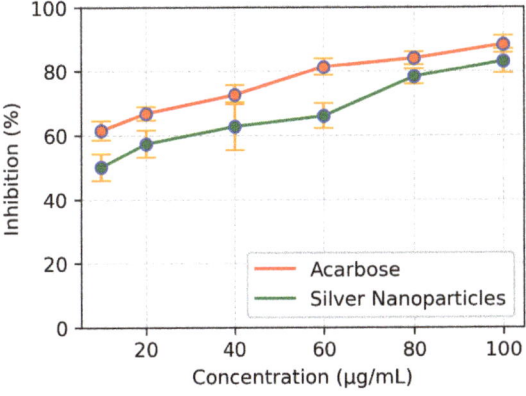

Figure 9. Comparison of α-amylase inhibition of various concentrations of AgNPs and acarbose.

3.3. Estimation of AgNP Treatment on Body Weight

Before the treatment, the animals were tested for blood glucose levels for confirmation of STZ-induced hyperglycaemia. The STZ group showed significantly higher levels of fasting blood glucose compared to normal (p = <0.001) (Figure 10). The effect on body weight of mice due to exposure to AgNPs was studied over the span of 4 weeks. The body weight of the diabetic group significantly increased since it did not receive any treatment after HFD, while in those of the treated groups in comparison to the control showed no significant effect on body weight (Figure 11).

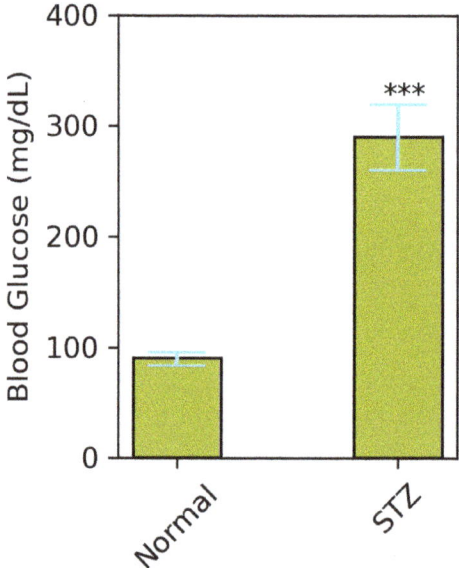

Figure 10. Average FBG levels (mmol/L) in control mice and mice fed with HFD from age of 3 to 9 weeks of age and rendered 2 doses of 100 mg/kg STZ (*** p < 0.001).

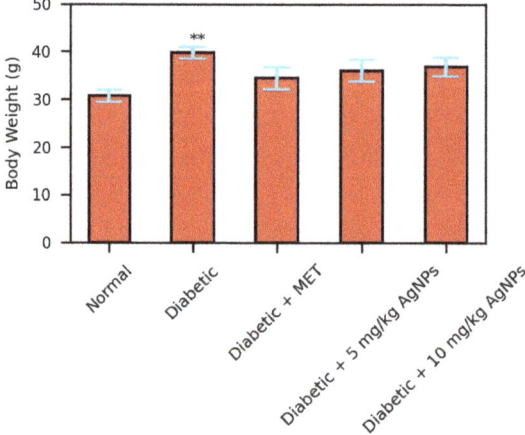

Figure 11. Body weights of treated and non-treated mice after 28 days treatment (** p < 0.01).

3.4. Comparative Analysis of Fasting Blood Glucose Levels in Treated Mice Groups

During 28 days of treatment, the blood glucose levels of mice of all groups were checked every week after overnight fasting. Due to HFD and STZ administration, the blood glucose levels of the untreated group noticeably raised to a significant level from the normal group. During the initial two weeks of treatment, no significant reduction in the blood glucose levels of the treated groups was observed.

Comparatively, the 10 mg/kg AgNPs treated group after 21 days of treatment significantly reduced the blood glucose levels. At the end of treatment (after 28 days), the FBG levels of AgNPs treated groups were slightly higher than the normal, but in comparison with the untreated diabetic mice models, there was a significant reduction in their glucose levels (Figure 12). Moreover, the metformin and 5 mg/kg AgNP treatment had almost similar effects, however, the 10 mg/kg AgNP treated group showed a better reduction in blood glucose levels, comparable to the control group.

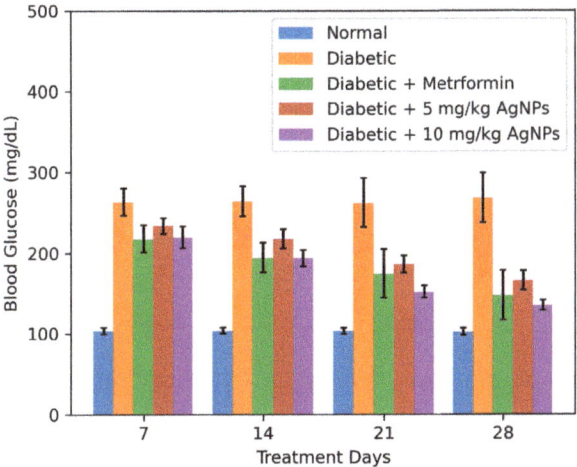

Figure 12. Histograms represent the effect of AgNPS treatment on fasting blood glucose (FBG levels (mmol/L) measured from Day 1 to Day 28 in comparison to the diabetic group. The data was analyzed using Two-way ANOVA following Bonferroni Post Hoc test and is shown as mean ± SEM.

Estimation of Improvement in Glucose Tolerance and Insulin Release

One of the main features of IR is an inability to tolerate the glucose. The untreated diabetic mice showed glucose intolerance while the metformin-treated 5 mg/kg and 10 mg/kg AgNP-treated group improved the glucose tolerance to a noticeable level (Figure 13). The glycaemic index in each group was expressed and monitored as Area under the Curve (AUC) (Figure 14).

An insulin tolerance test was carried out to determine the effect of AgNP treatment on the action of insulin. The untreated diabetic mice showed a greater extent of insulin intolerance, so the glucose levels remained significantly higher from the normal range, while the metformin-treated, 5 mg/kg and 10 mg/kg AgNP-treated group showed sensitivity towards the insulin and utilized the insulin to lower the blood glucose levels (Figure 15). The Area under the Curve in the untreated diabetic group was much higher after insulin administration as compared to the normal (Figure 16).

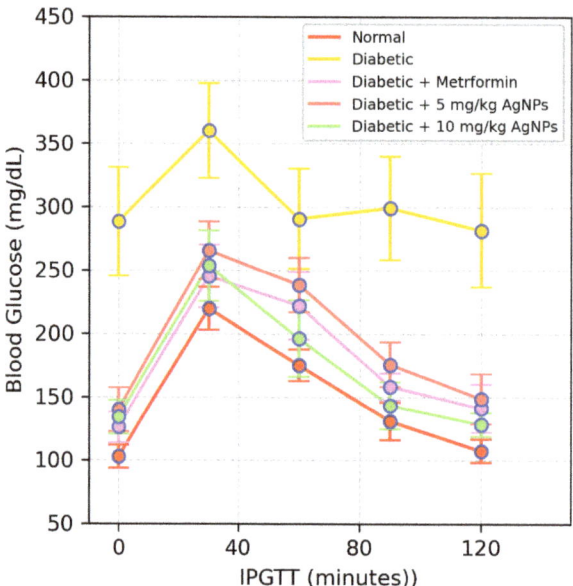

Figure 13. IPGTT results of the healthy, diabetic, metformin-treated and AgNP-treated low (5 mg/kg) and high dose (10 mg/kg) mice post treatment. The statistical significance of differences among different groups was analyzed using Two Way ANOVA Bonferroni test as implemented in Graph pad prism 5.0 software.

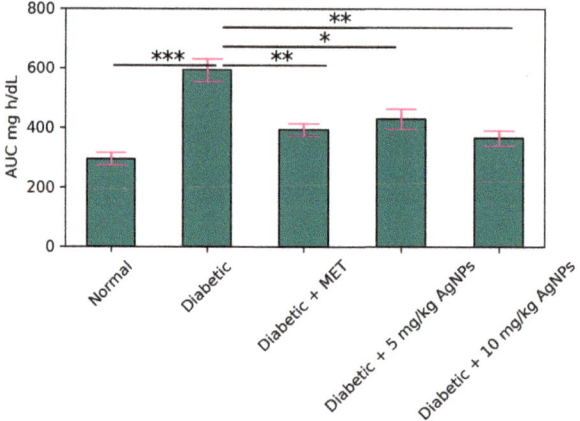

Figure 14. Area under the Curves (AUC) for IPGTT. The statistical significance of differences among different groups was analyzed using Two Way ANOVA Bonferroni test as implemented in Graph pad prism 5.0 software (* $p < 0.05$, ** $p < 0.01$, *** $p < 0.001$).

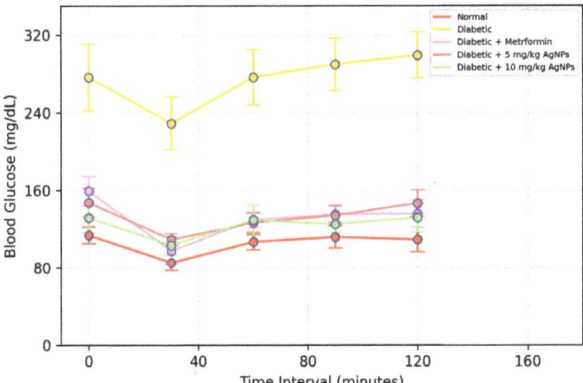

Figure 15. Effect of AgNPs on insulin tolerance (day 31). Results of the healthy, diabetic, metformin-treated and AgNP-treated low (5 mg/kg) and high dose (10 mg/kg) mice post treatment.

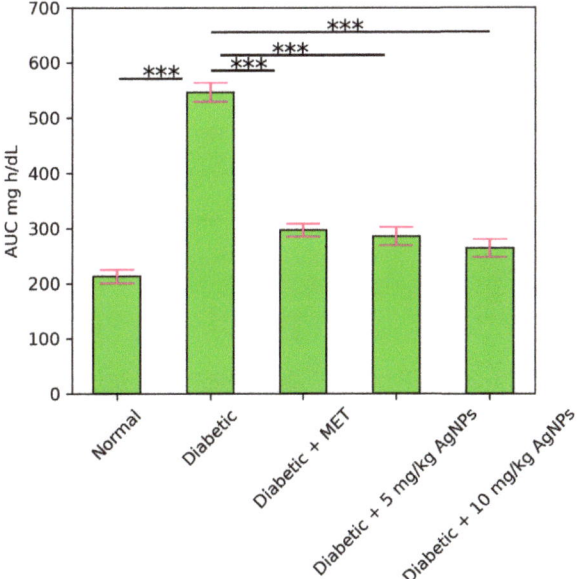

Figure 16. The Area under the Curve (AUC) for Insulin Tolerance Test (ITT). Analyzed Two Way ANOVA followed by Bonferroni Post Hoc test (*** $p < 0.001$).

3.5. AMPK and IRS1 Gene Expression Analysis through Real Time PCR

Quantitative real time PCR (qRT-PCR) was carried out to observe the expression of IRS1 gene in the liver of normal, diabetic, metformin treated, and AgNP (low and high dose) treated groups. It was observed that the expression level of IRS1 in the diabetic group was significantly decreased (* $p < 0.05$) as compared to the control (Figure 17). Upon comparison between the diabetic and metformin-treated group, a significant increase was observed (* $p < 0.05$) in the IRS1 expression in liver of the metformin-treated group. IRS1 expression in the 5 mg/kg AgNP-treated group was upregulated to almost 2.5-fold in comparison to the untreated diabetic group. Ten mg/kg AgNP treated mice depicted an approximately 4.5-fold increase in the expression of IRS1 in comparison to the untreated diabetic mice ($p < 0.001$). AMPK is the main target of the standard drug metformin, so metformin treatment in diabetic mice increased its expression to almost 3-fold (Figure 18). However, AMPK expression in the low dose treated silver nanoparticles group increased to almost 1.5-fold while in the 10 mg/kg AgNPs treated diabetic mice the expression fold increased almost 3 times, comparable to the metformin treated group with a significance of (*** $p < 0.001$).

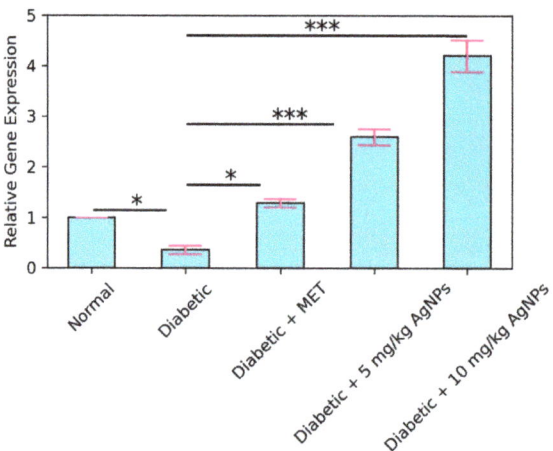

Figure 17. Relative expression of IRS1 in control, diabetic, and metformin- and AgNPs-treated (28 days) groups. Significance was determined by One Way ANOVA following Bonferroni's Multiple comparison Post Hoc Test. Error bars represent the SEM (standard error of the mean) (* $p < 0.05$, *** $p < 0.001$).

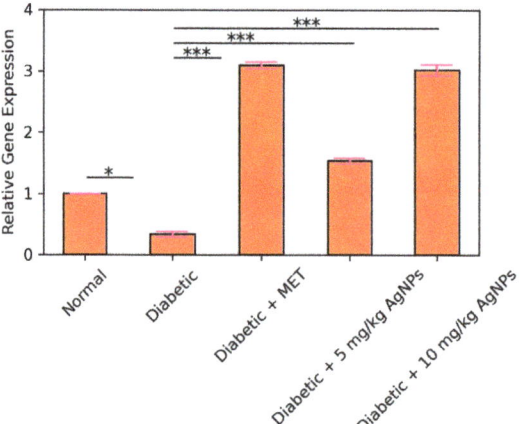

Figure 18. Relative expression of AMPK in control, diabetic untreated, and diabetic metformin- and AgNPs-treated (28 days) groups. Significance was determined by One Way ANOVA following Bonferroni's Multiple comparison Post Hoc Test. Error bars represent the SEM (standard error of the mean) (* $p < 0.05$, *** $p < 0.001$).

3.6. Histopathological Analysis of Control, Diabetic and NP Treated Balb/c Mice

The liver, kidney, and pancreas from all groups were obtained for histological analysis after the required treatment was carried out. The histological examination overall showed that high dose silver nanoparticles restored the damaged and necrotic tissues to a greater extent. Haematoxylin and Eosin stained sections of liver at 40× depicts the portal vein and the portal triad in normal, diabetic, metformin-treated, and silver nanoparticles-treated groups of BALB/c mice (Figure 19).

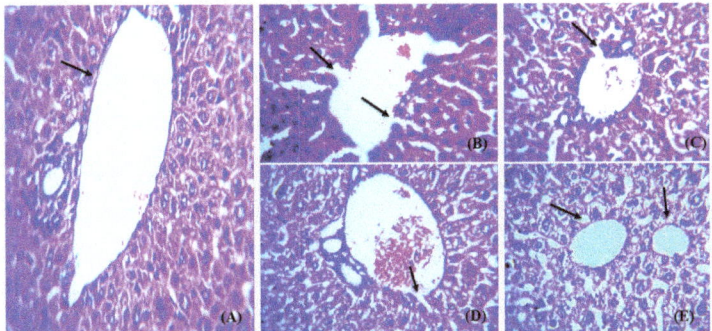

Figure 19. Representative images showing Hematoxylin and Eosin stained (H&E) (×40) sections of liver. (**A**) Control, normal liver histology with portal triad and portal vein (arrow) with hepatocytes radiating from the portal vein. (**B**) Diabetic group showing complete distortion of portal vein architecture and congestion of portal triad. (**C**) Diabetic metformin-treated group showing somewhat restoration of hepatic architecture but dilated vascular channels. (**D**) Diabetic mice liver treated with 5 mg/kg AgNPs showing greater restoration of portal vein boundary with little dilation (arrow) and normal portal triad. (**E**) Diabetic AgNP (10 mg/kg) treated mice liver showing restored architecture comparable to that of normal with intact portal triad and normal physiology of kupffer cells (arrow).

H & E stained sections of kidney from all groups were analyzed focusing on glomerulus and renal tubules arrangement and boundary (Figure 20). The cellular morphology and density is better restored by high dose silver nanoparticles, i.e., 10 mg/kg.

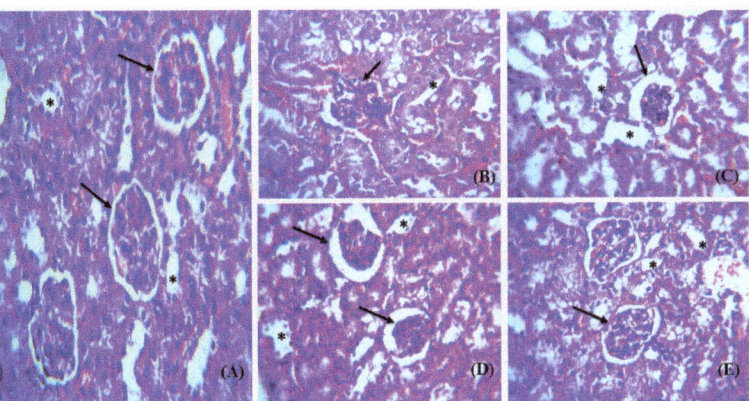

Figure 20. Photomicrograph showing Hematoxylin and Eosin stained (H&E) (×40) sections of kidney with asterisk showing kidney tubules and arrow showing glomerulus. (**A**) Control group with regular shape of glomerulus and distinguishable collecting duct and renal tubule. (**B**) Untreated diabetic group showing irregular renal cells distribution, tubules dilation, and glomerulosclerosis. (**C**) The diabetic metformin-treated group shows somewhat normal appearance of tubules and intact glomerular boundary when compared to the diabetic group. (**D**) Diabetic-AgNPs treated (5 mg/kg) group showing restoration of glomerular boundary and intact kidney tubules. (**E**) Diabetic AgNP (10 mg/kg) treated group show more reno-protective activity as compared to the standard metformin with complete restoration of glomerulus and renal tubules.

Pancreas from all the groups were H & E stained and histological examination was carried out on various magnifications to compare the morphology of all treated mice versus diabetic groups. Islets of Langerhans and pancreatic acini were focused on 40× in all groups (Figure 21).

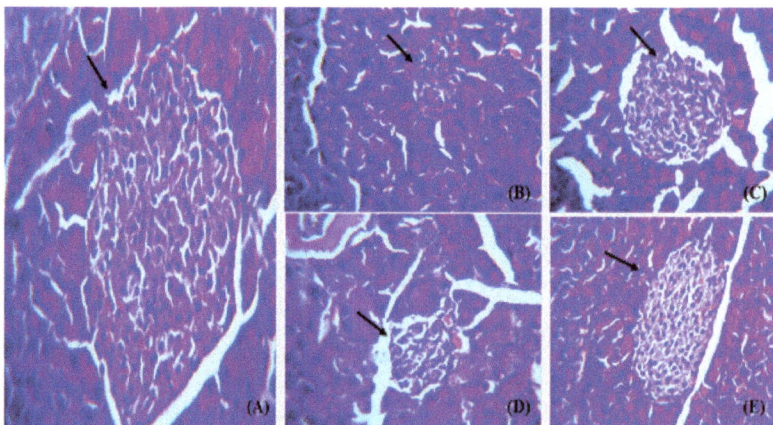

Figure 21. Representative images showing Hematoxylin and Eosin stained (H&E) (×40) sections of pancreas. (**A**) Control, normal pancreas showing normal islets of Langerhans in between normal pancreatic β-cells (arrow). (**B**) Diabetic pancreas showing ruptured and destructed islets of Langerhans with damage in β-cells. (**C**) Diabetic and Metformin group pancreas showing some normal islets of Langerhans. (**D**) Diabetic and AgNP (5 mg/kg) mice pancreas showing some restoration in islets of Langerhans with damage in β-cells (arrow). (**E**) Diabetic and AgNP (10 mg/kg) pancreas showing complete restoration of islets of Langerhans.

4. Conclusions

The main objective of this study was the formulation of a nanodrug for diabetes using green route, which could be safer and more biocompatible than synthetic drugs, with fewer side effects. The medicinal plant "*Thymus serpyllum*" has an aqueous extract rich in bioactive compounds, which served as a reducing agent to synthesize and then as a capping agent to stabilize the nanoparticles. The newly biosynthesized silver nanoparticles were confirmed via characterization techniques such as SEM, XRD, EDS, and FTIR, depicting spherical 42 nm size nanoparticles with JCPDS Card No. 00-04-0783. Smaller size nanoparticles show better antioxidant and alpha amylase inhibition activity, and increasing the dose of AgNPs increased the scavenging ability as well as amylase inhibition. The synthesized nanoparticles improved the insulin sensitivity and glucose tolerance in treated mice and reduced the fasting glucose levels significantly. We conclude that the synthesized nanoparticles are more powerful in reducing the hyperglycemia and enhances the IRS1 and AMPK expression for treatment of diabetes, which aid in glucose uptake and insulin sensitization. Our results suggest that silver nanoparticles potentially restore the cellular morphology of liver, kidney, and pancreas specifically. However, further research is needed to confirm these findings. There is a need for further in vivo pharmacological investigations to elucidate the mechanism of action of AgNPs by targeting other genes involved in diabetes. Furthermore, functionalization of these AgNPs with various drugs could be done to provide sustained release of the drug at the site of action in the body.

Author Contributions: Conceptualization, M.W. and A.B.; methodology, M.W. and A.B.; investigation, M.W.; writing—original draft preparation, M.W.; writing—in house reviewing and editing, A.B. and P.J.; supervision, A.B. and P.J.; project administration, A.B. and P.J. All authors have read and agreed to the published version of the manuscript.

Funding: This research received no external funding.

Institutional Review Board Statement: The approval for all the protocols carried out during research was obtained from the internal review board (IRB) of Attaur Rehman School of Applied Biosciences, National University of Sciences and Technology, (NUST). All the tests and experiments performed were according to the guidelines provided by the Institute of Laboratory Animal Research, Division on Earth and Life Sciences, National Institute of Health, USA (Guide for the Care and Use of Laboratory Animals: Eighth Edition, 2011).

Informed Consent Statement: Not applicable.

Data Availability Statement: Available on request to corresponding author.

Acknowledgments: The authors acknowledge the Atta-ur-Rahman School of Applied Biosciences, National University of Sciences and Technology for providing the necessary infrastructure and financial support to carry out this work.

Conflicts of Interest: The authors declare no conflict of interest.

References

1. Dixon, J.B.; Zimmet, P.; Alberti, K.G.; Rubino, F. Bariatric surgery: An IDF statement for obese type 2 diabetes. *Surg. Obes. Relat. Dis.* **2011**, *7*, 433–447. [CrossRef]
2. Brownlee, M. The pathobiology of diabetic complications: A unifying mechanism. *Diabetes* **2005**, *54*, 1615–1625. [CrossRef]
3. Marín-Peñalver, J.J.; Martín-Timón, I.; Sevillano-Collantes, C.; del Cañizo-Gómez, F.J. Update on the treatment of type 2 diabetes mellitus. *World J. Diabetes* **2016**, *7*, 354. [CrossRef]
4. Pradhan, A.D.; Manson, J.E.; Rifai, N.; Buring, J.E.; Ridker, P.M. C-reactive protein, interleukin 6, and risk of developing type 2 diabetes mellitus. *JAMA* **2001**, *286*, 327–334. [CrossRef]
5. National Diabetes Data Group. Classification and diagnosis of diabetes mellitus and other categories of glucose intolerance. *Diabetes* **1979**, *28*, 1039–1057. [CrossRef] [PubMed]
6. Drivsholm, T.; de Fine Olivarius, N.; Nielsen, A.B.S.; Siersma, V. Symptoms, signs and complications in newly diagnosed type 2 diabetic patients, and their relationship to glycaemia, blood pressure and weight. *Diabetologia* **2005**, *48*, 210–214. [CrossRef]
7. Steyn, N.P.; Mann, J.; Bennett, P.H.; Temple, N.; Zimmet, P.; Tuomilehto, J.; Lindström, J.; Louheranta, A. Diet, nutrition and the prevention of type 2 diabetes. *Public Health Nutr.* **2004**, *7*, 147–165. [CrossRef]
8. Jackson, R.A.; Hawa, M.I.; Jaspan, J.B.; Sim, B.M.; DiSilvio, L.; Featherbe, D.; Kurtz, A.B. Mechanism of metformin action in non-insulin-dependent diabetes. *Diabetes* **1987**, *36*, 632–640. [CrossRef]
9. Saini, V. Molecular mechanisms of insulin resistance in type 2 diabetes mellitus. *World J. Diabetes* **2010**, *1*, 68. [CrossRef]
10. Röder, P.V.; Wu, B.; Liu, Y.; Han, W. Pancreatic regulation of glucose homeostasis. *Exp. Mol. Med.* **2016**, *48*, e219. [CrossRef]
11. Prasad, V.S.S.; Adapa, D.; Vana, D.R.; Choudhury, A.; Asadullah, J.; Chatterjee, A. Nutritional components relevant to type-2-diabetes: Dietary sources, metabolic functions and glycaemic effects. *J. Res. Med. Dent. Sci.* **2018**, *6*, 52–75.
12. Cnop, M.; Vidal, J.; Hull, R.L.; Utzschneider, K.M.; Carr, D.B.; Schraw, T.; Scherer, P.E.; Boyko, E.J.; Fujimoto, W.Y.; Kahn, S.E. Progressive loss of β-cell function leads to worsening glucose tolerance in first-degree relatives of subjects with type 2 diabetes. *Diabetes Care* **2007**, *30*, 677–682. [CrossRef]
13. Kahn, S.E.; Hull, R.L.; Utzschneider, K.M. Mechanisms linking obesity to insulin resistance and type 2 diabetes. *Nature* **2006**, *444*, 840–846. [CrossRef]
14. Shackelford, D.B.; Shaw, R.J. The LKB1–AMPK pathway: Metabolism and growth control in tumour suppression. *Nat. Rev. Cancer* **2009**, *9*, 563–575. [CrossRef]
15. Garcia, D.; Shaw, R.J. AMPK: Mechanisms of cellular energy sensing and restoration of metabolic balance. *Mol. Cell* **2017**, *66*, 789–800. [CrossRef]
16. Hardie, D.G. AMPK: A target for drugs and natural products with effects on both diabetes and cancer. *Diabetes* **2013**, *62*, 2164–2172. [CrossRef]
17. Dong, H.; Wang, N.; Zhao, L.; Lu, F. Berberine in the treatment of type 2 diabetes mellitus: A systemic review and meta-analysis. *Evid.-Based Complement. Altern. Med.* **2012**, *2012*, 591654. [CrossRef]
18. Li, L.; Li, L.; Zhou, X.; Yu, Y.; Li, Z.; Zuo, D.; Wu, Y. Silver nanoparticles induce protective autophagy via Ca^{2+}/CaMKKβ/AMPK/mTOR pathway in SH-SY5Y cells and rat brains. *Nanotoxicology* **2019**, *13*, 369–391. [CrossRef]
19. Zhang, R.; Qin, X.; Zhang, T.; Li, Q.; Zhang, J.; Zhao, J. Astragalus polysaccharide improves insulin sensitivity via AMPK activation in 3T3-L1 adipocytes. *Molecules* **2018**, *23*, 2711. [CrossRef]
20. Fisher, J.S. Potential role of the AMP-activated protein kinase in regulation of insulin action. *Cellscience* **2006**, *2*, 68.
21. Lavin, D.P.; White, M.F.; Brazil, D.P. IRS proteins and diabetic complications. *Diabetologia* **2016**, *59*, 2280–2291. [CrossRef] [PubMed]
22. Alkaladi, A.; Abdelazim, A.M.; Afifi, M. Antidiabetic activity of zinc oxide and silver nanoparticles on streptozotocin-induced diabetic rats. *Int. J. Mol. Sci.* **2014**, *15*, 2015–2023. [CrossRef] [PubMed]

23. Bai, L.; Gao, J.; Wei, F.; Zhao, J.; Wang, D.; Wei, J. Therapeutic potential of ginsenosides as an adjuvant treatment for diabetes. *Front. Pharmacol.* **2018**, *9*, 423. [CrossRef] [PubMed]
24. Scalbert, A.; Johnson, I.T.; Saltmarsh, M. Polyphenols: Antioxidants and beyond. *Am. J. Clin. Nutr.* **2005**, *81*, 215S–217S. [CrossRef]
25. Johnson, P.; Krishnan, V.; Loganathan, C.; Govindhan, K.; Raji, V.; Sakayanathan, P.; Vijayan, S.; Sathishkumar, P.; Palvannan, T. Rapid biosynthesis of Bauhinia variegata flower extract-mediated silver nanoparticles: An effective antioxidant scavenger and α-amylase inhibitor. *Artif. Cells Nanomed. Biotechnol.* **2018**, *46*, 1488–1494. [CrossRef] [PubMed]
26. Seralathan, J.; Stevenson, P.; Subramaniam, S.; Raghavan, R.; Pemaiah, B.; Sivasubramanian, A.; Veerappan, A. Spectroscopy investigation on chemo-catalytic, free radical scavenging and bactericidal properties of biogenic silver nanoparticles synthesized using Salicornia brachiata aqueous extract. *Spectrochim. Acta Part A Mol. Biomol. Spectrosc.* **2014**, *118*, 349–355. [CrossRef] [PubMed]
27. Goodarzi, V.; Zamani, H.; Bajuli, L.; Moradshahi, A. Evaluation of antioxidant potential and reduction capacity of some plant extracts in silver nanoparticles' synthesis. *Mol. Biol. Res. Commun.* **2014**, *3*, 165.
28. Ghaffari-Moghaddam, M.; Hadi-Dabanlou, R.; Khajeh, M.; Rakhshanipour, M.; Shameli, K. Green synthesis of silver nanoparticles using plant extracts. *Korean J. Chem. Eng.* **2014**, *31*, 548–557. [CrossRef]
29. Iravani, S.; Zolfaghari, B. Green synthesis of silver nanoparticles using Pinus eldarica bark extract. *BioMed Res. Int.* **2013**, *2013*, 639725. [CrossRef]
30. Aritonang, H.F.; Koleangan, H.; Wuntu, A.D. Synthesis of silver nanoparticles using aqueous extract of medicinal plants' (*Impatiens balsamina* and *Lantana camara*) fresh leaves and analysis of antimicrobial activity. *Int. J. Microbiol.* **2019**, *2019*, 8642303. [CrossRef]
31. Mushtaq, M.N.; Bashir, S.; Ullah, I.; Karim, S.; Rashid, M.; Hayat Malik, M.N.; Rashid, H. Comparative hypoglycemic activity of different fractions of Thymus serpyllum L. in alloxan induced diabetic rabbits. *Pak. J. Pharm. Sci.* **2016**, *29*, 1483–1488.
32. Hussain, W.; Badshah, L.; Ullah, M.; Ali, M.; Ali, A.; Hussain, F. Quantitative study of medicinal plants used by the communities residing in Koh-e-Safaid Range, northern Pakistani-Afghan borders. *J. Ethnobiol. Ethnomed.* **2018**, *14*, 1–18. [CrossRef]
33. Ahamed, M.; AlSalhi, M.S.; Siddiqui, M. Silver nanoparticle applications and human health. *Clin. Chim. Acta* **2010**, *411*, 1841–1848. [CrossRef] [PubMed]
34. Tippayawat, P.; Phromviyo, N.; Boueroy, P.; Chompoosor, A. Green synthesis of silver nanoparticles in aloe vera plant extract prepared by a hydrothermal method and their synergistic antibacterial activity. *PeerJ* **2016**, *4*, e2589. [CrossRef]
35. Sanganna, B.; Chitme, H.R.; Vrunda, K.; Jamadar, M.J. Antiproliferative and antioxidant activity of leaves extracts of Moringa oleifera. *Int. J. Curr. Pharm. Rev. Res.* **2016**, *8*, 54–56. [CrossRef]
36. Haritha, H. Green Synthesis and Characterization of Silver Nanoparticles Using Pterocarpus Marsupium Roxb. and Assessment of Its In-Vitro Anti Diabetic Activity. Ph.D. Thesis, Sri Ramakrishna Institute of Paramedical Sciences, Coimbatore, India, 2017.
37. Noor, A.; Zahid, S. Alterations in adult hippocampal neurogenesis, aberrant protein s-nitrosylation, and associated spatial memory loss in streptozotocin-induced diabetes mellitus type 2 mice. *Iran. J. Basic Med. Sci.* **2017**, *20*, 1159.
38. Dong, J.; Xu, H.; Wang, P.F.; Cai, G.J.; Song, H.F.; Wang, C.C.; Dong, Z.T.; Ju, Y.J.; Jiang, Z.Y. Nesfatin-1 stimulates fatty-acid oxidation by activating AMP-activated protein kinase in STZ-induced type 2 diabetic mice. *PLoS ONE* **2013**, *8*, e83397. [CrossRef]
39. Lanje, A.S.; Sharma, S.J.; Pode, R.B. Synthesis of silver nanoparticles: A safer alternative to conventional antimicrobial and antibacterial agents. *J. Chem. Pharm. Res.* **2010**, *2*, 478–483.

Article

Effect of Solvent on Superhydrophobicity Behavior of Tiles Coated with Epoxy/PDMS/SS

Srimala Sreekantan [1,*], Ang Xue Yong [1], Norfatehah Basiron [1], Fauziah Ahmad [2,*] and Fatimah De'nan [2]

[1] School of Materials and Mineral Resources Engineering, Engineering Campus, Universiti Sains Malaysia, Nibong Tebal 14300, Pulau Pinang, Malaysia; xueyong95@hotmail.com (A.X.Y.); fatehahbasiron31@gmail.com (N.B.)
[2] School of Civil Engineering, Engineering Campus, Universiti Sains Malaysia, Nibong Tebal 14300, Pulau Pinang, Malaysia; cefatimah@usm.my
* Correspondence: srimala@usm.my (S.S.); cefahmad@usm.my (F.A.); Tel.: +604-599-5255 (S.S.); +604-599-6268 (F.A.)

Abstract: Superhydrophobic coatings are widely applied in various applications due to their water-repelling characteristics. However, producing a durable superhydrophobic coating with less harmful low surface materials and solvents remains a challenge. Therefore, the aim of this work is to study the effects of three different solvents in preparing a durable and less toxic superhydrophobic coating containing polydimethylsiloxane (PDMS), silica solution (SS), and epoxy resin (DGEBA). A simple sol-gel method was used to prepare a superhydrophobic coating, and a spray-coating technique was employed to apply the superhydrophobic coating on tile substrates. The coated tile substrates were characterized for water contact angle (WCA) and tilting angle (TA) measurements, Field-Emission Scanning Electron Microscopy (FESEM), Atomic Force Microscopy (AFM), and Fourier Transform Infrared Spectroscopy (FTIR). Among 3 types of solvent (acetone, hexane, and isopropanol), a tile sample coated with isopropanol-added solution acquires the highest water contact angle of $152 \pm 2°$ with a tilting angle of $7 \pm 2°$ and a surface roughness of 21.80 nm after UV curing for 24 h. The peel off test showed very good adherence of the isopropanol-added solution coating on tiles. A mechanism for reactions that occur in the best optimized solvent is proposed.

Keywords: superhydrophobic; spray-coating; polydimethylsiloxane; silica solution; epoxy resin; acetone; hexane; isopropanol

1. Introduction

A material is known to be superhydrophobic when it has a water contact angle of more than 150° and a negligible tilting angle of less than 10° [1,2]. Superhydrophobic coating has gained more attention and lead to many applications such as anti-icing [3,4], anti-fogging [5,6], oil-water separation [7,8], anti-corrosion [9–11], self-cleaning [12,13], antibacterial [14,15] and biomedical applications [16]. Various methods that have been used for preparation of rough surfaces are layer by layer assembly [17,18], spray-coating [19–21], lithography [22,23], sol gel processing [24,25], electrochemical deposition [26,27] and chemical vapour deposition [28]. Among them, spray-coating is a fabrication process used for industrial applications due to its availability in commercial form and a simple procedure that uses inexpensive materials [2,29]. In regard to this, the increase in surface roughness and the reduction of surface energy are vital in forming a superhydrophobic surface [30–32].

A summary of findings on the choice of materials used for spray-coating is shown in Table 1. As seen, the main precursors needed for superhydrophobic spraying techniques are solvent, low surface energy material and nanoparticles. The function of low surface energy material is to reduce the wettability of the surface up to 120°. Nanoparticles are to achieve appropriate roughness to trap the air to further reduce the wettability to achieve a water contact angle greater than 150° and a tilting angle smaller than 10°. The function of solvent is to reduce the viscosity of low surface energy material for easy application of superhydrophobic coating [33] on material surface. The emphasis of this work was on the solvent utilized to generate the superhydrophobic coating, as there have been many studies on nanoparticles and low surface energy materials.

Table 1. Main precursors for spray-coating fabrication method with their respective water contact angle, titling angle and surface roughness.

Low Surface Energy Material	Solvent	Nanoparticle	Substrate	WCA (°)	TA (°)	Surface Roughness (μm)	Reference
PU		Hexadecyl polysiloxane modified SiO_2	Glass	163.9	3.7		[3]
PDMS	Ethyl acetate	SiO_2	Glass	157		5.357	[10]
PDMS	Hexane	SiO_2		164	2	0.023	[12]
OCTES	Ethanol, water	SiO_2	Metal	155	<5		[18]
PU	THF	SiO_2	Glass	160	<2	2.550	[20]
Fluorosilicone	Ethyl acetate, butyl acetate	SiO_2	Glass	153	2.5		[34]
PDMS, PMMS	Hexane	SiO_2	Polyurethane acrylate	160	5		[35]
PDMS	Hexane	SiO_2	Glass	156	1		[36]
Fluorinated polysiloxane	Butyl acetate	ZnO	Steel	166	4		[37]
Glass resin	Isopropanol	SiO_2	Aluminium	155	4		[38]
PDMS	THF	Camphor soot particles	Glass	171	5	1.491	[39]
HMDS	γ-Aminopropyltri ethoxysilane	SiO_2	Glass, wood, filter paper, cotton, plastic, stone, fabric and aluminium foil	161	6.5		[40]
Hexadecyltrimetho xysilane (HDTMS)	Ethanol	SiO_2	White rice husks and cabbage	155	-		[41]
Hydrophobic silica	Hexane	SiO_2	Body of motorcycle, building wall	160	6		[42]
PDMS	Hexane	SiO_2	Glass, paper, and plastic	156.4	5		[43]
Paper mulch	Anhydrous ethanol	SiO_2		160.6	4.2		[44]

According to Luo et al. the dispersibility of nanoparticles in solvent influences the generation of even dispersion or aggregation, which controls the formation of a smooth or rough coating on the surface of a substrate [45]. To produce a well-dispersed or aggregated superhydrophobic coating, the polarity and relative permittivity of the solvent are the important criteria to be considered. For instance, inorganic components such as SiO_2 nanoparticles aggregate when non-polar solvents are used but disperse evenly in polar solvents. This happens because polar solvents stabilize silica dispersion through strong hydrogen bonding to silanol groups on the silica surface. Conversely, low-polarity solvents result in destabilization and gelation of silica particles via hydrogen bonding between adjacent silica particles [46,47]. The relative permittivity of the solvent is also a measure of solvent polarity; the lower the relative permittivity of the solvent, the lower the polarity of the mixture; the polarity of the mixture may eventually decrease to the point where it is no longer sufficient to sustain the dispersion of the polar silica nanoparticles, resulting in particle aggregation. Nonetheless, high aggregation conditions prior to the phase separation limit (gelation of SiO_2 nanoparticles) are desirable as they help to create multi-scale roughness [47]. Therefore, an appropriate solvent needs to be selected to form the required roughness to reduce the wettability behavior.

A variety of solvents used by researchers to prepare superhydrophobic coating is tabulated in Table 2. In 2018, Zhang et al. used xylene as solvent for superhydrophobic epoxy/PDMS nanocomposite coating fabrication [6]. Tetrahydrofuran was also reported as solvent to synthesize superhydrophobic wood surfaces, hydrophobic sol-gel coating and UV-cured superhydrophobic cotton fabric surfaces [25,48,49]. Saleem used toluene as solvent in the development of superhydrophobic surfaces [33]. Among these solvents, hexane is the most popular solvent used to produce superhydrophobic coating due to low permittivity nature as it helps to create multi-scale roughness surface to trap air and increase the WCA. [12,35,50–52] are among the author researchers fabricated superhydrophobic coating by using hexane as solvent. However, these solvents are toxic and hazardous towards organs through prolonged exposure [53]. Other solvents such as isopropanol and acetone that has less hazardous effect towards users are not explored probably due to the relatively high permittivity value, 17.9 and 20.7, respectively as compared to hexane (1.9) 54. Due to safety concern, these two solvent are still explored in this work because the use of hydroxyl terminated PDMS (ε = 2.30 − 2.80) would react with mild polar hydroxyl and carbonyl group of those solvents because of the like-dissolve-like concept [54,55]. This would sustain the dispersion of the polar silica nanoparticles, with certain degree of aggregation. Besides, to reduce the toxicity of the superhydrophobic coating produced in this work, a low surface energy material, PDMS was used. To value add, the SiO_2 nanoparticles that is required to increase the surface roughness for the superhydrophobic coating was extracted from palm oil fuel ash waste (POFA). A suitable mechanism was also proposed for the development of super-hydrophobic coating on the tiles.

Table 2. Various solvents with respective relative permittivity and water contact angle.

Solvent	Relative Permittivity, ε	WCA (°)	Reference
Xylene	2.30	154	[6]
		156.8	[13]
Tetrahydrofuran	7.60	152	[48]
		153	[49]
Toluene	2.38	154	[33]
Hexane	1.90	160	[7]
		164	[12]
		160	[35]
		160	[42]
		156.4	[43]
		163	[50]
		160	[51]
		158	[52]
		172	[56]
Ethyl	25	149	[57]

2. Materials and Methods

2.1. Materials

In this work, the tiles with dimension size of 3 cm × 3 cm were used as substrate were obtained from Ceramic Research Company Sdn Bhd. Palm Oil Fuel Ash (POFA) and Hydroxyl-terminated Polydimethylsiloxane (PDMS-OH) act as precursor were obtained from Malpom Industries Berhad (Pulau Pinang, Malaysia) and Sigma Aldrich (Darmstadt, Germany), respectively. Sodium Hydroxide (NaOH) and Sulfuric Acid (H_2SO_4) with purity of 98% were used as POFA extraction and also Hexane (purity of 98.5%) and Isopropanol (purity of 99.8%) was used as solvent were purchased from Merck (Darmstadt, Germany). For solvent and cleaning agent, acetone; with purity of 99.9% was obtained from J.T. Baker (Avantor, Radnor, PA, USA). Meanwhile, epoxy resin crystal clear (diglycidyl ether of bisphenol A) were used as binder and epoxy hardener crystal clear (trimethyl hexamethylene diamine) were used as curing agent. Both materials were obtained from Euro Chemo-Pharma (Pulau Pinang, Malaysia). For the cross-linking agent, 3-aminopropyltrimethoxysilane (AMPS) with purity of 98% and dibutyltin dilaurate (DBTL) with purity of 98% was used as catalyst. Both materials were obtained from Sigma Aldrich, (Darmstadt, Germany). All the chemicals used were used as received without further purification.

2.2. Extraction of SS from POFA

Alkali extraction method was used to extract SS from POFA by mixing 10 g of POFA and 100 mL of 1 M NaOH at 80 °C for 1 h. After that, the mixture was allowed to cool to room temperature and filtered using a Whatman filter paper with pore size of 11 μm. Then, the filtrate was titrated against 1 M H_2SO_4, to adjust its pH to 3 and the solution produced was silica solution (SS) [36].

2.3. Preparation of the Hydrophobic Solution

The procedure of epoxy/PDMS/SS preparation is shown in Scheme 1 and mainly consists of 3 mixtures (mixture A, B and C). Mixture A was prepared by stirring 100 mL of solvent and 10 mL of SS solution vigorously for 1 h. Next, 50 mL of solvent, 10 mL of PDMS, 5 mL of AMPS, and 1 mL of DBTL were stirred moderately and heated at 60 °C for 20 min to obtain Mixture B. Then, mixtures A and B were mixed and stirred vigorously for 1 h. Simultaneously, Mixture C was prepared by mixing 2 mL of epoxy resin with 1 mL of hardener. After that, mixture C was added into mixture A + B and stirred vigorously for 2 h. The same procedure is repeated for different types of solvent (acetone, hexane, isopropanol).

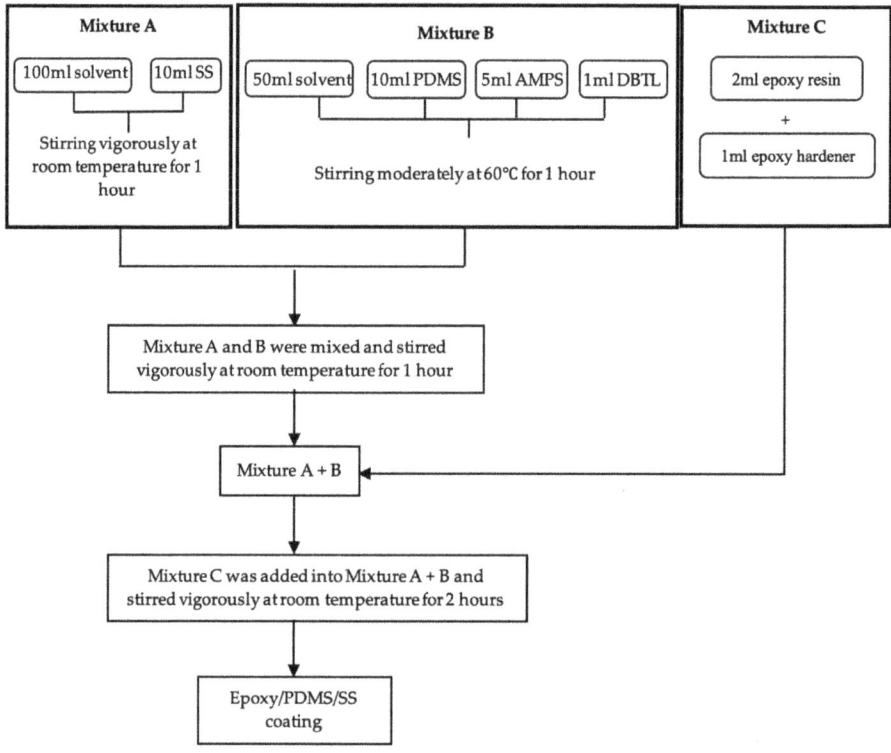

Scheme 1. Illustration of the Epoxy/PDMS/SS coating preparation.

2.4. Fabrication of Hydrophobic Coating on Tiles Substrate

For the fabrication of film on tiles, the epoxy/PDMS/SS coating was sprayed on tile from 20 cm away with the aid of a spraying gun (air pressure: 40 psi), followed by 5 min of drying at 80 °C in an oven. The spraying and drying processes were repeated subsequently until a 3-layer Epoxy/PDMS/SS coating was obtained, and the coated substrate was cured overnight at 80 °C in an oven [36]. Super hydrophobicity characterization (WCA and TA) was used to analyze the coated substrate, and the best solvent was fixed for the Epoxy/PDMS/SS coating preparation for the rest of the experiment. Samples that were coated with acetone, hexane, and isopropanol-added solutions were labelled as S1, S2 and S3. The effect of UV curing was also studied on the respective samples and labelled as S1-UV, S2-UV and S3-UV.

2.5. Characterizations and Analysis

2.5.1. Water Contact Angle (WCA) and Tilting Angle (TA) Measurement

In this work, water contact angle and tilting angle of the samples were measured by a contact angle goniometer (Model 250-F1, Rame-Hart Instruments Co., Mountain Lakes, NJ, USA). The measurement was carried out by placing the samples in the goniometer that is attached to an Image analyzer. A drop of water with a volume of 5 μm was used to determine the water contact angle. Each sample was subjected to 10 measurements in each 4-angle position, including vertical left, vertical right, horizontal left, horizontal right. The WCA and TA were obtained by using DROPimage Advanced software [36]. The measurements were also carried out on the samples before and after durability test to investigate the super hydrophobicity of the sample. In addition, surface energy of coated samples was determined for a more thorough analysis on water contact angle measurement.

2.5.2. Atomic Force Microscopy (AFM)

The surface topology was characterized using an atomic force microscope (AFM, Nano Navi, SPA400, Seiko Instruments, Chiba, Japan) operated in contact mode. The surface roughness of samples was measured, to study the effect of different solvent in PDMS/SS coating on water contact angle. Root-mean-square roughness of the samples was determined by operating at the contact mode of 5 μm × 5 μm. Surface roughness of the samples was analyzed by using NanoNavi software [58]. AFM images subjected to polynomial background subtraction.

2.5.3. Field-Emission Scanning Electron Microscopy (FESEM)

Field-Emission Scanning Electron Microscope (FESEM-EDX, Supra 35VP, Zeiss, Oberkochen, Germany) was used to study the surface morphology of samples at an acceleration voltage of 5 kV. As the substrate was non-conductive, a thin layer of gold was sputtered onto the sample surface to make it conductive in order to obtain a clear FESEM image of the surface morphology [59].

2.5.4. Fourier Transform Infrared Spectroscopy (FTIR)

In this work, transmission, and absorbance mode Fourier Transform Infrared (FTIR, Perkin Elmer, Ohio, USA) was used to investigate the functional groups present on the coated samples surface and the interface between coating and substrate, in order to study the formation of bonding after the application of PDMS/SS coating. Besides, the samples before and after immersion test were analyzed by using FTIR to evaluate the changes in the functional groups of the sample. The samples were tested in absorbance and transmittance mode from 4000 to 550 cm^{-1} wavenumber [35].

2.5.5. Peel-Off Test

Peel-off test was carried out to investigate adhesion between PDMS/SS coating and substrate surface. It was carried out by using double-sided foam tape in which the tape was adhered to the surface of coating and pressed to confirm the adhesion was tight where no gap was found at the interface of tape and sample. Lastly, the tape was peeled off. This test was repeated for 5 cycles. The value of water contact angle before and after peel-off test was measured to evaluate the durability of the coating [28].

3. Results and Discussion

3.1. Effect of Solvent on Wettability

Table 3 shows the water contact angle, tilting angle, surface energy and the roughness of tiles coated with S1, S2 and S3 that were prepared with different solvents. Water contact angles for S1 and S2 coated with acetone and hexane added solution are $88 \pm 1°$ and $85° \pm 1°$, respectively. Therefore, S1 and S2 exhibit hydrophilic behavior. On the other hand, S3 that was coated with isopropanol-added solution has water contact angles of $149 \pm 2°$, showing hydrophobic behavior. The high surface energy of the S1 and S2 coatings, which is 15 times higher than the S3 with 2 J/m^2, contributes for the difference in water contact angle. S1 and S2 may indeed be affected by the dispersion of PDMS and SS particles in different solvents, which will be explained in more detail later. Furthermore, as compared to S1 and S2, sample S3 has a higher roughness, which could be another aspect that improves the sample's hydrophobicity. As for tilting angle, it was found S1 with 7.09 nm has a lower tilting angle ($20°$) as compared to S2 that has a roughness of 13.01 with a tilting angle of $46°$. The findings are consistent with [29,60], which demonstrated that for hydrophilic samples, higher surface roughness results in a higher tilting angle. Besides, higher surface roughness will also result in a stronger water pining effect due to the absence of air pockets. This situation will result in the penetration of water droplets into the grooves. Thus, S1 and S2 are predicted to be in a Wenzel state, as the Wenzel equation states that roughness emphasizes the effect of surface chemistry. In other words, for hydrophilic surfaces, the higher the surface roughness, the more hydrophilic the surface is while for a hydrophobic surface, the higher the surface roughness, the more hydrophobic the surface will be. The wetting mechanisms of hydrophobic surfaces (S3) can also be determined based on their respective tilting angles. The tilting angle of $10 \pm 2°$ for S3 suggests the surface is in a Cassie-Baxter state, which is further supported by the surface morphology and high surface roughness (18.57 nm), which render a hierarchical structure. Such morphology leads to the formation of air voids that help in water droplet suspension, which results in a low tilting angle.

Table 3. Water contact angle, tilting angle, surface energy and the roughness of tiles coated with S1, S2 and S3 that were prepared with different solvents before and after UV curing.

Sample	Type of Solvent	WCA (°)	TA (°)	RMS (nm)	Surface Energy (J/m^2)
S1	Acetone	88 ± 2	20 ± 2	7.09	31 ± 1
S2	Hexane	85 ± 1	46 ± 1	13.01	33 ± 1
S3	Isopropanol	149 ± 2	10 ± 2	18.57	2 ± 0
S1-UV	Acetone	82 ± 2	25 ± 2	9.934	34 ± 1
S2-UV	Hexane	88 ± 1	38 ± 1	11.57	31 ± 1
S3-UV	Isopropanol	152 ± 2	7 ± 2	21.80	1 ± 0

Apart from that, slight changes in water contact angle were also observed after the samples were cured under UV for 24 h (Table 3). After UV-curing, the water contact angles of samples that were coated with hexane-added solution and isopropanol-added solution were increased by ~$3°$, while the water contact angle of sample that was coated with acetone-added solution was decreased by ~$6°$. The increase in water contact angle of S2-UV and S3-UV after UV curing may be attributed to the increase in grafting density of the PDMS polymer chain to form a stronger 3D network 36. Besides, UV-curing helps to cure the polymer phase that was disrupted by aggregated particles in oven curing, producing a coating with strong adhesion, mechanical reliability, and chemical resistance [51,61,62]. Even though the difference in WCA was not statistically significant, the effect of UV was still considered crucial as the WCA was increased to above $150°$, causing S3-UV to be superhydrophobic. However, the decrease in water contact angle of S1-UV after UV curing may be attributed to surface oxidation, which is triggered by the functional carbonyl group (C=O) presence in acetone. This accelerates the degradation of long alkyl chains into smaller chains [63,64], thus increasing the wettability.

3.2. Effect of Solvent on Surface Morphology and Surface Roughness

Surface roughness is another key contributor to the superhydrophobic characteristics of coated samples other than surface energy 35. Figure 1 shows the 3D AFM topographical images (Figure 1a–f), AFM line profile (Figure 1(a^a–f^a), Figure 1(c^c,f^c)) and FESEM images (Figure 1(a^b–f^b)) of S1, S2 and S3 before and after UV treatment. From Figure 1(a^b,b^b), it is observed that the surfaces of S1 and S2 before UV curing do not have obvious micro papillae structure. However, under AFM, the surface possesses certain roughness as indicated by the peaks and valleys in (Figure 1(a^a)) and (Figure 1(b^a)) corresponding to RMS of 7.09 nm and 13.01 nm, respectively. In comparison to S1 and S2, FESEM images of S3 before UV curing show that the surfaces are relatively rough (Figure 1(c^b)). S3 that was coated with isopropanol-added solution has an RMS value of 18.57 nm. The high surface roughness in S3 is due to the dispersion of the coating solution, in which multiscale roughness is created. This sample has an appropriate degree of particle aggregation in the isopropanol-added solution, which induces hierarchical structure [47].

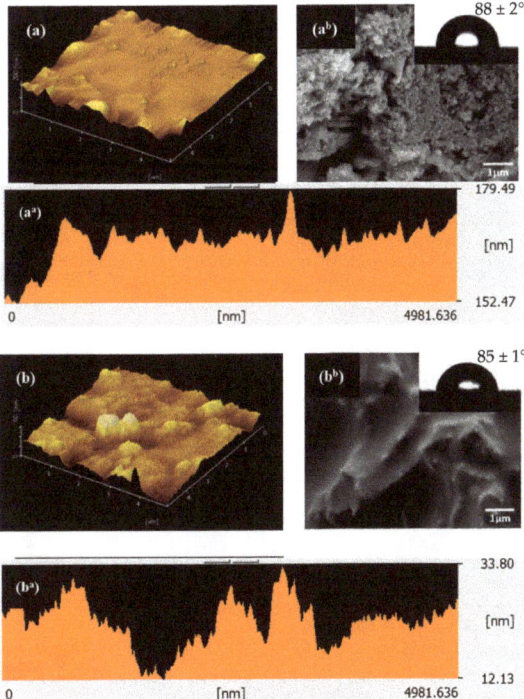

Figure 1. *Cont.*

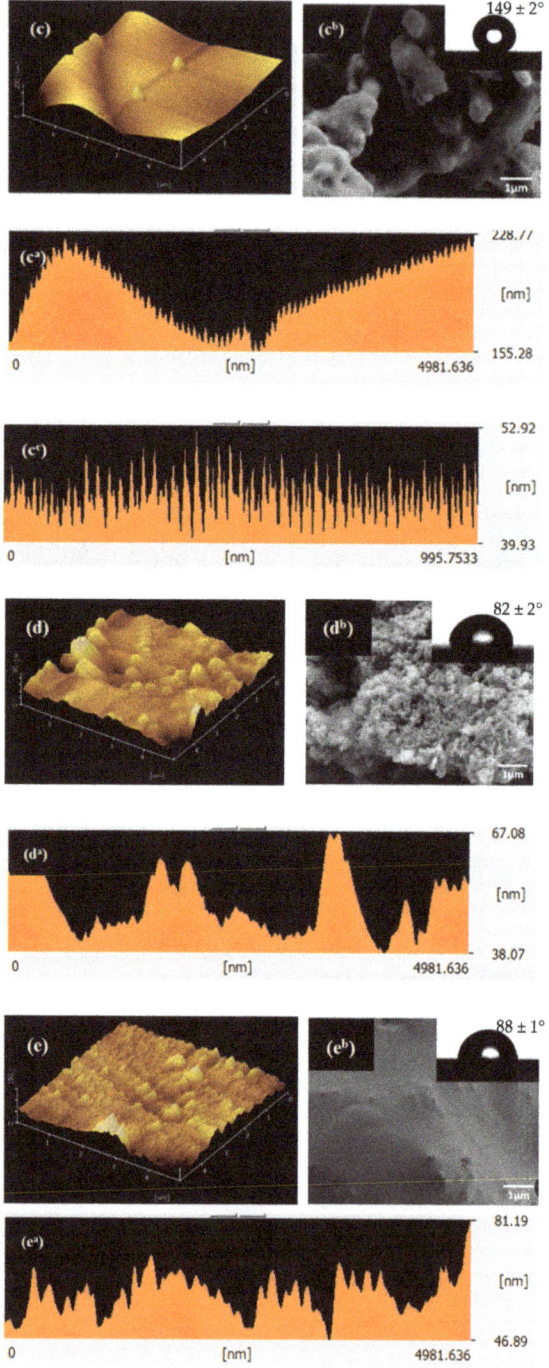

Figure 1. *Cont.*

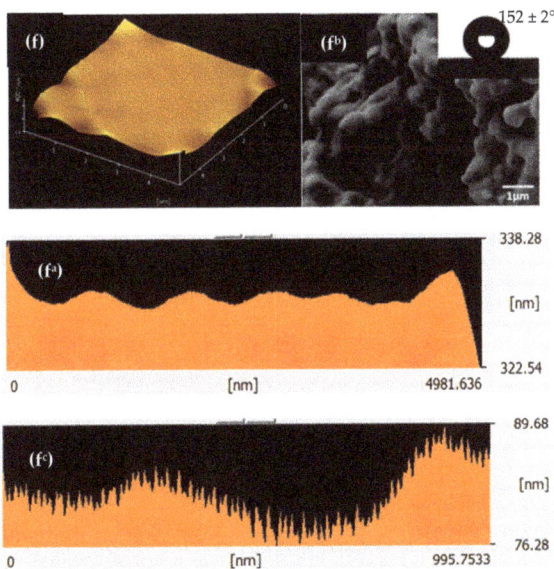

Figure 1. 3D AFM topographical images of (**a**) S1, (**b**) S2, (**c**) S3, (**d**) S1-UV, (**e**) S2-UV and (**f**) S3-UV on 5 μm × 5 μm area; Image of AFM line profile of (**a^a**) S1, (**b^a**) S2, (**c^a**) S3, (**d^a**) S1-UV, (**e^a**) S2-UV and (**f^a**) S3-UV. FESEM images and images of water droplets on (**a^b**) S1, (**b^b**) S2, (**c^b**) S3, (**d^b**) S1-UV, (**e^b**) S2-UV and (**f^b**) S3-UV. Image of AFM line profile of (**e^c**) S2-UV and (**f^c**) S3-UV on 1 μm × 1 μm area.

For samples cured with UV (Figure 1(d^b,e^b,f^b)), a similar trend was observed as before UV curing (Figure 1(a^b,b^b,c^b)), respectively. However, the distance between the hills for S1-UV after UV curing (Figure 1(d^a)) is wider than that before UV curing (Figure 1(a^a)), leading to a lower water contact angle after UV curing. Figure 1(e^a) shows that S2-UV after UV curing has peaks and valleys with a lower contrast as compared to after UV curing, leading to a lower RMS roughness value of 11.57 nm. The S3-UV that was coated with isopropanol-added solution and UV-cured has the highest RMS roughness value of 21.80 nm. This is probably ascribed to the even distribution of peaks and valleys with spikes that renders a hierarchical structure in S3 after UV curing (Figure 1(f^a,f^c)) as compared to before curing (Figure 1(c^a,c^c)). In summary, the formation of hierarchical structures with appropriate roughness and distance between hills improves hydrophobicity. Those characteristics were achieved in the isopropanol-added solution that has been UV cured.

3.3. Effect of Solvent on Dispersion of Coating Solution

The effects of solvents such as hexane, acetone, and isopropanol on the dispersion of superhydrophobic coating solutions were investigated. Figure 2 shows that agglomeration was evident in all three solutions, but that after shaking, the isopropanol-added solution became well-dispersed whereas the agglomeration in the other two solutions remained. The relative permittivity of the solvent utilized causes such circumstance. Hexane is a non-polar solvent with a relative permittivity of 1.90, making it the least polar. As a result, it failed to disperse SS particles, which are polar in nature with OH groups. Hexane's relative permittivity is too low, resulting in SS particle gelation, rather than particle aggregation. It was proven by the fact that samples S2 coated with hexane-added solution have low water contact angles before and after UV curing due to the absence of hierarchical structure on their surface (Figure 1(b^b,e^b)).

(a)  (b)

Figure 2. Images of (i) hexane-added solution, (ii) acetone-added solution and (iii) isopropanol-added solution (**a**) before and (**b**) after they were shaken.

On the other hand, acetone and isopropanol are polar solvents with a relative permittivity of 20.7 and 17.9, respectively. Hence, both solvents were able to disperse SS particles with a certain degree of aggregation, resulting in surface roughness with a maximum RMS value of 21.80 nm (Table 3). In addition, the agglomeration in the isopropanol-added solution could be dispersed after it was shaken due to the like-dissolves-like concept. This dissolves-like concept indicates that solutes, which are PDMS, SS and epoxy in this work, will dissolve best in a solvent that has a similar chemical structure to them [65]. From Figure 3A-a, OH-bonds were found in isopropanol but not in hexane (Figure 3B-a) and acetone (Figure 3C-a) Therefore, PDMS, SS, and epoxy dissolved best in isopropanol instead of acetone or hexane.

3.4. FTIR of Different Types of Solvent and Coating Solution

FTIR spectroscopy was used to investigate functional groups in coating solutions of different types of solvent by analyzing their respective precursors. For isopropanol (Figure 3(Aa)), it can be seen that O-H stretching, and O-H bending are indicated at 3318 cm^{-1} and 951 cm^{-1}, respectively. At 2970 cm^{-1}, C-H stretching of CH$_3$ group is indicated. In addition, C-H bending vibration of CH$_2$ and CH$_3$ are shown at 1467 cm^{-1} and 1379 cm^{-1}, correspondingly. A peak noticed at 3338 cm^{-1} corresponds to O-H stretching, which is attributed to its hydroxyl-terminated structure of PDMS (Figure 3(Ab)). At 1640 cm^{-1}, O-C = O stretching is detected. Peaks that are noticed at 1259 cm^{-1}, 1022 cm^{-1} and 795 cm^{-1} in Figure 3(Ab), representing Si-CH$_3$, O-Si-O and Si-C bonds, respectively [36,66,67]. The finding of functional groups in isopropanol and PDMS matches with their respective chemical formulae of C$_3$H$_8$O and (C$_2$H$_6$OSi)$_n$. For Mixture B (Figure 3(Ac)), PDMS, AMPS and DBTL were added into isopropanol. The reaction occurred in the preparation of Mixture B can be proven by the reduced intensity of O-H stretching at 3325 cm^{-1} and presence of C-N at 1200 cm^{-1}, as a result of the AMPS is grafted at the hydroxyl-end of PDMS [6]. C-H stretching and bending of CH$_3$ groups and O-H bending that were originally absent in PDMS spectra, were observed at 2969 cm^{-1}, 1378 cm^{-1} and 951 cm^{-1}, respectively, ascertaining the formation of methanol (CH$_3$OH) as side-product in the proposed mechanism (Figure 4). Figure 3(Ad) shows FTIR spectrum of SS, in which O-H, Si-O and O-Si-O bonds are detected at 3318 cm^{-1}, 1261 cm^{-1} and 1081 cm^{-1}, respectively. A peak observed at 1640 cm^{-1} belongs to O-C = O stretching. In Figure 3(Ae), O-H stretching is also found in epoxy (DGEBA) at 3319 cm^{-1}. C-H stretching and rocking vibrations are noticed at 2967 cm^{-1} and 794 cm^{-1}. In addition, C-O-C stretching and C-O stretching of oxirane ring are also found at 1022 cm^{-1} and 1259 cm^{-1} [6]. Lastly, Figure 3(Af) shows the FTIR spectrum of isopropanol-coating. Both O-H stretching, and O-H out-of-plane bending are detected at 3338 cm^{-1} and 951 cm^{-1}, respectively. Besides, C-H stretching of CH$_3$, C-H bending vibrations of CH$_2$ and CH$_3$ groups are also detected at 2970 cm^{-1}, 1467 cm^{-1} and 1379 cm^{-1}, correspondingly. At 1261 cm^{-1}, Si-CH$_3$ bond is indicated. Peaks that are observed at 1128 cm^{-1} and 815 cm^{-1} correspond to O-Si-O and Si-C bonds, respectively. From Figure 3(Ac) to Figure 3(Af), intensity of O-H stretching at

~3300 cm^{-1} increases, which may be attributed to the reaction occurred between Mixture B, epoxy and SS. Other than this, a strong and sharp peak observed at 951 cm^{-1} corresponds to high intensity of O-H out-of-plane bending may be caused by the presence of methanol as side-product [66].

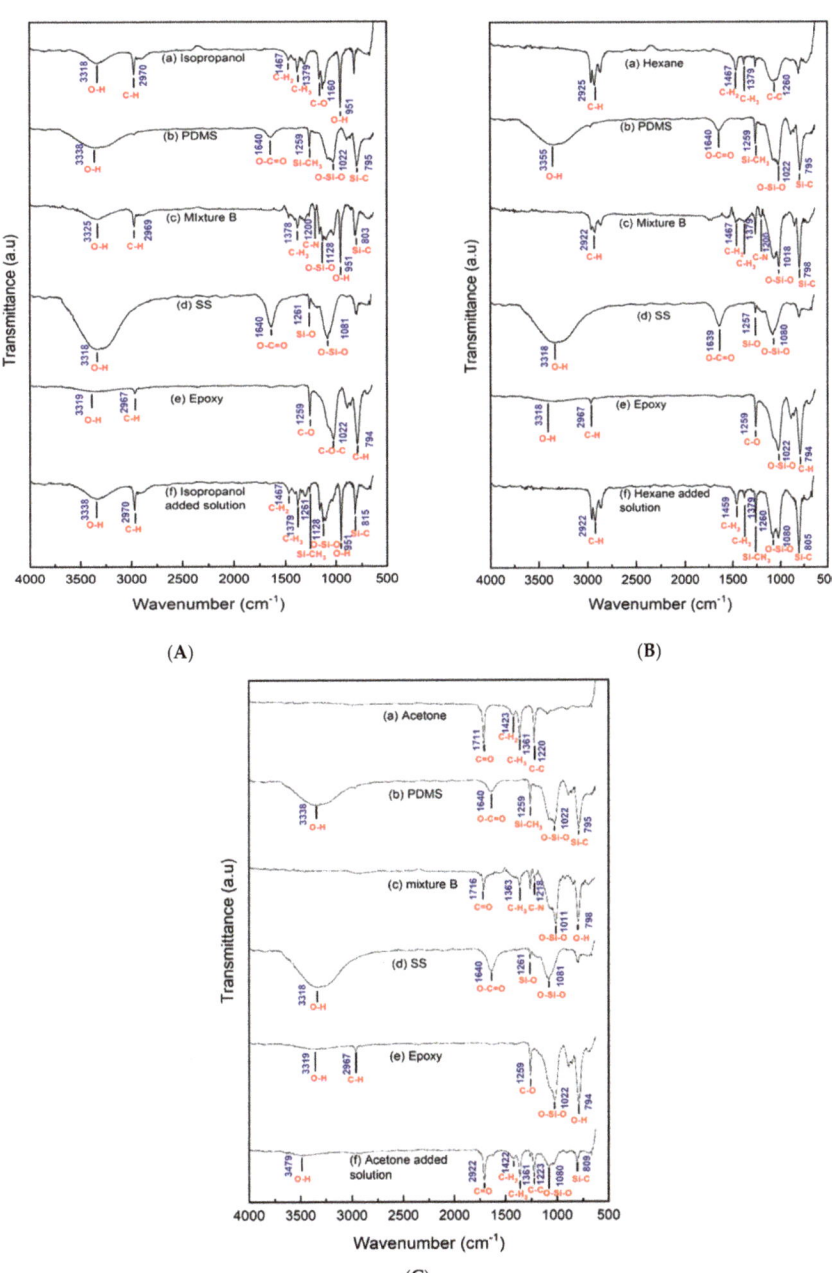

Figure 3. FTIR spectra for (**A**) isopropanol-added solution, (**B**) hexane-added solution, (**C**) acetone-added solution and their respective precursors.

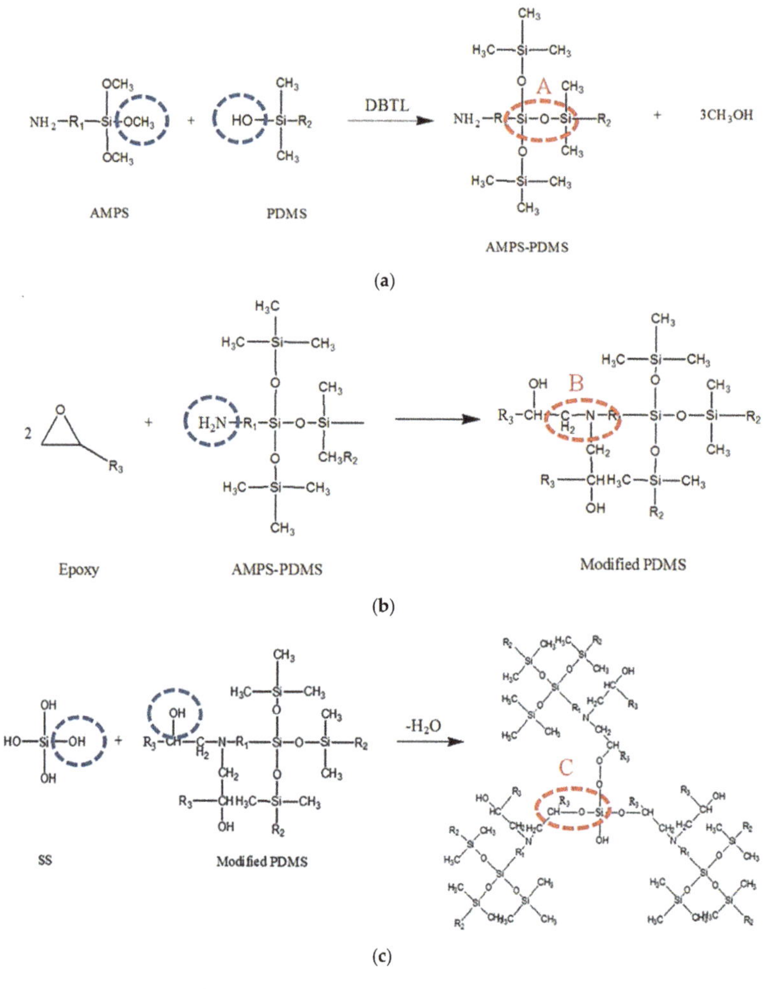

Figure 4. The proposed mechanisms of S3 solution (**a**) reaction between AMPS and PDMS with DBTL as catalyst, (**b**) reaction between epoxy resin and AMPS-PDMS molecules, (**c**) reaction between SS and modified PDMS. R1, R2 and R3 symbolize alkyl groups.

Next, Figure 3B shows FTIR spectra of hexane solution and its precursors. The functional groups of PDMS (Figure 3(Bb)), SS (Figure 3(Bd)) and epoxy (Figure 3(Be)) and are similar to those in Figure 3A. The precursor was included in Figure 3B for comparison purpose. For hexane (Figure 3(Ba)), with chemical formula of C_6H_{14}, O-H bond is absent, while C-H stretching of CH_3 (2925 cm^{-1}), C-H bending of CH_2 and CH_3 groups (1467 cm^{-1} and 1379 cm^{-1}), as well as C-C stretching (1260 cm^{-1}) are detected. In this case, reaction between PDMS and AMPS is verified as O-H bonds from PDMS are fully contributed in this reaction, causing depletion of O-H bonds and formation of C-N groups as observed at 1218 cm^{-1} in Mixture B (Figure 3(Bc)). Based on Figure 3(Bf), (OH) bond is not detected after the addition of epoxy and SS, which may be due to the polarity of hexane that results in agglomeration of SS particles; thus, the solution is not well-dispersed [46].

FTIR spectra of acetone solution and its precursors are shown in Figure 3C. The functional groups of PDMS (Figure 3(Cb)), SS (Figure 3(Cd)) and epoxy (Figure 3(Ce)) are similar to those in Figure 3A. The precursor was included in Figure 3C for comparison purpose. For acetone (Figure 3(Ca)), shows peaks correspond to C = O (1711 cm^{-1}), C-H stretching of CH_2 (1423 cm^{-1}) and CH_3 groups (1361 cm^{-1}), together with C-C stretching (1220 cm^{-1}), which are identical to its chemical formula of C_3H_6O [68]. For Mixture B (Figure 3(Cc)) of acetone-added solution, the presence of C-N groups at 1200 cm^{-1} and O-H stretching is also absent as O-H groups from PDMS and are utilized for bonding between AMPS and PDMS. In acetone-added coating solution (Figure 3(Cf)), intensity of O-H stretching noticed at 3479 cm^{-1} is low. This may be attributed to "like-dissolves-like" concept, leading to a higher degree of dispersion as compared to hexane-added solution but lower than that of isopropanol-added solution, as acetone is a polar solvent, but hydroxyl group is absent [65].

As seen, compared to S1 solution, S2 solution has similar peaks expect for peaks that correspond to O-H stretching and bending vibration. When S3 solution compared S1, an extra peak of C=O was observed while the Si-CH_3 and O-H bending vibration were absence. Absence of C-H stretching (~2900 cm^{-1}) and Si-CH_3 bond (~1260 cm^{-1}) in S3 is probably due to oxidation of C-H to form C=O bonds [69,70]. The oxidation is further affirmed by color of acetone-added solution, which is brownish in color as compared to other solutions (Figure 2).

3.5. The Mechanism of Reaction of S3 Solution

Since S3 with isopropanol solvent shows the best hydrophobic behavior, the mechanism of this solution is proposed. A mechanism for reactions occurred in the preparation of isopropanol-added solution is illustrated schematically in Figure 4a–c, based on FTIR analysis. In the preparation of Mixture B (Figure 3A-c), Si-O-Si (refer A) bonds were formed upon the mixing of AMPS and PDMS with the presence of DBTL as catalyst, by hydrogen bonding with PDMS at the –OCH_3 ends of AMPS and methanol is formed as by-product (Figure 4a). Then the ring opening reaction of epoxy occur with AMPS-PDMS at the amine end and forming modified PDMS (Figure 4b, refer B). Then, the modified PDMS is attached to a central Si atom of SS particles, by forming hydrogen bonding and covalent bond after heating and release water (Figure 4c, refer C).

3.6. Peel of Test

A peel-off test was performed on S3 coated tiles to determine the durability of the coating on the substrate. This test was repeated five times. In Figure 5, the appearance of the tape surface and the water contact angle after the peel off test are presented. As seen, the surface of the tape is clear from any debris, indicating the coating's resistance to separation from the substrate. This characteristic implies that the coating has a good adhered to the surface and forms a strong bond. As a result, there are no significant changes in the topography and roughness of the S3 sample before and after peeling, resulting in a WCA that is identical.

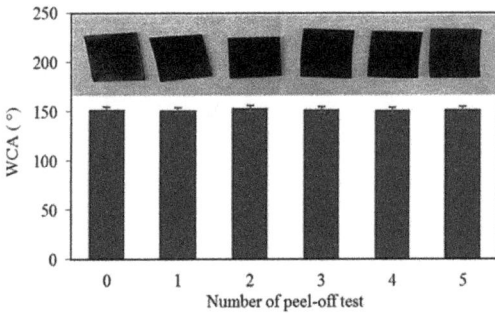

Figure 5. Changes in water contact angle of after peel-off test and appearance of surface of initial and peeled off tape after each peel-off test.

4. Conclusions

In this work, the best superhydrophobic coating has been synthesized using PDMS, SS, and epoxy resin (DGEBA) with isopropanol as the solvent rather than hexane and acetone. Isopropanol, which is a polar solvent, poses dissolving-like behavior towards PDMS, SS, and epoxy and aids in the creation of multiscale roughness. The S3 coated sample has a water contact angle of $152 \pm 2°$ after UV irradiation and showed good durability. Such a superhydrophobic coating can be utilized on tiles to act as a self-cleaning surface.

Author Contributions: Conceptualization, S.S. and A.X.Y.; methodology, A.X.Y.; validation, S.S., A.X.Y. and N.B.; formal analysis, A.X.Y. and N.B.; investigation, A.X.Y., N.B. and S.S.; resources, F.A. and F.D.; data curation, A.X.Y.; writing—original draft preparation, S.S., A.X.Y. and F.A.; writing—review and editing, S.S. and F.A.; visualization, A.X.Y.; supervision, S.S. and F.A.; project administration, F.A. and F.D.; funding acquisition, F.A., S.S. and F.D. All authors have read and agreed to the published version of the manuscript.

Funding: This research was funded by the Ministry of Higher Education under the Fundamental Research Grant Scheme (FRGS) with grant number FRGS/1/2019/TK06/USM/01/1.

Institutional Review Board Statement: Not applicable.

Informed Consent Statement: Not applicable.

Data Availability Statement: Not applicable.

Acknowledgments: The author would like to thank the Ministry of Higher Education for funding this project under the Fundamental Research Grant Scheme (FRGS) with grant number FRGS/1/2019/TK06/USM/01/1. The authors are also grateful to Universiti Sains Malaysia (USM) for providing necessary facilities for this research work.

Conflicts of Interest: The authors declare no conflict of interest.

References

1. Varshney, P.; Mohapatra, S.S. Fabrication of durable and regenerable superhydrophobic coatings on metallic surfaces for potential industrial applications. In Proceedings of the 2018 8th International Conference on Mechanical and Intelligent Manufacturing Technologies (ICMIMT), Cape Town, South Africa, 3–6 February 2017; IEEE: Piscataway, NJ, USA, 2018; pp. 16–20.
2. Barati Darband, G.; Aliofkhazraei, M.; Khorsand, S.; Sokhanvar, S.; Kaboli, A. Science and engineering of superhydrophobic surfaces: Review of corrosion resistance, Chemical and Mechanical Stability. *Arab. J. Chem.* **2018**, *13*, 1763–1802. [CrossRef]
3. Li, Y.; Li, B.; Zhao, X.; Tian, N.; Zhang, J. Totally waterborne, nonfluorinated, mechanically robust, and self-healing superhydrophobic coatings for actual anti-icing. *ACS Appl. Mater. Interfaces* **2018**, *10*, 39391–39399. [CrossRef] [PubMed]
4. Zhuo, Y.; Hakonsen, V.; He, Z.; Xiao, S.; He, J.; Zhang, Z. Enhancing the mechanical durability of icephobic surfaces by introducing autonomous self-healing function. *ACS Appl. Mater. Interfaces* **2018**, *10*, 11972–11978. [CrossRef] [PubMed]
5. Zhang, D.; Wang, L.; Qian, H.; Li, X. Superhydrophobic surfaces for corrosion protection: A review of recent progresses and future directions. *J. Coat. Technol. Res.* **2015**, *13*, 11–29. [CrossRef]
6. Zhang, Y.; Ren, F.; Liu, Y. A superhydrophobic EP/PDMS nanocomposite coating with high gamma radiation stability. *Appl. Surf. Sci.* **2018**, *436*, 405–410. [CrossRef]
7. Su, X.; Li, H.; Lai, X.; Zhang, L.; Liang, T.; Feng, Y.; Zeng, X. Polydimethylsiloxane-based superhydrophobic surfaces on steel substrate: Fabrication, reversibly extreme wettability and oil–water separation. *ACS Appl. Mater. Interfaces* **2017**, *9*, 3131–3141. [CrossRef]
8. Gao, S.; Dong, X.; Huang, J.; Li, S.; Li, Y.; Chen, Z.; Lai, Y. Rational construction of highly transparent superhydrophobic coatings based on a non-particle, fluorine-free and water-rich system for versatile oil-water separation. *Chem. Eng. J.* **2018**, *333*, 621–629. [CrossRef]
9. Qing, Y.; Yang, C.; Sun, Y.; Zheng, Y.; Wang, X.; Shang, Y.; Wang, L.; Liu, C. Facile fabrication of superhydrophobic surfaces with corrosion resistance by nanocomposite coating of TiO_2 and polydimethylsiloxane. *Colloids Surf. A Physicochem. Eng. Asp.* **2015**, *484*, 471–477. [CrossRef]
10. Zhang, Z.; Ge, B.; Men, X.; Li, Y. Mechanically durable, superhydrophobic coatings prepared by dual-layer method for anti-corrosion and self-cleaning. *Colloids Surf. A Physicochem. Eng. Asp.* **2016**, *490*, 182–188. [CrossRef]
11. Zhang, B.; Yan, J.; Xu, W.; Yu, T.; Chen, Z.; Duan, J. Eco-friendly anticorrosion superhydrophobic Al_2O_3@PDMS coating with salt deliquescence self-coalescence behaviors under high atmospheric humidity. *Front. Mater.* **2022**, *9*, 839948. [CrossRef]
12. Wang, P.; Liu, J.; Chang, W.; Fan, X.; Li, C.; Shi, Y. A facile cost-effective method for preparing robust self-cleaning transparent superhydrophobic coating. *Appl. Phys. A* **2016**, *122*, 916. [CrossRef]

13. Zhang, Y.; Zhou, S.; Lv, Z.; Fan, L.; Huang, Y.; Liu, X. A facile method to prepare superhydrophobic coatings for various substrates. *Appl. Sci.* **2022**, *12*, 1–9. [CrossRef]
14. Acikbas, G.; Acikbas, N.C. Nanoarchitectonics for polymer-ceramic hybrid coated ceramic tiles for antibacterial activity and wettability. *Appl. Phys. A* **2021**, *127*, 794. [CrossRef]
15. Acikbas, G.; Acikbas, N.C. The effect of sintering regime on superhydrophobicity of silicon nitride modified ceramic surfaces. *J. Asian Ceram. Soc.* **2021**, *9*, 733–743. [CrossRef]
16. Abbas, A.; Zhang, C.; Asad, M.; Waqas, A.; Khatoon, A.; Hussain, S.; Husain Mir, S. Recent developments in artificial super-wettable surfaces based on bioinspired polymeric materials for biomedical applications. *Polymers* **2022**, *14*, 238. [CrossRef]
17. Wu, M.; An, N.; Li, Y.; Sun, J. Layer-by-layer assembly of fluorine-free polyelectrolyte–surfactant complexes for the fabrication of self-healing superhydrophobic films. *Langmuir* **2016**, *32*, 12361–12369. [CrossRef]
18. Guo, X.-J.; Xue, C.-H.; Li, M.; Li, X.; Ma, J.-Z. Fabrication of robust, superhydrophobic, electrically conductive and UV-blocking fabrics via layer-by-layer assembly of carbon nanotubes. *RSC Adv.* **2017**, *7*, 25560–25565. [CrossRef]
19. Liu, H.; Huang, J.; Chen, Z.; Chen, G.; Zhang, K.-Q.; Al-Deyab, S.S.; Lai, Y. Robust translucent superhydrophobic PDMS/PMMA film by facile one-step spray for self-cleaning and efficient emulsion separation. *Chem. Eng. J.* **2017**, *330*, 26–35. [CrossRef]
20. El Dessouky, W.I.; Abbas, R.; Sadik, W.A.; El Demerdash, A.G.M.; Hefnawy, A. Improved adhesion of superhydrophobic layer on metal surfaces via one step spraying method. *Arab. J. Chem.* **2015**, *10*, 368–377. [CrossRef]
21. Hejazi, I.; Seyfi, J.; Sadeghi GM, M.; Jafari, S.H.; Khonakdar, H.A.; Drechsler, A.; Davachi, S.M. Investigating the interrelationship of superhydrophobicity with surface morphology, topography and chemical composition in spray-coated polyurethane/silica nanocomposites. *Polymer* **2017**, *128*, 108–118. [CrossRef]
22. Sung, Y.H.; Kim, Y.D.; Choi, H.-J.; Shin, R.; Kang, S.; Lee, H. Fabrication of superhydrophobic surfaces with nano-in-micro structures using UV-nanoimprint lithography and thermal shrinkage films. *Appl. Surf. Sci.* **2015**, *349*, 169–173. [CrossRef]
23. Kothary, P.; Dou, X.; Fang, Y.; Gu, Z.; Leo, S.-Y.; Jiang, P. Superhydrophobic hierarchical arrays fabricated by a scalable colloidal lithography approach. *J. Colloid Interface Sci.* **2017**, *487*, 484–492. [CrossRef] [PubMed]
24. Latthe, S.S.; Terashima, C.; Nakata, K.; Sakai, M.; Fujishima, A. Development of sol–gel processed semi-transparent and self-cleaning superhydrophobic coatings. *J. Mater. Chem. A* **2014**, *2*, 5548–5553. [CrossRef]
25. Kumar, D.; Wu, X.; Fu, Q.; Ho JW, C.; Kanhere, P.D.; Li, L.; Chen, Z. Hydrophobic sol–gel coatings based on polydimethylsiloxane for self-cleaning applications. *Mater. Des.* **2015**, *86*, 855–862. [CrossRef]
26. Liu, Y.; Li, S.; Zhang, J.; Liu, J.; Han, Z.; Ren, L. Corrosion inhibition of biomimetic super-hydrophobic electrodeposition coatings on copper substrate. *Corros. Sci.* **2015**, *94*, 190–196. [CrossRef]
27. Ashoka, S.; Saleema, N.; Sarkar, D. Tuning of superhydrophobic to hydrophilic surface: A facile one step electrochemical approach. *J. Alloy. Compd.* **2017**, *695*, 1528–1531. [CrossRef]
28. Xu, L.; Zhu, D.; Lu, X.; Lu, Q. Transparent, thermally and mechanically stable superhydrophobic coating prepared by an electrochemical template strategy. *J. Mater. Chem. A* **2015**, *3*, 3801–3807. [CrossRef]
29. Mohamed AM, A.; Abdullah, A.M.; Younan, N.A. Corrosion behavior of superhydrophobic surfaces: A review. *Arab. J. Chem.* **2015**, *8*, 749–765. [CrossRef]
30. Fihri, A.; Bovero, E.; Al-Shahrani, A.; Al-Ghamdi, A.; Alabedi, G. Recent progress in superhydrophobic coatings used for steel protection: A review. *Colloids Surf. A Physicochem. Eng. Asp.* **2017**, *520*, 378–390. [CrossRef]
31. Khaskhoussi, A.; Calabrese, L.; Patané, S.; Proverbio, E. Effect of chemical surface texturing on the superhydrophobic behavior of micro–nano-roughened AA6082 surfaces. *Materials* **2021**, *14*, 7161. [CrossRef]
32. Wang, X.L.; Wang, W.K.; Qu, Z.G.; Ren, G.F.; Wang, H.C. Surface roughness dominated wettability of carbon fiber in gas diffusion layer materials revealed by molecular dynamics simulations. *Int. J. Hydrogen Energy* **2021**, *46*, 26489–26498. [CrossRef]
33. Saleem, M.S. Development of Super Hydrophobic Surfaces Using Silica Nanoparticles. Master's Thesis, CUNY City College, New York, NY, USA, 2015.
34. Wang, X.; Li, X.; Lei, Q.; Wu, Y.; Li, W. Fabrication of superhydrophobic composite coating based on fluorosilicone resin and silica nanoparticles. *R. Soc. Open Sci.* **2018**, *5*, 180598. [CrossRef] [PubMed]
35. Li, Y.; Shao, H.; Lv, P.; Tang, C.; He, Z.; Zhou, Y.; Shuai, M.; Mei, J.; Lau, W.-M. Fast preparation of mechanically stable superhydrophobic surface by UV cross-linking of coating onto oxygen-inhibited layer of substrate. *Chem. Eng. J.* **2018**, *338*, 440–449. [CrossRef]
36. Saharudin, K.A.; Sreekantan, S.; Basiron, N.; Chun, L.K.; Kumaravel, V.; Abdullah, T.K.; Ahmad, Z.A. Improved super-hydrophobicity of eco-friendly coating from palm oil fuel ash (POFA) waste. *Surf. Coat. Technol.* **2018**, *337*, 126–135. [CrossRef]
37. Qing, Y.; Yang, C.; Hu, C.; Zheng, Y.; Liu, C. A facile method to prepare superhydrophobic fluorinated polysiloxane/ZnO nanocomposite coatings with corrosion resistance. *Appl. Surf. Sci.* **2015**, *326*, 48–54. [CrossRef]
38. Zhang, Y.; Ge, D.; Yang, S. Spray-coating of superhydrophobic aluminum alloys with enhanced mechanical robustness. *J. Colloid Interface Sci.* **2014**, *423*, 101–107. [CrossRef]
39. Sahoo, B.N.; Nanda, S.; Kozinski, J.A.; Mitra, S.K. PDMS/camphor soot composite coating: Towards a self-healing and a self-cleaning superhydrophobic surface. *RSC Adv.* **2017**, *7*, 15027–15040. [CrossRef]
40. Subramanian, B.T.; Alla, J.P.; Essomba, J.S.; Nishter, N.F. Non-fluorinated superhydrophobic spray coatings for oil-water separation applications: An eco-friendly approach. *J. Clean. Prod.* **2020**, *256*, 120693. [CrossRef]

41. Caldona, E.B.; Sibaen, J.W.; Tactay, C.B.D.; Mendiola, S.L.; Abance, C.B.; Añes, M.P.; DSerrano, F.D.; SDe Guzman, M.M. Preparation of spray coated surfaces from green formulated superhydrophobic coatings. *SN Appl. Sci.* **2019**, *1*, 1657. [CrossRef]
42. Latthe, S.S.; Sutar, R.S.; Kodag, V.S.; Bhosale, A.K.; Madhan Kumarc, A.; Sadasivuni, K.K.; Xing, R.; Liu, S. Self–cleaning superhydrophobic coatings: Potential industrial applications. *Prog. Org. Coat.* **2019**, *128*, 52–58. [CrossRef]
43. Gong, X.; He, S. Highly Durable Superhydrophobic Polydimethylsiloxane/Silica Nanocomposite Surfaces with Good Self-Cleaning Ability. *ACS Omega* **2020**, *5*, 4100–4108. [CrossRef] [PubMed]
44. Zhang, F.; Li, A.; Zhao, W. Analysis of the Acid and Alkali Resistance of Superhydrophobic Paper Mulch. *Cellulose* **2021**, *28*, 8705–8718. [CrossRef]
45. Luo, X.; Hu, W.; Cao, M.; Ren, H.; Feng, J.; Wei, M. An environmentally friendly approach for the fabrication of conductive superhydrophobic coatings with sandwich-like structures. *Polymers* **2018**, *10*, 378. [CrossRef] [PubMed]
46. Raghavan, S.R.; Walls, H.; Khan, S.A. Rheology of silica dispersions in organic liquids: New evidence for solvation forces dictated by hydrogen bonding. *Langmuir* **2000**, *16*, 7920–7930. [CrossRef]
47. Lee, D.H.; Jeong, J.; Han, S.W.; Kang, D.P. Superhydrophobic surfaces with near-zero sliding angles realized from solvent relative permittivity mediated silica nanoparticle aggregation. *J. Mater. Chem. A* **2014**, *2*, 17165–17173. [CrossRef]
48. Chang, H.; Tu, K.; Wang, X.; Liu, J. Fabrication of mechanically durable superhydrophobic wood surfaces using polydimethyl-siloxane and silica nanoparticles. *RSC Adv.* **2015**, *5*, 30647–30653. [CrossRef]
49. Qiang, S.; Chen, K.; Yin, Y.; Wang, C. Robust UV-cured superhydrophobic cotton fabric surfaces with self-healing ability. *Mater. Des.* **2017**, *116*, 395–402. [CrossRef]
50. Liu, X.; Xu, Y.; Ben, K.; Chen, Z.; Wang, Y.; Guan, Z. Transparent, durable and thermally stable PDMS-derived superhydrophobic surfaces. *Appl. Surf. Sci.* **2015**, *339*, 94–101. [CrossRef]
51. Schaeffer, D.A.; Polizos, G.; Smith, D.B.; Lee, D.F.; Hunter, S.R.; Datskos, P.G. Optically transparent and environmentally durable superhydrophobic coating based on functionalized SiO_2 nanoparticles. *Nanotechnology* **2015**, *26*, 055602. [CrossRef]
52. Long, M.; Peng, S.; Deng, W.; Yang, X.; Miao, K.; Wen, N.; Miao, X.; Deng, W. Robust and thermal-healing superhydrophobic surfaces by spin-coating of polydimethylsiloxane. *J. Colloid Interface Sci.* **2017**, *508*, 18–27. [CrossRef]
53. ECHA. *Substance Information: N-Hexane*; ECHA: Helsinki, Finland, 2018. Available online: https://echa.europa.eu/substance-information/-/substanceinfo/100.003.435 (accessed on 13 November 2018).
54. Smallwood, I. *Handbook of Organic Solvent Properties*; Butterworth-Heinemann: Oxford, UK, 2012.
55. Chen, I.-J.; Lindner, E. The stability of radio-frequency plasma-treated polydimethylsiloxane surfaces. *Langmuir* **2007**, *23*, 3118–3122. [CrossRef] [PubMed]
56. Zhang, T.; Zhou, A.G.; Sun, B.R.; Chen, K.S.; Yu, H.-Z. Functional and versatile superhydrophobic coatings via stoichiometric silanization. *Nat. Commun.* **2021**, *12*, 982. [CrossRef] [PubMed]
57. Liu, X.; Chen, K.; Zhang, D.; Guo, Z. Stable and Durable Conductive Superhydrophobic Coatings Prepared by Double-Layer Spray Coating Method. *Nanomaterials* **2021**, *11*, 1506. [CrossRef] [PubMed]
58. Chuah, W.H. *Investigation on Hydrophobic Coating for Improved Functionality and Durability*; Universiti Sains Malaysia: Pulau Pinang, Malaysia, 2015.
59. Tu, K.; Kong, L.; Wang, X.; Liu, J. Semitransparent, durable superhydrophobic polydimethylsiloxane/SiO_2 nanocomposite coatings on varnished wood. *Holzforschung* **2016**, *70*, 1039–1045. [CrossRef]
60. Celia, E.; Darmanin, T.; de Givenchy, E.T.; Amigoni, S.; Guittard, F. Recent advances in designing superhydrophobic surfaces. *J. Colloid Interface Sci.* **2013**, *402*, 1–18. [CrossRef]
61. Polizos, G.; Tuncer, E.; Sauers, I.; More, K.L. Physical properties of epoxy resin/titanium dioxide nanocomposites. *Polym. Eng. Sci.* **2011**, *51*, 87–93. [CrossRef]
62. Hollande, L.; Do Marcolino, I.; Balaguer, P.; Domenek, S.; Gross, R.A.; Allais, F. Preparation of renewable epoxy-amine resins with tunable thermo-mechanical properties, wettability and degradation abilities from lignocellulose-and plant oils-derived components. *Front. Chem.* **2019**, *7*, 159. [CrossRef]
63. Gindl, M.; Sinn, G.; Stanzl-Tschegg, S.E. The effects of ultraviolet light exposure on the wetting properties of wood. *J. Adhes. Sci. Technol.* **2006**, *20*, 817–828. [CrossRef]
64. Tsuda, Y. Surface wettability controllable polyimides by UV light irradiation for printed electronics. *J. Photopolym. Sci. Technol.* **2016**, *29*, 383–390. [CrossRef]
65. Chaban, V.V.; Maciel, C.; Fileti, E.E. Does the like dissolves like rule hold for fullerene and ionic liquids? *J. Solut. Chem.* **2014**, *43*, 1019–1031. [CrossRef]
66. Coates, J. Interpretation of infrared spectra, a practical approach. In *Encyclopedia of Analytical Chemistry: Applications, Theory and Instrumentation*; John Wiley & Sons Ltd.: New York, NJ, USA, 2006.
67. Li, K.-M.; Jiang, J.-G.; Tian, S.-C.; Chen, X.-J.; Yan, F. Influence of Silica Types on Synthesis and Performance of Amine−Silica Hybrid Materials Used for CO_2 Capture. *J. Phys. Chem. C* **2014**, *118*, 2454–2462. [CrossRef]
68. Chen, K.; Zhou, S.; Wu, L. Facile fabrication of self-repairing superhydrophobic coatings. *Chem. Commun.* **2014**, *50*, 11891–11894. [CrossRef] [PubMed]

69. Ernault, E.; Richaud, E.; Fayolle, B. Thermal oxidation of epoxies: Influence of diamine hardener. *Polym. Degrad. Stab.* **2016**, *134*, 76–86. [CrossRef]
70. Krauklis, A.E.; Echtermeyer, A.T. Mechanism of yellowing: Carbonyl formation during hygrothermal aging in a common amine epoxy. *Polymers* **2018**, *10*, 1017. [CrossRef] [PubMed]

Article

Stabilization of Silver Nanoparticles on Polyester Fabric Using Organo-Matrices for Controlled Antimicrobial Performance

Ana Isabel Ribeiro [1], Vasyl Shvalya [2], Uroš Cvelbar [2,3], Renata Silva [4,5], Rita Marques-Oliveira [4,5], Fernando Remião [4,5], Helena P. Felgueiras [1], Jorge Padrão [1] and Andrea Zille [1,*]

1. Centre for Textile Science and Technology (2C2T), Department of Textile Engineering, University of Minho, Campus de Azurém, 4800-058 Guimaraes, Portugal; afr@2c2t.uminho.pt (A.I.R.); helena.felgueiras@2c2t.uminho.pt (H.P.F.); padraoj@2c2t.uminho.pt (J.P.)
2. Department of Gaseous Electronics (F6), Jožef Stefan Institute, SI-1000 Ljubljana, Slovenia; vasyl.shvalya@ijs.si (V.S.); uros.cvelbar@ijs.si (U.C.)
3. Faculty of Mathematics and Physics, University of Ljubljana, SI-1000 Ljubljana, Slovenia
4. Associate Laboratory i4HB—Institute for Health and Bioeconomy, Faculty of Pharmacy, University of Porto, 4051-401 Porto, Portugal; rsilva@ff.up.pt (R.S.); ritoliveira.m@gmail.com (R.M.-O.); remiao@ff.up.pt (F.R.)
5. UCIBIO—Applied Molecular Biosciences Unit, REQUIMTE, Laboratory of Toxicology, Department of Biological Sciences, Faculty of Pharmacy, University of Porto, 4051-401 Porto, Portugal
* Correspondence: azille@2c2t.uminho.pt

Abstract: Antimicrobial textiles are helpful tools to fight against multidrug-resistant pathogens and nosocomial infections. The deposition of silver nanoparticles (AgNPs) onto textiles has been studied to achieve antimicrobial properties. Yet, due to health and environmental safety concerns associated with such formulations, processing optimizations have been introduced: biocompatible materials, environmentally friendly agents, and delivery platforms that ensure a controlled release. In particular, the functionalization of polyester (PES) fabric with antimicrobial agents is a formulation in high demand in medical textiles. However, the lack of functional groups on PES fabric hinders the development of cost-effective, durable systems that allow a controlled release of antimicrobial agents. In this work, PES fabric was functionalized with AgNPs using one or two biocompatible layers of chitosan or hexamethyldisiloxane (HMDSO). The addition of organo-matrices stabilized the AgNPs onto the fabrics, protected AgNPs from further oxidation, and controlled their release. In addition, the layered samples were efficient against *Staphylococcus aureus* and *Escherichia coli*. The sample with two layers of chitosan showed the highest efficacy against S. aureus (log reduction of 2.15 ± 1.08 after 3 h of contact). Against *E. coli*, the sample with two layers of chitosan showed the best properties. Chitosan allowed to control the antimicrobial activity of AgNPs, avoid the complete loss of AgNPs after washings and act in synergy with AgNPs. After 3 h of incubation, this sample presented a log reduction of 4.81, and 7.27 of log reduction after 5 h of incubation. The antimicrobial results after washing showed a log reduction of 3.47 and 4.88 after 3 h and 5 h of contact, respectively. Furthermore, the sample with a final layer of HMDSO also presented a controlled antimicrobial effect. The antimicrobial effect was slower than the sample with just an initial layer of HMDSO, with a log reduction of 4.40 after 3 h of incubation (instead of 7.22) and 7.27 after 5 h. The biocompatibility of the composites was confirmed through the evaluation of their cytotoxicity towards HaCaT cells (cells viability > 96% in all samples). Therefore, the produced nanocomposites could have interesting applications in medical textiles once they present controlled antimicrobial properties, high biocompatibility and avoid the complete release of AgNPs to the environment.

Keywords: silver nanoparticles; chitosan; hexamethyldisiloxane; antimicrobial textiles; spray deposition

Citation: Ribeiro, A.I.; Shvalya, V.; Cvelbar, U.; Silva, R.; Marques-Oliveira, R.; Remião, F.; Felgueiras, H.P.; Padrão, J.; Zille, A. Stabilization of Silver Nanoparticles on Polyester Fabric Using Organo-Matrices for Controlled Antimicrobial Performance. *Polymers* **2022**, *14*, 1138. https://doi.org/10.3390/polym14061138

Academic Editor: Suguna Perumal

Received: 17 February 2022
Accepted: 10 March 2022
Published: 12 March 2022

Publisher's Note: MDPI stays neutral with regard to jurisdictional claims in published maps and institutional affiliations.

Copyright: © 2022 by the authors. Licensee MDPI, Basel, Switzerland. This article is an open access article distributed under the terms and conditions of the Creative Commons Attribution (CC BY) license (https://creativecommons.org/licenses/by/4.0/).

1. Introduction

Textiles may provide an excellent environment for microorganisms to thrive, presenting a suitable availability of nutrients, moisture, oxygen, and favorable temperature ranges. The functionalization of textiles with antimicrobial finishing agents has been widely applied to achieve technical materials and avoid the deterioration caused by microorganisms [1]. Increased healthcare awareness about hygiene and health issues has extended the global market to the antimicrobial textiles field. These textiles may be employed in several applications (e.g., wounds, sutures, and tissue engineering products) to prevent microbial proliferation and, hence, bad odors, stains, infections, a reduction in the textile's mechanical properties, and cross-contamination [2].

The usage of nanocomposite-based coatings has opened several possibilities in functional and high-performance textiles. Different metal-oxide (e.g., copper oxide, zinc oxide, and titanium dioxide) and metal (e.g., gold, zinc, copper, and silver) nanoparticles (NPs) have received significant attention as promising antimicrobial agents. These NPs possess superior action due to the higher surface-area-to-volume ratio, inducing their antimicrobial action via multiple mechanisms, namely by the direct interaction with the bacterial cell wall, inhibition of the biofilm formation, activation of the intrinsic and adaptive host immune responses, generation of reactive oxygen species (ROS) and interaction with intracellular components (e.g., DNA and proteins) [3–6]. In this respect, silver nanoparticles (AgNPs) have presented interesting antimicrobial properties, even in low concentrations [7,8].

Several techniques have been used to formulate textiles with AgNPs: the in situ thermal reduction, sonication, padding, dip-coating, spray, exhaustion, layer-by-layer, and electrospinning. However, numerous studies reported the potential uncontrolled leaching behavior of AgNPs, becoming a relevant environmental and health problem [9–11]. There is a need for new strategies to increase the stability of AgNPs on the fabrics.

Polyester (PES) fabrics have been largely used in various industries owing to their excellent strength, chemical resistance, processability, quick-drying and dimensional stability. Nevertheless, it presents a hydrophobic surface, where microorganisms can proliferate due to the abundant adsorption of metabolic products from the skin sweat/sebaceous glands [12]. The functionalization of PES with AgNPs can avoid the problems related to the microorganism's proliferation, but the strategies for AgNPs deposition rarely promote acceptable adhesion of the AgNPs due to the absence of functional groups on the PES structure [13,14]. Surface modification techniques have been applied to introduce other chemical groups onto the PES surface, namely photo-induced irradiation, electron beam irradiation, enzymatic modification, alkaline hydrolysis, aminolysis, alcoholysis, and plasma treatments [12,15,16]. However, most of the accessible methods for stabilizing the AgNPs on PES require various functionalization steps, final treatments, drying, and/or curing processes. Each step increases time and cost, hindering large-scale production. Embedding NPs on the fiber polymeric matrix or reducing metallic salts to NPs in the bulk polymeric matrix are presented as high-performance methods. However, enveloping the particles in the fiber core significantly compromises their antimicrobial performance. Another developed strategy is the application of a binder to improve the adhesion of NPs onto textile substrates. Though, few reports were found in the literature using PES [17,18]. Enhancing the adhesion strength between the NPs and the PES fibers' surface is imperative to ensure an efficient antimicrobial action, durability and avoid the undesirable release of metal NPs and ions [19].

Chitosan is an interesting biopolymer due to its inherent antimicrobial properties, biodegradability, non-toxicity, blood coagulating efficiency, antistatic features, and biocompatibility. It has been commonly studied for textile functionalization, namely as a binder for pigment printing, cationization of cotton, antimicrobial, anti-odor, and crease-resistant finishing [20–22]. In particular, metals easily interact with chitosan through electrostatic and chemical forces due to the presence of hydroxyl and amine groups [23]. Additionally, the combined antimicrobial effect of chitosan and metals have been explored to prepare novel nanocomposite materials with improved antimicrobial properties [24].

Organosilicon compounds have received more attention as coating agents due to their lack of toxicity, environmental friendliness, abrasion resistance, and physiological inertia [25]. Additionally, organosilicon compounds, including hexamethyldisiloxane

(HMDSO), have been shown to tune the adhesion properties of a surface by exhibiting methyl groups within the silicon-organic matrix [26,27].

Herein, a fast and cost-effective method was developed to functionalize PES fabric with AgNPs via spray coating. Chitosan or HMDSO were sprayed in different layers, before and after the AgNPs deposition, to promote the adhesion of NPs onto the textile, to protect AgNPs from further oxidation, and to control the AgNPs release. In the AgNPs dispersion preparation, commercial polyvinylpyrrolidone-AgNPs (20.0–30.0 nm in size) were used. They were redispersed in ethanol and characterized by dynamic light scattering (DLS) and zeta potential. The textile samples were characterized by scanning electron microscopy (SEM) and X-ray photoelectron spectroscopy (XPS). The antimicrobial activity of textiles was evaluated against *Staphylococcus aureus* and *Escherichia coli*. Finally, the cytotoxicity of PES composites was assessed in HaCaT cells by the neutral red uptake assay.

2. Materials and Methods

2.1. Materials

Commercial pre-washed PES fabric (weight per unit area of 100 g·m^{-2}) was used. The fabric was washed using a non-ionic detergent (1.0 g·L^{-1}) at 60 °C for 60 min., rinsed with distilled water, and dried at 40 °C. Commercial spherical polyvinylpyrrolidone-coated (PVP) AgNPs 99.95%, with sizes of 20–30 nm, were purchased from SkySpring Nanomaterials Inc, Houston, TX, USA. Chitosan, Chito Clear 42,030– 800 CPS, was purchased from Primex, Siglufjordur, Iceland. Ethanol, acetic acid, nitric acid, HMDSO, Neutral red (NR) solution, and Triton™ X-100 detergent solution were purchased from Sigma-Aldrich, Taufkirchen, Germany. Dulbecco's modified Eagle's medium (DMEM) with 4.5 g·L^{-1} glucose and GlutaMAX™, fetal bovine serum (FBS), antibiotic (10,000 U·mL^{-1} penicillin, 10,000 µg·mL^{-1} streptomycin), Hanks' balanced salt solution (HBSS) without calcium and magnesium [HBSS (-/-)] and 0.25% trypsin·1 mM^{-1} EDTA were obtained from Gibco™, Thermo Fisher Scientific, Waltham, MA, USA. All the reagents used were of analytical or of the highest purity grade available.

2.2. Preparation of AgNPs Dispersions

All materials were previously cleaned with nitric acid (10% (v/v)) and rinsed with distilled water. Then, PVP-AgNPs were dispersed in ethanol (1.0 mg·mL^{-1}) using an ultrasonic bath (30 min, 40 Hz) and ultrasound tip (15 min, 20 Hz).

2.3. Formulation of PES Composites by Spray

The AgNPs, chitosan solution (0.25% (w/v) of chitosan in 1% (v/v) acetic acid) and HMDSO layers were applied in both sides of the PES samples (10 × 10 cm^2) via spray system, pressurized at 1.5 bar with a distance of 5 cm. Samples with different formulations were prepared: (i) only with AgNPs; (ii) HMDSO + AgNPs; (iii) HMDSO + AgNPs + HMDSO; (iv) chitosan + AgNPs; (v) chitosan + AgNPs + chitosan; vi) only HMDSO; (vii) only chitosan (Figure 1).

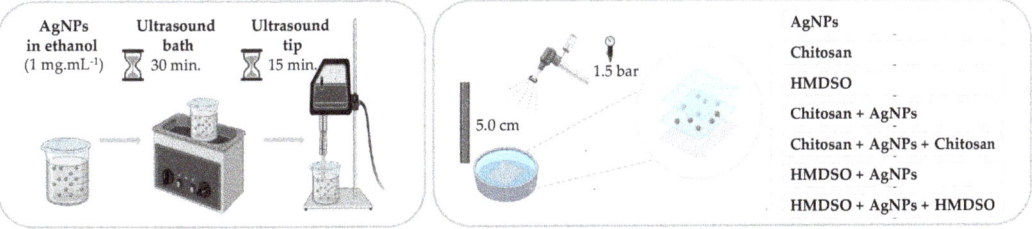

Figure 1. Schematic representation of the methodology adopted (images adapted from https://smart.servier.com (3 February 2022). Servier Medical Art by Servier is licensed under a Creative Commons Attribution 3.0 Unported License).

2.4. Washing Fastness

The washing fastness was assessed by performing 5 washing cycles (WC) according to EN ISO 15797 in a Datacolor Ahiba Lab Dyeing Machine (Lawrenceville, NJ, USA) at 75 °C, 40 rpm, for 15 min using a non-ionic surfactant (1.0 g·L^{-1}) in a liquor bath ratio of 1/30 (v/v) [28].

2.5. Dynamic Light Scattering (DLS) Analysis

The size and zeta potential of PVP-AgNPs in the dispersion were measured using a Zeta Sizer-Nano (Malvern Instruments, Malvern, UK). Data were collected after 30 scans at 25 ± 1 °C, and zeta potential was measured in a moderate electrolytic concentrated solution.

2.6. Scanning Electron Microscopy (SEM)

Morphological analyses were carried out with an ultra-high-resolution FEG-SEM, NOVA 200 Nano, FEI Company (Hillsboro, OR, USA). Secondary electron images were performed with an acceleration voltage of 5 kV. Backscattering electron images were realized with an acceleration voltage of 15 kV. Samples were coated with an Au-Pd (20–80 weight %) film using a high-resolution sputter coater, 208 HR Cressington Company (Watford, UK), coupled to an MTM-20 Cressington High-Resolution Thickness Controller.

2.7. X-ray Photoelectron Spectroscopy (XPS)

Detailed surface atomic composition and bonding environment research was conducted employing XPS PHI-TFA spectrometer (Physical Electronics Inc., Chanhassen, MN, USA) equipped with an Al- monochromatic (7 mm) X-ray source operating at pass energy equal 1486.6 eV, with active surface charge neutralization. Data acquisition was performed with a vacuum better than 1×10^{-8} Pa. Spectra have been corrected to give the adventitious C 1s spectral component (C–C, C–H) binding energy of 284.5 eV. Spectra were analyzed for elemental composition using Multipack software.

Deconvolution into sub-peaks was performed by OriginLab software, using the Gaussian fitting function and Shirley-type background subtraction. No tailing function was considered in the peak fitting procedure.

2.8. Evaluation of Antibacterial Properties of PES Samples

Antibacterial testing was performed according to the ASTM-E2149 standard for the determination of the antimicrobial activity of antimicrobial agents under dynamic contact conditions. The tests were performed immediately after sample preparation with slight modifications. Both Gram-positive and Gram-negative bacteria were used, respectively *Staphylococcus aureus* (American Type Culture Collection (ATCC 25923) and *Escherichia coli* (*E. coli*, ATCC 25922). The pre-inoculum of each bacterium was prepared in tryptic soy broth (Merck) and after 12 h of incubation at 37 °C and 120 rpm, the inoculum of each bacterium was centrifuged, the supernatant was eliminated, and the bacteria washed with sterile phosphate buffer saline (PBS). Then, the concentration of each bacterium was adjusted to 2×10^7 CFU·mL^{-1}. PES samples (1 × 1.5 cm) were inoculated in 5 mL of bacterial suspension for 3 h and 5 h at 37 °C and 120 rpm. Afterward, aliquots of these suspensions were collected and used to prepare 10-fold serial dilutions, which were cultured on agar plates for the determination of viable cells. The number of colony-forming units (CFUs)/mL was established before (0 h) and after (3 h and 5 h) contact with the fabrics. The results were expressed as log reductions, calculated as the ratio between the number of surviving bacteria colonies present on the tryptic soy agar (TSA) plates, before and after contact with the fabric. Antibacterial studies were performed in triplicate in two independent experiments.

2.9. Cytotoxicity

The cytotoxicity of the PES composites was evaluated in HaCaT cells, an immortalized human keratinocyte cell line, after 24 h after exposure, by the NR uptake assay.

HaCaT cells were routinely cultured in 75 cm^2 flasks using DMEM with 4.5 g·L^{-1} glucose and GlutaMAX™, supplemented with 10% of FBS, 100 U·mL^{-1} of penicillin, and 100 µg·mL^{-1} of streptomycin. Cells were grown at 37 °C, in a 5% CO_2-95% air atmosphere, and the medium was changed every 2 days. At 80–90% of confluence, cells were detached from the culture flasks via trypsinization (0.25% trypsin·mM^{-1} EDTA). The cells were seeded in 96-well plates at a density of 20,000 cells/well. Freshly prepared extracts of each sample (previously hermetically sealed and sterilized in an autoclave at 121 °C, 1.2 bar for 20 min.) were used in the evaluation of cytotoxicity, accordingly with ISO 1993-5 (Biological evaluation of medical devices—Part 5: Tests for in vitro cytotoxicity) [29]. Briefly, the extraction was performed in a complete cell culture medium, at a proportion of 0.1 g·mL^{-1} (ratio recommended for textiles), in a sterile, chemically inert, and closed container, for 24 ± 2 h, at 37 ± 1 °C, and under agitation. The extract was then directly used (100% concentration) or diluted in fresh cell culture medium at different concentrations (2, 4, 8, 16, and 32-times dilutions leading to 50, 25, 12,5, 6.25, and 3.125% concentrations). After 24 h seeding, the cells were exposed to the extracts of the medical devices (0–100%) for another 24 h. Extraction cell culture medium (without the test material) was also submitted to the same extraction conditions and used as a control. Triton™ X-100 (1% (v/v)) was used as positive control. The cells used in all experiments were taken between the 45th and 50th passages.

2.10. Neutral Red (NR) Uptake Assay

The cytotoxicity of the PES samples was assessed by the NR uptake assay that, based on the capacity of living cells to incorporate and retain the supravital dye NR within the lysosomes, provides a quantitative estimation of the number of viable cells in the culture. After 24 h exposure to the extracts of the samples, the cell culture medium was aspirated and a fresh cell culture medium containing NR (50 µg·mL^{-1}) was added. Cells were then incubated for 90 min, at 37 °C, in a humidified 5% CO_2-95% air atmosphere. After incubation, the cell culture medium was removed, and the NR dye retained only by viable cells was extracted (absolute ethyl alcohol/distilled water (1:1) with 5% (v/v) acetic acid). The absorbance was subsequently measured, at 540 nm, in a multi-well plate reader (PowerWaveX BioTek Instruments, VT, Santa Clara, CA, USA). The percentage of NR uptake relative to that of the control cells (0%) was used as the cytotoxicity measure. Four independent experiments were performed in triplicate.

2.11. Statistical Analysis

All statistical calculations were performed using the GraphPad Prism 8 for Windows (GraphPad Software, San Diego, CA, USA). The normality of the data distribution was assessed using the KS, D'Agostino & Pearson omnibus, and Shapiro–Wilk normality tests. One-way analysis of variance (ANOVA) was used to perform the statistical comparisons, followed by Dunnett's multiple comparisons test. Details of the performed statistical analysis are described in the figure captions. Differences were considered significant for p values lower than 0.05.

3. Results and Discussion

3.1. PES Fabrics Functionalization and Characterization

The present research focuses on the stabilization of AgNPs on PES fabrics using a biopolymer, the chitosan, and an organosilicon compound, HMDSO, to improve the stability of the AgNPs and control antibacterial efficacy. PES nanocomposites were prepared by spray deposition of AgNPs and chitosan or HMDSO layers. The following structures were prepared: (i) chitosan + AgNPs, (ii) chitosan + AgNPs + chitosan, (iii) HMDSO + AgNPs, and (iv) HMDSO + AgNPs + HMDSO. Control samples were also prepared by loading the fabric only with AgNPs, HMDSO, or chitosan (Figure 1). The AgNPs dispersion was characterized by DLS and zeta potential. The textile samples were characterized by SEM and XPS. Furthermore, the nanocomposites were washed 5 times to predict the release of AgNPs and the effect of the chitosan or HMDSO layers. The differences between washed and unwashed samples were detected by XPS analysis and confirmed via antimicrobial testing. The HMDSO and chitosan layers were meant to delay the AgNPs oxidation to a more controlled and durable release of the Ag ions. Especially after several washing cycles.

In the first step, commercially available AgNPs were redispersed using an ultrasound bath and ultrasound tip in absolute ethanol. The AgNPs dispersion was assessed by DLS measurements showing an average size of 281.0 ± 1.5 nm, a polydispersity index of 0.10 ± 0.02, and suitable colloidal stability with a zeta potential value of -31.0 ± 1.2 mV.

In the second step, the PES formulations were obtained by a spray method, where the AgNPs distribution onto the fabric and corresponding morphology were evaluated by SEM (Figure 2). The SEM images confirm the presence of AgNPs in all samples with no significant differences in their distribution or morphology, even in samples with an initial layer of chitosan or HMDSO, which may be explained by the successful utilization of the spray method for NPs deposition. The distance and the velocity of the spray flow were the same in all samples. The AgNPs onto PES showed uniform distribution, some agglomeration in all samples, and a quasi-spherical structure. In samples with a final layer of chitosan, it was possible to observe that the available AgNPs linked to the fabrics' surface were mostly covered by chitosan (some internal areas were uncovered). It was not possible to detect the distribution of HMDSO by SEM. No visual differences were found in samples containing HMDSO in the performed magnifications.

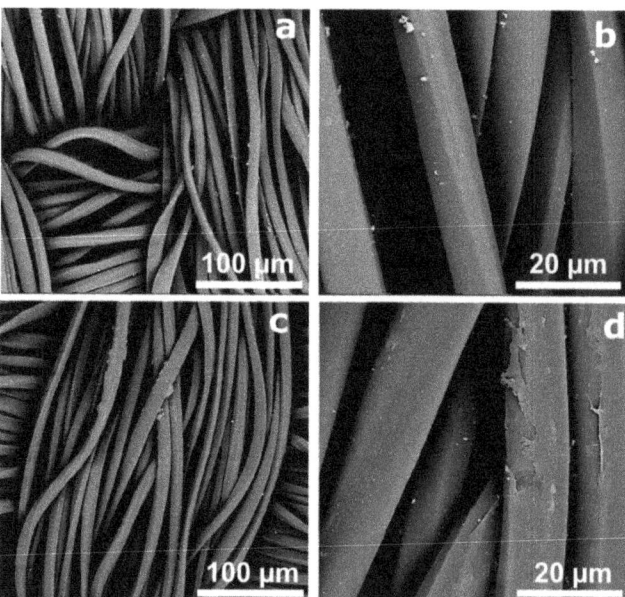

Figure 2. SEM micrographs of PES fabric with AgNPs (**a**,**b**) and with a final chitosan layer (chitosan + AgNPs + chitosan, (**c**,**d**) at magnifications of ×1000 and ×5000.

Chemical composition analysis of the unprotected AgNPs distributed over a control PES substrate surface was investigated before and after washing (Figure 3a). An evident spin-orbit doublet between 366–376 eV is observed, and it is attributed to Ag 3d core levels confirming the presence of noble nanoparticles on both PES substrates, before and after washing, respectively. However, after repetitive washing cycles, the intensity of silver-related peaks undergoes a relative decrease (Figure 3a). Deconvolution also confirms that the area under a final fit (black dashed curves) before washing is about 1.79 times larger than after washing (Figure 3b). It can mean the only fact that an unprotected nanosilver has been partially rinsed off during laundry. The oxidation states configuration of the remaining silver differs from that observed for the non-washed sample. By using the formula:

$$Ag(0) = \frac{Ag^0}{Ag^0 + Ag^{1+}} \quad (1)$$

and considering peak areas, it can be estimated that a portion of metallic silver of non-treated sample is larger (59%) than the washed one (52%). Note that neither carbon C 1s nor oxygen O 1s peaks related to PES substrate have not been considerably modified, proving its stability under multiple laundry cycles (Figure 3c). The oxygen/carbon atomic ratio remains unchanged as well, being equal to 0.30 in both cases.

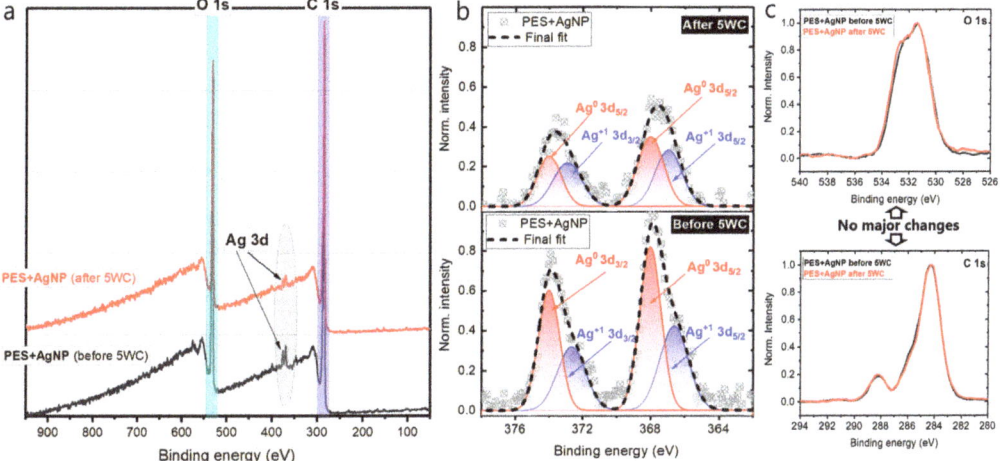

Figure 3. (a) Survey spectra of control PES loaded with AgNPs before and after five consequent washing procedures. (b) Gaussian deconvolution of high-resolution XPS Ag 3d core levels peaks. (c) O 1s and C 1s high-resolution spectra.

Aiming to protect the chemical environment of AgNPs, and especially to persist their population on PES substrate, two different sandwich-like spray-coated protective polymer layers were investigated (HMDSO and Chitosan). C 1s and O 1s spectra were elaborated to monitor the presence of HMDSO and chitosan protections after spraying. Considering PES oxygen/carbon ratio stability before and after washing, this feature can serve as a "presence indicator" of HMDSO and chitosan on top of PES. This parameter is different for each case and was found to be 0.30, 0.32, and 0.34 for PES, HMDSO, and chitosan, respectively. Slightly higher oxygen content is observed for chitosan coating, and the lowest is attributed to silver decorated PES sample. This finding denotes a good correlation with the Gaussian deconvolution, where a contribution of the C-O surface component, defined by a peak area, increases following the oxygen/carbon atomic ratio. A calculated area enclosed under the "blue" peak given in arbitrary units is as follows: PES = 0.37, HMDSO = 0.43, chitosan = 0.49 (Figure 4a).

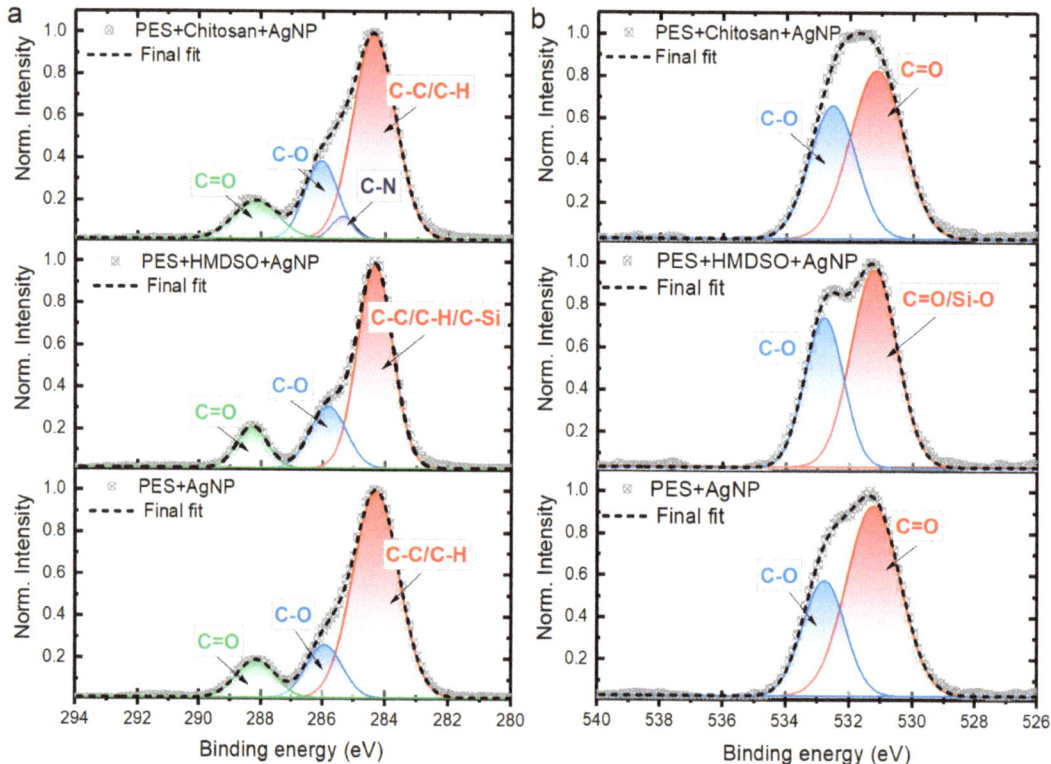

Figure 4. (**a**) High-resolution C 1s peak accompanied by Gaussian components deconvolution. (**b**) The same elaboration for O 1s spectra of the indicated samples.

Similarly, the C-O component behaves within recorded O 1s spectra. The integrated area of the peak located around 532.7 eV (blue peak) increases its contribution into a final fitted curve (Figure 4b). The numbers expressed in shares (%) are 32%, 38%, and 41% for PES, HMDSO, and chitosan, respectively. These observations confirm a successful coating of PES by HDMSO and chitosan protective layers.

Further, a new portion of nanosilver was attached to freshly created PES + HDMSO and PES + chitosan samples. Collected XPS results related to Ag $3d_{5/2}$ core levels revealed a high level of similarity in shape for all three unwashed samples, indicating no major alterations in oxidation shares between Ag(0) and Ag(+1) (Figure 5a). The presence of Ag(+1) can be related to the oxidation of the metallic silver in contact with air to AgOH, which consequently decomposes to Ag_2O [30]. After completing a sandwich-like structure by depositing a second protective layer made of the same polymer, noble metal NPs were not detected by XPS, additionally proving the efficiency of a spray-coating technique. The influence of washing cycles was tested again for PES + HDMSO + AgNP + HMDSO and PES + chitosan + AgNP + chitosan. Following the XPS data, neither the shape nor peak intensity of C 1s was altered. Additionally, the oxygen/carbon atomic ratio remained unchanged (Figure 5b). Additionally, a typical small Si 2p peak for HMDSO and N 1s peak for chitosan remained with similar features before and after laundry, indicating superior stability of sprayed protective layers.

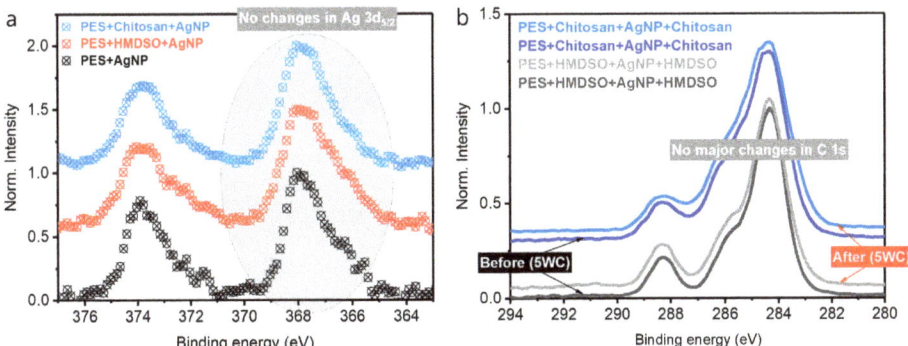

Figure 5. (**a**) XPS cut-off representing Ag 3d electronic orbital binding energy range. (**b**) The effect of washing cycles on high-resolution C 1s spectra of protective polymer coatings.

3.2. Evaluation of Antibacterial Properties of PES Samples

The antibacterial activity of PES samples functionalized with AgNPs and chitosan or HMDSO, as well as the corresponding control samples, were tested against *S. aureus* and *E. coli* by shake flask method (Figure 6). The tests were performed at different time points, after 3 h and 5 h of contact, to evaluate the time-kill kinetics of the AgNPs activity and the effect of the different layers of each configuration. The antimicrobial effect of the samples, before and after 5 WC, was also tested to assess the AgNP's endurance.

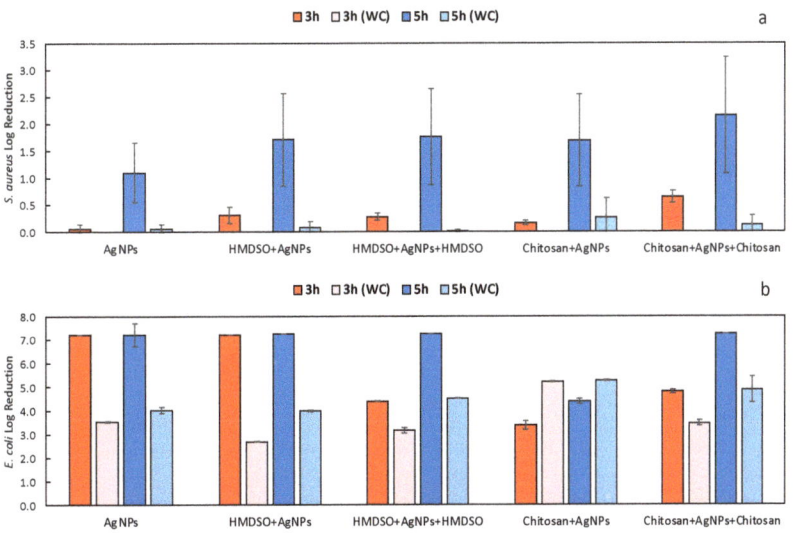

Figure 6. Antimicrobial action of samples against *S. aureus* (**a**) and *E. coli* (**b**) after 3 h and 5 h of contact, before and after 5 WC.

The composites exhibited low antimicrobial efficacy against *S. aureus* and strong activity against *E. coli*. This can be justified by the differences in the structure of the cell walls of the two types of bacteria, once *S. aureus* presents a thicker peptidoglycan layer (30 nm thickness) than the thinner structure of the *E. coli* cell wall (~3–4 nm thickness). Thus, the probability of the positively charged AgNPs being immobilized in the negative and thicker peptidoglycan layer of *S. aureus* bacteria is much higher than in *E. coli*. This suggests that the antimicrobial effect is controlled by the capability of the silver ions and AgNPs to disrupt the bacterial cell wall [31–33].

Generally, the results after 5 h of contact are equal or higher than the results after 3 h of contact but the layers showed to promote different behaviors in the AgNPs action. Starting from the results against *S. aureus*, after 5 h of incubation, the sample with 2 layers of chitosan (chitosan + AgNPs + chitosan) showed the highest log reduction (2.15 ± 1.08). Despite the higher exposition of the AgNPs in other samples, data suggests a potential synergistic effect to have taken place between chitosan and AgNPs. Some studies have been performed using chitosan to improve the adhesion of metal NPs onto textiles, while improving their inherent features, such as antimicrobial activity (chitosan and NPs) [34]. However, these studies mostly focus on cotton fabrics and no examples were found using PES. Chitosan has been described as a bacteriostatic polymer. Chitosan can interact with the cell membrane of pathogens once it presents a negative surface charge, reducing cell permeability to important environmental factors necessary to their viability. It can also interact with DNA, form chelates with microorganisms' nutrients, and form an intense film on the cell's surface, inhibiting their growth [35]. The remainder tested samples presented similar results after 5 h of contact with log reduction between 1.10 and 1.76. The results after 3 h of contact and after 5 WC did not show any relevant antibacterial efficacy against *S. aureus*.

The results against *E. coli* provided more information about the layer's efficacy. The control sample with just AgNPs showed an analogous log reduction after 3 h and 5 h of contact (7.22). A similar log reduction was obtained in the sample with an initial layer of HMDSO (HMDSO + Ag), 7.22 and 7.27 after 3 h and 5 h of contact. After 5 WC, the control sample exhibited 3.53 of log reduction and the HMDSO + AgNPs sample presented 2.69 after 3 h of incubation, and 4.09 and 3.99 after 5 h, respectively. These results suggest that only one layer of HMDSO does not provide any improvement in AgNPs adhesion over the control. However, when a final layer of HMDSO was added (HMDSO + AgNPs + HMDSO), the sample presented a slower antimicrobial effect, with a log reduction of 4.40 after 3 h of incubation (instead of 7.22) and 7.27 after 5 h. Again, after washing a slower antimicrobial effect was observed. This sample showed a log reduction of 3.16 after 3 h and a log reduction of 4.53 after 5 h, showing a superior result than the control and the HMDSO + AgNPs samples.

Since the amount of AgNPs in the unwashed samples is the same (spray application), the reduction in activity in the first hours of contact when a final layer of HMDSO is present is due to the greater stabilization of the AgNPs onto the fabric but also to the superior protection against the oxidation. The mechanism of action of AgNPs against *E. coli* have shown to be closely related to the release of silver ions by (i) oxidative stress caused by ROS, (ii) interaction of silver ions with thiol groups in proteins, and (iii) the destruction of the bacteria cells via strong affinity between silver ions and cell membrane. The release of silver ions and the ROS generation have been widely induced when Ag_2O is present [36,37]. In this work, the used AgNPs have an average diameter of 20–30 nm, and after immobilization form bigger agglomerated clusters that cannot enter inside the bacteria. Thus, the antibacterial effect seems mainly to be promoted by the ions release.

When using chitosan different effects were observed. The adhesion of AgNPs onto the fabric was superior using an initial layer of chitosan. The sample chitosan + AgNPs displayed a log reduction of 3.37 and 4.40 after 3 h and 5 h of contact, respectively. These results demonstrated the protective effect of chitosan over AgNPs and their superior adhesion to the substrate. However, after 5 WC, the same sample presented a superior antimicrobial effect, 5.22 and 5.27 of log reduction after 3 h and 5 h of contact, respectively. This can be attributed to a dual effect: a better adhesion of AgNPs and to an increased oxidation after washings as proved in the XPS results. Moreover, when the AgNPs layer was deposited over the first layer of chitosan, the chitosan was dissolved and the AgNPs were wrapped in the chitosan, justifying the inferior antimicrobial effect before washings. After washings, some chitosan was removed, decreasing the protective effect under AgNPs, the exposition of AgNPs increased, and consequently, the antimicrobial action also increased. Following this evidence, the inferior but suitable antimicrobial data of the chitosan + AgNPs sample before washing and the superior antimicrobial effect after washing should be attributed to the superior adhesion and protection of AgNPs using the initial layer of chitosan. Lastly, when a final layer of chitosan was added, a synergistic antimicrobial effect was also observed between chitosan and AgNPs. Despite the higher AgNPs exposition in the chitosan + AgNPs sample, the samples chitosan + AgNPs + chitosan showed a higher antimicrobial effect after 3 h of incubation (4.81 of log reduction) and 5 h of incubation (7.27 of log reduction). After WC, the antimicrobial results were also relevant, but the synergism was not that evident reporting log reductions of 3.47 and 4.88 after 3 h and 5 h of contact, respectively.

Although the results with chitosan layers after washing were comparable to the control samples just with AgNPs, by using chitosan it was possible to guarantee the washing fastness of the AgNPs (the AgNPs remained on the fabric). This strategy may prevent environmental contamination with heavy metals during the use and washing of antimicrobial textiles, enhance the durability of the antimicrobial effects, and protects the users from unnecessary exposure to AgNPs and, consequently, their cytotoxicity.

3.3. Evaluation of the Cytotoxicity of PES Samples Extracts

The cytotoxicity of the extracts of the PES samples was assessed after 24 h exposure, by the NR uptake assay. For that purpose, HaCaT keratinocyte-like cells were used as an in vitro model. No significant effects on NR uptake were detected after 24 h exposure to the extracts of AgNPs, HMDSO + AgNPs, HMDSO + AgNPs + HMDSO, chitosan + AgNPs, chitosan + AgNPs + chitosan, PES control + HDMSO and PES control + chitosan, at all the tested concentrations (0–100%) (Figure 7). For PES control extract, a small but significant reduction in NR uptake was observed for the highest tested concentrations (NR uptake significantly decreased to 96.74 and 96.75, 24 h after exposure to 50 and 100% of PES control extract, respectively, and when compared to control cells (0%)). Noteworthy, and accordingly with the ISO 1993-5, the medical devices under study are considered non-cytotoxic as the relative cell viability observed for the highest concentration of the sample extract (100% extract) was always higher than 70% when compared to the control cells (0%) [29].

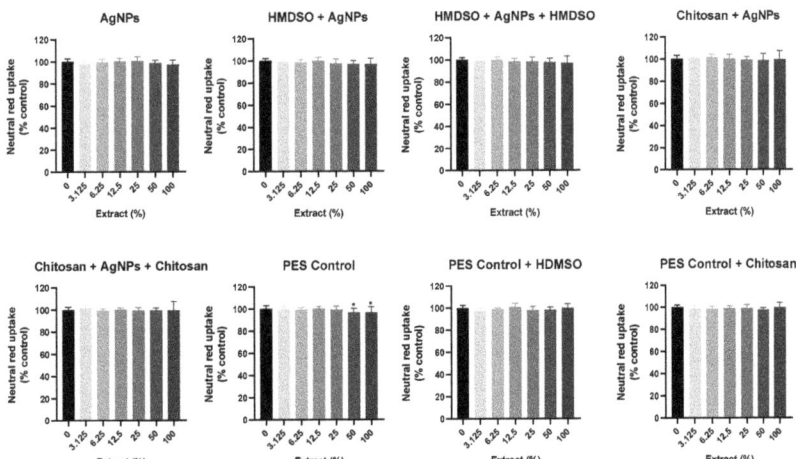

Figure 7. Cytotoxicity of the PES samples extracts (0–100%) evaluated in HaCat cells by the NR uptake assay, after 24 h exposure. Results are expressed as mean ± standard deviation (SD) from 4 independent experiences, performed in triplicate. Statistical comparisons were made using one-way ANOVA followed by Dunnett's multiple comparisons tests (* $p < 0.05$ vs. 0%).

4. Conclusions

This research envisaged the development of PES nanocomposites with controlled antimicrobial performance against *S. aureus* and *E. coli*, using AgNPs and chitosan or HMDSO layers. The chitosan or HMDSO layers were applied on PES fabric before and/or after AgNPs deposition. The samples were successfully prepared by spray coating and 5 WC were conducted after deposition. Successful PES functionalization and AgNPs content were verified by XPS and SEM analyses. The antimicrobial results showed that just an initial layer of HMDSO does not improve the AgNPs adhesion. However, when an initial and final layer of HMDSO was applied, AgNPs were stabilized onto PES fabric and the treatment prevented the complete loss of AgNPs during the washings. When chitosan was used, different results were obtained. With only one layer of chitosan, the adhesion of AgNPs to the PES fabric was significantly improved. Moreover, when an initial and final layer of chitosan was added, a controlled antimicrobial action was attained, and synergistic antimicrobial effects were evidenced between chitosan and AgNPs. Here, too, superior washing fastness was observed. Lastly, cytotoxicity studies showed the biocompatibility of the prepared PES nanocomposites.

These nanocomposites will open new perspectives for the use of AgNPs to PES functionalization in the healthcare sector with minimal environmental contamination during the use and washing of antimicrobial textiles, superior durability, and controlled antimicrobial effect.

Author Contributions: A.I.R.—investigation, data collection, interpretation, original draft, and editing; V.S. and U.C.—XPS analysis, XPS deconvolutions, and XPS interpretation; F.R., R.S. and R.M.-O.—cytotoxicity tests and interpretation; H.P.F.—antimicrobial tests and interpretation; J.P.—antimicrobial tests, interpretation, and review; A.Z.—supervision, interpretation, review, and editing. All authors have read and agreed to the published version of the manuscript.

Funding: This research was funded by the projects PLASMAMED-PTDC/CTM-TEX/28295/2017 and MEDCOR-PTDC/CTM-TEX/1213/2020, the project UID/CTM/00264/2021 of 2C2T under the COMPETE and FCT/MCTES (PIDDAC) co-financed by FEDER through the PT2020 program.

Institutional Review Board Statement: Not applicable.

Informed Consent Statement: Not applicable.

Data Availability Statement: Not applicable.

Acknowledgments: Ana Isabel Ribeiro acknowledges FCT the scholarship SFRH/BD/137668/2018.

Conflicts of Interest: The authors declare no conflict of interest.

References

1. Buşilă, M.; Muşat, V.; Textor, T.; Mahltig, B. Synthesis and characterization of antimicrobial textile finishing based on Ag:ZnO nanoparticles/chitosan biocomposites. *RSC Adv.* **2015**, *5*, 21562–21571. [CrossRef]
2. Abid, S.; Hussain, T.; Nazir, A. Antimicrobial textiles for skin and wound infection management. In *Antimicrobial Textiles from Natural Resources*; Mondal, M.I.H., Ed.; Woodhead Publishing: Sawston, UK, 2021; pp. 313–347. [CrossRef]
3. Saleem, H.; Zaidi, S. Sustainable Use of Nanomaterials in Textiles and Their Environmental Impact. *Materials* **2020**, *13*, 5134. [CrossRef]
4. Ribeiro, A.I.; Senturk, D.; Silva, K.S.; Modic, M.; Cvelbar, U.; Dinescu, G.; Mitu, B.; Nikiforov, A.; Leys, C.; Kuchakova, I.; et al. Efficient silver nanoparticles deposition method on DBD plasma-treated polyamide 6,6 for antimicrobial textiles. *IOP Conf. Ser. Mater. Sci. Eng.* **2018**, *460*, 012007. [CrossRef]
5. Mehravani, B.; Ribeiro, A.; Zille, A. Gold Nanoparticles Synthesis and Antimicrobial Effect on Fibrous Materials. *Nanomaterials* **2021**, *11*, 1067. [CrossRef]
6. Baptista, P.V.; McCusker, M.P.; Carvalho, A.; Ferreira, D.A.; Mohan, N.M.; Martins, M.; Fernandes, A.R. Nano-Strategies to Fight Multidrug Resistant Bacteria—"A Battle of the Titans". *Front. Microbiol.* **2018**, *9*, 1441. [CrossRef]
7. Ribeiro, A.I.; Senturk, D.; Silva, K.K.; Modic, M.; Cvelbar, U.; Dinescu, G.; Mitu, B.; Nikiforov, A.; Leys, C.; Kuchakova, I.; et al. Antimicrobial Efficacy of Low Concentration PVP-Silver Nanoparticles Deposited on DBD Plasma-Treated Polyamide 6,6 Fabric. *Coatings* **2019**, *9*, 581. [CrossRef]
8. Ribeiro, A.I.; Dias, A.M.; Zille, A. Synergistic Effects Between Metal Nanoparticles and Commercial Antimicrobial Agents: A Review. *ACS Applied Nano Materials* **2022**. [CrossRef]

9. Salama, A.; Abouzeid, R.E.; Owda, M.E.; Cruz-Maya, I.; Guarino, V. Cellulose–Silver Composites Materials: Preparation and Applications. *Biomolecules* **2021**, *11*, 1684. [CrossRef]
10. Pollard, Z.A.; Karod, M.; Goldfarb, J.L. Metal leaching from antimicrobial cloth face masks intended to slow the spread of COVID-19. *Sci. Rep.* **2021**, *11*, 19216. [CrossRef]
11. Shah, M.A.; Pirzada, B.M.; Price, G.; Shibiru, A.L.; Qurashi, A. Applications of nanotechnology in smart textile industry: A critical review. *J. Adv. Res.* **2022**, in press. [CrossRef]
12. Chen, S.; Zhang, S.; Galluzzi, M.; Li, F.; Zhang, X.; Yang, X.; Liu, X.; Cai, X.; Zhu, X.; Du, B.; et al. Insight into multifunctional polyester fabrics finished by one-step eco-friendly strategy. *Chem. Eng. J.* **2019**, *358*, 634–642. [CrossRef]
13. Lin, S.; Chen, L.; Huang, L.; Cao, S.; Luo, X.; Liu, K. Novel antimicrobial chitosan–cellulose composite films bioconjugated with silver nanoparticles. *Ind. Crops Prod.* **2015**, *70*, 395–403. [CrossRef]
14. Lorusso, E.; Ali, W.; Hildebrandt, M.; Mayer-Gall, T.; Gutmann, J.S. Hydrogel Functionalized Polyester Fabrics by UV-Induced Photopolymerization. *Polymers* **2019**, *11*, 1329. [CrossRef] [PubMed]
15. Morshed, M.N.; Behary, N.; Bouazizi, N.; Guan, J.; Chen, G.; Nierstrasz, V. Surface modification of polyester fabric using plasma-dendrimer for robust immobilization of glucose oxidase enzyme. *Sci. Rep.* **2019**, *9*, 15730. [CrossRef] [PubMed]
16. Baghriche, O.; Rtimi, S.; Pulgarin, C.; Roussel, C.; Kiwi, J. RF-plasma pretreatment of surfaces leading to TiO_2 coatings with improved optical absorption and OH-radical production. *Appl. Catal. B Environ.* **2013**, *130–131*, 65–72. [CrossRef]
17. Xu, Q.; Xie, L.; Diao, H.; Li, F.; Zhang, Y.; Fu, F.; Liu, X. Antibacterial cotton fabric with enhanced durability prepared using silver nanoparticles and carboxymethyl chitosan. *Carbohydr. Polym.* **2017**, *177*, 187–193. [CrossRef]
18. Dastjerdi, R.; Montazer, M.; Shahsavan, S. A new method to stabilize nanoparticles on textile surfaces. *Colloids Surf. A Physicochem. Eng. Asp.* **2009**, *345*, 202–210. [CrossRef]
19. Zhang, Y.; Xu, Q.; Fu, F.; Liu, X. Durable antimicrobial cotton textiles modified with inorganic nanoparticles. *Cellulose* **2016**, *23*, 2791–2808. [CrossRef]
20. Abate, M.T.; Ferri, A.; Guan, J.; Chen, G.; Ferreira, J.A.; Nierstrasz, V. Single-step disperse dyeing and antimicrobial functionalization of polyester fabric with chitosan and derivative in supercritical carbon dioxide. *J. Supercrit. Fluids* **2019**, *147*, 231–240. [CrossRef]
21. Rouhani Shirvan, A.; Shakeri, M.; Bashari, A. Recent advances in application of chitosan and its derivatives in functional finishing of textiles. In *The Impact and Prospects of Green Chemistry for Textile Technology*; Shahid-ul-Islam, B.S.B., Ed.; Woodhead Publishing: Sawston, UK, 2019; pp. 107–133. [CrossRef]
22. Said, M.M.; Rehan, M.; El-Sheikh, S.M.; Zahran, M.K.; Abdel-Aziz, M.S.; Bechelany, M.; Barhoum, A. Multifunctional Hydroxyapatite/Silver Nanoparticles/Cotton Gauze for Antimicrobial and Biomedical Applications. *Nanomaterials* **2021**, *11*, 429. [CrossRef]
23. Ali, F.; Khan, S.B.; Kamal, T.; Alamry, K.A.; Asiri, A.M. Chitosan-titanium oxide fibers supported zero-valent nanoparticles: Highly efficient and easily retrievable catalyst for the removal of organic pollutants. *Sci. Rep.* **2018**, *8*, 6260. [CrossRef] [PubMed]
24. Perinelli, D.R.; Fagioli, L.; Campana, R.; Lam, J.K.W.; Baffone, W.; Palmieri, G.F.; Casettari, L.; Bonacucina, G. Chitosan-based nanosystems and their exploited antimicrobial activity. *Eur. J. Pharm. Sci.* **2018**, *117*, 8–20. [CrossRef] [PubMed]
25. Sun, Z.; Wen, J.; Wang, W.; Fan, H.; Chen, Y.; Yan, J.; Xiang, J. Polyurethane covalently modified polydimethylsiloxane (PDMS) coating with increased surface energy and re-coatability. *Prog. Org. Coat.* **2020**, *146*, 105744. [CrossRef]
26. Beier, O.; Pfuch, A.; Horn, K.; Weisser, J.; Schnabelrauch, M.; Schimanski, A. Low Temperature Deposition of Antibacterially Active Silicon Oxide Layers Containing Silver Nanoparticles, Prepared by Atmospheric Pressure Plasma Chemical Vapor Deposition. *Plasma Processes Polym.* **2013**, *10*, 77–87. [CrossRef]
27. Ribeiro, A.I.; Modic, M.; Cvelbar, U.; Dinescu, G.; Mitu, B.; Nikiforov, A.; Leys, C.; Kuchakova, I.; Souto, A.P.; Zille, A. Atmospheric-Pressure Plasma Spray Deposition of Silver/HMDSO Nanocomposite on Polyamide 6,6 with Controllable Antibacterial Activity. *AATCC J. Res.* **2020**, *7*, 1–6. [CrossRef]
28. ISO 15797:2017; Textiles—Industrial Washing and Finishing Procedures for Testing of Workwear. In 2. International Organization for Standardization: Geneva, Switzerland, 2017; 14.
29. ISO 10993-5:2009; Biological Evaluation of Medical Devices—Part 5: Tests for In Vitro Cytotoxicity. In 3. International Organization for Standardization: Geneva, Switzerland, 2009; 34.
30. Rtimi, S.; Baghriche, O.; Sanjines, R.; Pulgarin, C.; Bensimon, M.; Kiwi, J. TiON and TiON-Ag sputtered surfaces leading to bacterial inactivation under indoor actinic light. *J. Photochem. Photobiol. A Chem.* **2013**, *256*, 52–63. [CrossRef]
31. Ribeiro, A.I.; Modic, M.; Cvelbar, U.; Dinescu, G.; Mitu, B.; Nikiforov, A.; Leys, C.; Kuchakova, I.; De Vrieze, M.; Felgueiras, H.P.; et al. Effect of Dispersion Solvent on the Deposition of PVP-Silver Nanoparticles onto DBD Plasma-Treated Polyamide 6,6 Fabric and Its Antimicrobial Efficiency. *Nanomaterials* **2020**, *10*, 607. [CrossRef] [PubMed]
32. Tavares, T.D.; Antunes, J.C.; Padrão, J.; Ribeiro, A.I.; Zille, A.; Amorim, M.T.P.; Ferreira, F.; Felgueiras, H.P. Activity of Specialized Biomolecules against Gram-Positive and Gram-Negative Bacteria. *Antibiotics* **2020**, *9*, 314. [CrossRef] [PubMed]
33. Dakal, T.C.; Kumar, A.; Majumdar, R.S.; Yadav, V. Mechanistic Basis of Antimicrobial Actions of Silver Nanoparticles. *Front. Microbiol.* **2016**, *7*, 1831. [CrossRef] [PubMed]
34. Gadkari, R.R.; Ali, S.W.; Joshi, M.; Rajendran, S.; Das, A.; Alagirusamy, R. Leveraging antibacterial efficacy of silver loaded chitosan nanoparticles on layer-by-layer self-assembled coated cotton fabric. *Int. J. Biol. Macromol.* **2020**, *162*, 548–560. [CrossRef]

35. Matica, M.A.; Aachmann, F.L.; Tøndervik, A.; Sletta, H.; Sletta, H.; Ostafe, V. Chitosan as a Wound Dressing Starting Material: Antimicrobial Properties and Mode of Action. *Int. J. Mol. Sci.* **2019**, *20*, 5889. [CrossRef] [PubMed]
36. Li, D.; Chen, S.; Zhang, K.; Gao, N.; Zhang, M.; Albasher, G.; Shi, J.; Wang, C. The interaction of Ag_2O nanoparticles with *Escherichia coli*: Inhibition–sterilization process. *Sci. Rep.* **2021**, *11*, 1703. [CrossRef] [PubMed]
37. Rtimi, S.; Nadtochenko, V.; Khmel, I.; Bensimon, M.; Kiwi, J. First unambiguous evidence for distinct ionic and surface-contact effects during photocatalytic bacterial inactivation on Cu–Ag films: Kinetics, mechanism and energetics. *Mater. Today Chem.* **2017**, *6*, 62–74. [CrossRef]

Article

Antibacterial Effect of Functionalized Polymeric Nanoparticles on Titanium Surfaces Using an In Vitro Subgingival Biofilm Model

Jaime Bueno [1], Leire Virto [1], Manuel Toledano-Osorio [2], Elena Figuero [1], Manuel Toledano [2], Antonio L. Medina-Castillo [3], Raquel Osorio [2,*], Mariano Sanz [1] and David Herrera [1]

1. ETEP (Etiology and Therapy of Periodontal and Peri-Implant Diseases) Research Group, University Complutense, Pza. Ramón y Cajal s/n, 28040 Madrid, Spain; jaimebue@ucm.es (J.B.); lvirto@ucm.es (L.V.); elfiguer@ucm.es (E.F.); marsan@ucm.es (M.S.); davidher@ucm.es (D.H.)
2. Faculty of Dentistry, University of Granada, Colegio Máximo de Cartuja s/n, 18071 Granada, Spain; mtoledano@correo.ugr.es (M.T.-O.); toledano@ugr.es (M.T.)
3. Faculty of Sciences, University of Granada, Colegio Máximo de Cartuja s/n, 18071 Granada, Spain; amedina@nanomyp.com
* Correspondence: rosorio@ugr.es

Abstract: This investigation aimed to evaluate the antibacterial effect of polymeric nanoparticles (NPs), functionalized with calcium, zinc, or doxycycline, using a subgingival biofilm model of six bacterial species (*Streptococcus oralis*, *Actinomyces naeslundii*, *Veillonela parvula*, *Fusobacterium nucleatum*, *Porphyromonas gingivalis*, and *Aggregatibacter actinomycetemcomitans*) on sandblasted, large grit, acid-etched titanium discs (TiDs). Undoped NPs (Un-NPs) or doped NPs with calcium (Ca-NPs), zinc (Zn-NPs), or doxycycline (Dox-NPs) were applied onto the TiD surfaces. Uncovered TiDs were used as negative controls. Discs were incubated under anaerobic conditions for 12, 24, 48, and 72 h. The obtained biofilm structure was studied by scanning electron microscopy (SEM) and its vitality and thickness by confocal laser scanning microscopy (CLSM). Quantitative polymerase chain reaction of samples was used to evaluate the bacterial load. Data were evaluated by analysis of variance ($p < 0.05$) and post hoc comparisons with Bonferroni adjustments ($p < 0.01$). As compared with uncovered TiDs, Dox-NPs induced higher biofilm mortality (47.21% and 85.87%, respectively) and reduced the bacterial load of the tested species, after 72 h. With SEM, scarce biofilm formation was observed in Dox-NPs TiDs. In summary, Dox-NPs on TiD reduced biofilm vitality, bacterial load, and altered biofilm formation dynamics.

Keywords: polymers; nanoparticles; doxycycline; antibacterial; zinc; calcium; biofilm

1. Introduction

Different strategies have been developed over the years to rehabilitate lost dentition. The current use of dental implants has demonstrated long-term survival and success [1], but also frequent complications [2], as peri-implant diseases, which are chronic inflammatory conditions of the peri-implant tissues. When the inflammatory lesion affects the peri-implant mucosa, without loss of supporting bone, the condition is diagnosed as peri-implant mucositis [3], whereas, when the inflammation also results in progressive loss of supporting bone, the diagnosis is peri-implantitis [4], and can eventually lead to the loss of the implant [5].

The prevalence of peri-implant diseases is high, ranging between 19% and 65% in the case of peri-implant mucositis, and between 1% and 47% for peri-implantitis [6–8], what makes prevention, early diagnosis, and treatment of these conditions very important.

Although the etiology of peri-implant diseases is multifactorial, the accumulation of bacterial biofilms on the implant crown, abutment, and implant surfaces is the most important factor [5]. A history of periodontitis has been shown to be a risk factor for peri-implantitis, and this inflammatory condition is usually more aggressive [9,10] and its treatment less predictable and efficacious as compared with the treatment of periodontitis [11,12]. Thus, distinct strategies have been proposed for the treatment of peri-implantitis, mainly through the decontamination of biofilm contaminated surfaces by means of debridement/instrumentation of these surfaces, with or without adjunctive systemic or local antimicrobial agents [13]. However, the predictability of these treatment protocols remains to be a challenge, and future research is seeking new approaches and adjunctive agents.

In recent years, nanotechnology has gained relevance in medicine and dentistry for its use in prevention, diagnosis, and treatment of different conditions [14–16]. Different nanostructured materials have been proposed for the treatment of periodontal and peri-implant diseases [17,18], such as polymeric nanoparticles due to their antimicrobial activity [19,20]. These polymeric nanoparticles (NPs) are non-resorbable and exhibit carboxyl groups on their external surface, which may be functionalized with different molecules, thus, enhancing their antibacterial properties. For example, they can effectively chelate and release calcium and zinc [19,21]. Similarly, doxycycline loaded NPs have been shown to release most of this antibacterial agent (around 70%) up to 28 d [21]. These functionalized NPs act as antibacterial agents not only on planktonic cultures, but also in subgingival biofilms when grown on hydroxyapatite discs [20,21], which makes their use a potentially effective tool adjunctive to the mechanical treatment of peri-implant diseases.

The objective of this in vitro investigation was to evaluate the antibacterial capacity of NPs functionalized with zinc, calcium, and doxycycline, in a validated in vitro oral biofilm model over titanium discs with sandblasted, large grit, and acid-etched (SLA) surface. As specific objectives, the bacterial load, the vitality and thickness of the biofilms were analyzed.

2. Materials and Methods

2.1. Bacterial Strains and Culture Conditions

The bacterial strains *Streptococcus oralis* CECT 907T, *Veillonella parvula* NCTC 11810, *Actinomyces naeslundii* ATCC 19039, *Fusobacterium nucleatum* DMSZ 20482, *Aggregatibacter actinomycetemcomitans* DSMZ 8324, and *Porphyromonas gingivalis* ATCC 33277 were selected for bacterial growth in blood agar plates (Blood Agar Oxoid N° 2, Oxoid, Basingstoke, UK), supplemented with 5% (v/v) sterile horse blood (Oxoid), 5.0 mg L-1 hemin (Sigma, St. Louis, MO, USA), and 1.0 mg L-1 menadione (Merck, Darmstadt, Germany), under anaerobic conditions (10% H_2, 10% CO_2, and balance N_2) at 37 °C for 24–72 h.

2.2. Nanoparticle Production

Four different NPs, developed as previously described by Osorio et al. [19], were studied: (a) Undoped NPs (Un-NPs), (b) NPs loaded with zinc (Zn-NPs), (c) NPs loaded with calcium (Ca-NPs), and (d) NPs doped with doxycycline (Dox-NPs). NPs were created through a polymerization/precipitation procedure with a composition of 2-hydroxyethyl methacrylate (backbone monomer), ethylene glycol dimethacrylate (cross-linker), and methacrylic acid (functional monomer), with a final diameter of approximately 150 nm [22,23]. For the functionalizing process, 30 mg of Zn-NPs and Ca-NPs were immersed for 3 days at room temperature, and under continuous agitation, in an aqueous solution of $ZnCl_2$ and $CaCl_2$ (containing zinc and calcium at 40 ppm, at pH 6.5) until reaching the adsorption equilibrium of metal ions. Next, the particles were removed from the supernatant and suspended in phosphate buffered saline (PBS). Subsequently, the suspensions were centrifuged for 20 min (6000× g), and the particles were detached from the supernatant. The same centrifugation technique, with the addition of PBS was used to wash the samples, and it was repeated twice. Ion complexation values were 2.15 ± 0.05 µg Zn/mg NPs and 0.96 ± 0.04 µg Ca/mg NPs, respectively [21]. For doping NPs with doxycycline, 30 mg of nanoparticles were submerged in a 40 mgL^{-1} aqueous solution of doxycycline hyclate (Sigma-Aldrich, Chemie Gmbh, Riedstr, Germany). NPs were maintained for 30 min under constant shaking. The achieved amount of doxycycline per gram of NPs was 70 µg [21].

As with the other NPs, the suspensions were centrifuged. The NPs were separated from the supernatant and re-suspended in PBS [21]. Doped NPs have been previously shown to effectively liberate zinc, calcium, and doxycycline [21].

2.3. Specimen Production

Sterile titanium discs (TiDs) (grade 2) of 5 mm of diameter (manufactured and donated by Straumann, Institut Straumann AG, Basel, Switzerland) with surfaces comparable to the commercially available SLActive® surface (Institut Straumann AG) were used. The different types of NPs were all diluted in PBS (10 mg/mL) and were applied onto the surfaces of the TiDs. Discs coated with PBS without NPs were used as the control.

2.4. Biofilm Development on the Prepared Specimens

The surfaces of TiDs covered with Un-NPs, Zn-NPs, Ca-NPs, Dox-NPs, and PBS were used for the establishment of the multispecies biofilms [24]. Pure cultures of each bacterium were grown anaerobically in a medium containing a high concentration of proteins formed by brain heart infusion (BHI) (Becton, Dickinson and Company, Franklin Lakes, NJ, USA) supplemented with 2.5 g L^{-1} mucin (Oxoid), 1.0 g L^{-1} yeast extract (Oxoid), 0.1 g L^{-1} cysteine (Sigma), 2.0 g L^{-1} sodium bicarbonate (Merck), 5.0 mg L^{-1} hemin (Sigma, St. Louis, MO, USA), 1.0 mg L^{-1} menadione (Merck, Darmstadt, Germany) and 0.25% (v/v) glutamic acid (Sigma). The bacteria were collected at the mid-exponential phase of bacterial growth (measured by spectrophotometry) and a mixed bacterial suspension was prepared in modified BHI medium containing 10^3 colony forming units (CFU) mL^{-1} of *S. oralis*, 10^5 CFU mL^{-1} of *V. parvula* and *A. naeslundii*, and 10^6 CFU mL^{-1} of *F. nucleatum, A. actinomycetemcomitans*, and *P. gingivalis*.

Then, 1.5 mL of mixed bacteria suspension was placed over TiDs coated with PBS or with the different tested products, in a multi-well plate of a 24-well tissue culture plate (Greiner Bio-one, Frickenhausen, Germany) and were incubated in anaerobic conditions (10% H_2, 10% CO_2, and balance N_2) at 37 °C for 12, 24, 48, and 72 h. Plates containing only culture medium were also cultured in order to be sure of the sterility of the culture medium.

2.5. Morphological Analysis of Biofilms by Scanning Electron Microscope (SEM)

Biofilms were analyzed by SEM at different times of growth (12, 24, 48, and 72 h). Specimen fixation was performed by immersion in a 4% paraformaldehyde and 2.5% glutaraldehyde solution for 4 h, at 4 °C. Then, the discs were washed with PBS and sterile water (10 min each) and submitted to critical point drying. Specimens were sputter-coated with gold and analyzed by SEM using a JSM 6400 (JSM6400, JEOL, Tokyo, Japan), with a back-scattered electron detector at an image resolution of 25 kV.

2.6. Analysis of Biofilms' Vitality and Thickness by Confocal Laser Scanning Microscopy (CLSM)

The non-invasive confocal imaging of fully hydrated biofilms was carried out by means of a fixed-stage Ix83 Olympus inverted microscope coupled to an Olympus FV1200 confocal system (Olympus, Shinjuku, Tokyo, Japan). The objective lens was a ×63 water-immersion lens (Olympus). Specimens were stained at room temperature with LIVE/DEAD® BacLight™ Bacterial Viability Kit solution (L7012, Molecular Probes B. V., Leiden, The Netherlands). A staining time of 8 ± 1 min, in a 1:1 fluorochrome ratio was used to obtain the best fluorescence signal at the corresponding wavelengths (Syto9, 515–530 nm and propidium iodide, PI > 600 nm). At least three different and demonstrative locations of the discs were selected for the study. A z-series of scans (xyz) of 1 µm thickness (8 bits, 1024 × 1024 pixels) were analyzed thanks to the configuration of the CLSM control software. Image stacks were analyzed by using the Olympus® software (Olympus®). To quantify the biomass and cell viability within the biofilm, total fluorescent staining of the confocal micrographs was analyzed using an open source image analysis software (Fiji ImageJ) by measuring voxel intensities from two-channel images and, thus, calculating the percentage of the biomass and cell viability within the stacks [25].

2.7. DNA Isolation and Quantitative Polymerase Chain Reaction (qPCR)

DNA of the biofilms at 12, 24, 48, and 72 h was obtained from all samples using a commercial kit (D-321-100, MolYsis Complete5, Molzym GmgH & CoKG, Bremen, Germany), following the manufacturer's instructions. The qPCR technique with hydrolysis probes was used to detect and quantify the bacterial DNA. The primers and probes were obtained by Life Technologies Invitrogen (Carlsbad, CA, USA), Applied Biosystems (Carlsbad, CA, USA) and Roche (Roche Diagnostic GmbH, Mannheim, Germany) and were fixed against the 16S rRNA gene. A total volume of 10 µL of the reaction mixture was used for the amplification of the qPCR. The reaction mixtures contained 5 µL Master Mix 2x (LC 480 Probes Master, Roche), optimal primers and probe concentrations [25] (900, 900, and 300 nM for *S. oralis*; 300, 300, and 300 nM for *A. naeslundii*; 750 and 400 nM for *V. parvula*; 300, 300, and 200 nM for *A. actinomycetemcomitans*; 300, 300, and 300 nM for *P. gingivalis;* and 600, 600, and 300 nM for *F. nucleatum*), as well as 2 µL of DNA of the corresponding samples. The negative control was 2 µL of sterile water (Water PCR grade, Roche). The samples were subjected to an initial amplification cycle at 95 °C for 10 min, followed by 45 cycles at 95 °C for 15 s and 60 °C for 1 min. The analyses were performed with a LightCycler® 480 II thermal cycler (Roche). Plates LightCycler® 480 Multiwell Plate 384 (Roche), sealed with qPCR Adhesive Clear Seals (4titude), were employed.

Each DNA sample was analyzed in duplicate. The value of the quantification cycle (Cq) was determined using a computer software (LC 480 Software 1.5, Roche Diagnostic GmbH, Mannheim, Germany) based on standard curves. The correlation between the Cq values and the CFU mL^{-1} was generated automatically through the software (LC 480 Software 1.5, Roche).

2.8. Data Analysis

The primary outcome variable to compare biofilms formed over TiD surfaces exposed to the different NPs was CFUs mL^{-1} of the six bacterial species presented in the biofilms at 12, 24, 48, and 72 h. The secondary outcomes were cell vitality (72 h), thickness (72 h), and morphological appearance of the biofilms (12, 24, 48, and 72 h).

Data were expressed as means and standard deviations (SD). The Kolmogorov–Smirnov test was used to assess data normality. The non-parametric analyses Kruskal-Wallis ANOVA and pairwise Mann–Whitney comparisons were used. Cell vitality and the thickness of the formed biofilms at 72 h ($n = 6$) on the different treated discs were compared to the control group and also the effects of NPs at different exposure times on CFU mL^{-1} ($n = 9$), in this case, the different time periods were analyzed separately. Statistical significance was set at $p < 0.05$, except for post hoc comparisons, where a Bonferroni correction was applied, and significance was set at $p < 0.01$. The morphological appearance of the biofilms, in the different groups, was described as a secondary outcome variable. The software package (IBM SPSS Statistics 24.0, IBM Corporation, Armonk, NY, USA) was used for all data analysis.

3. Results

3.1. Morphological Analysis of Biofilms by Scanning Electron Microscope (SEM)

After 12 h of biofilm growth, the SLA TiD surfaces in the negative control group showed the typical morphology of early biofilm formation, with the presence of individual bacterial cells, bacterial chains, and bacterial co-aggregates (Figure 1A,B). At this stage, *F. nucleatum* was easily identified through its characteristic fusiform bacillus appearance.

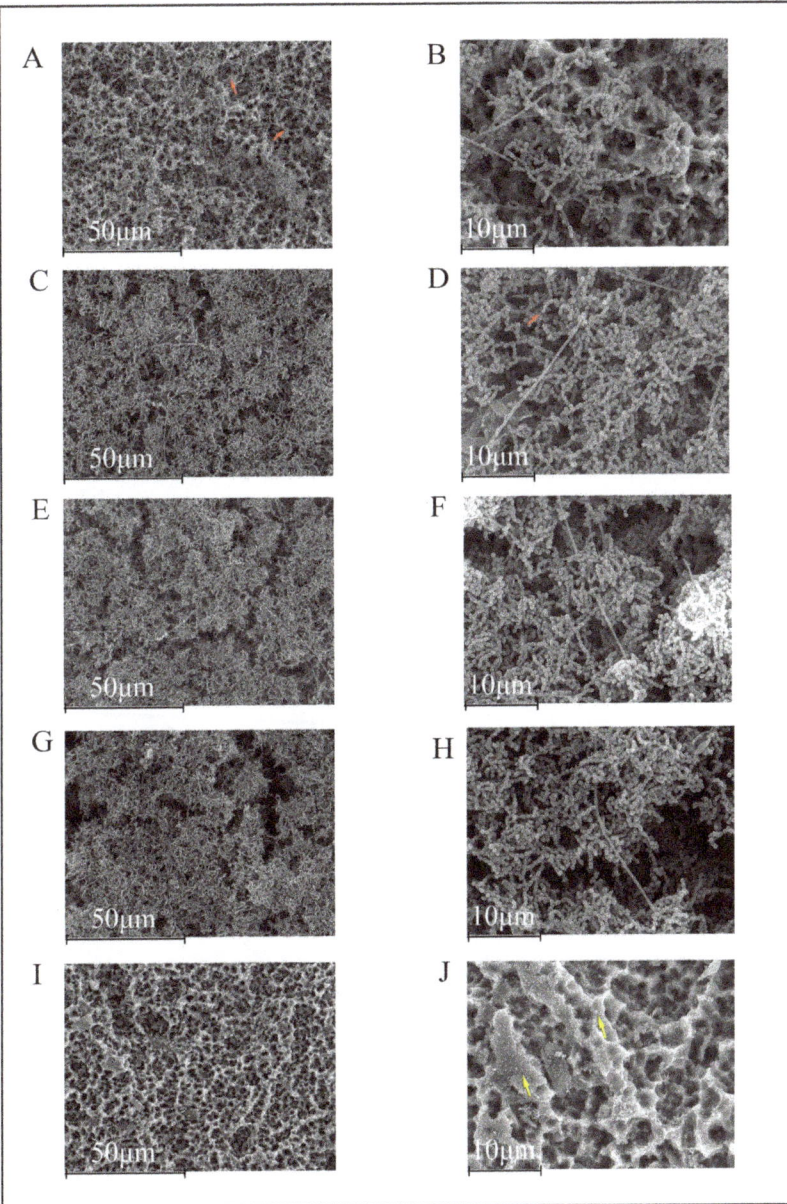

Figure 1. Scanning electron microscopy images of 12 h biofilms on titanium discs (TiDs). *F. nucleatum* is recognized due to its morphological fusiform bacillus appearance. Bacteria colonized the entire rough surface of the phosphate buffer saline (PBS, negative control) coated TiDs (**A**,**B**). TiDs covered with undoped nanoparticles (NPs) (**C**,**D**), calcium nanoparticles (**E**,**F**), and zinc nanoparticles (**G**,**H**) presented a similar pattern of bacterial presence and distribution. The surfaces of TiDs covered with doxycycline nanoparticles (**I**,**J**) did not present biofilm formation. Bacilli and cocci are marked with red arrows. The NPs are sometimes visible (yellow arrows). Magnification: (**A**,**C**,**E**,**G**,**I**) 1000×; (**B**,**D**,**F**,**H**,**J**) 3000×.

TiDs covered with Un-NPs, Ca-NPs, and Zn-NPs presented a similar pattern of biofilm formation, although with lesser bacterial biomass as compared with the negative control group (uncovered TiDs) (Figure 1C–H). However, as compared with TiDs covered with Dox-NPs, these TiDs lacked biofilm formation, with evidence of the presence of NPs coating the SLA titanium surfaces (Figure 1I,J).

As compared with 12-hour biofilms, the 24-hour biofilms in the negative control group presented thicker biomass and larger amounts of *F. nucleatum* (Figure 2A,B). In Un-NPs, Zn-NPs, and Ca-NPs, the structure was similar, but with a higher bacteria load (Figure 2C–H). In the Dox-NPs treated surfaces, there was no biofilm formation, only depicting some isolated bacterial cells on the disc surfaces (Figure 2I–J).

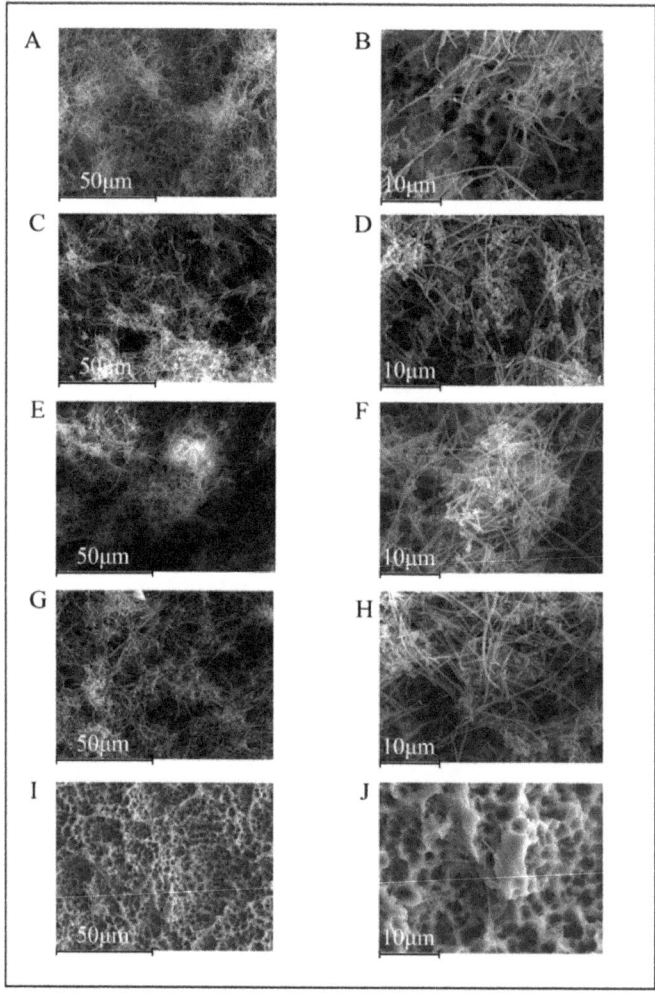

Figure 2. Scanning electron microscopy images of biofilms, after 24 h of development on titanium discs (TiDs). Discs with undoped nanoparticles (**C,D**), calcium nanoparticles (**E,F**), and zinc nanoparticles (**G,H**), demonstrated a similar biofilm structure (**A,B**), with higher amounts of bacteria, and covering the entire surface of the TiDs. However, the discs with doxycycline nanoparticles evidenced the surfaces of the TiDs free from biofilm formation, with only a few isolated bacterial cells (**I,J**). Magnification: (**A,C,E,G,I**) 1000×; (**B,D,F,H,J**) 3000×.

At 48 h, the morphological characteristics of the biofilms in the negative control discs were similar, with evidence of bacterial stacks and tunnel formation, the typical features of mature biofilms (Figure 3A,B). In the Un-NPs group, the entire surface of the TiD was covered with biofilm growth, in contrast with Ca-NPs and Zn-NPs TiDs, where some areas of titanium free of bacterial growth were observed (Figure 3C–H). The TiDs covered with Dox-NPs was again free of biofilm formation, with the presence of only a few scattered deposits and cells were visible (Figure 3I,J).

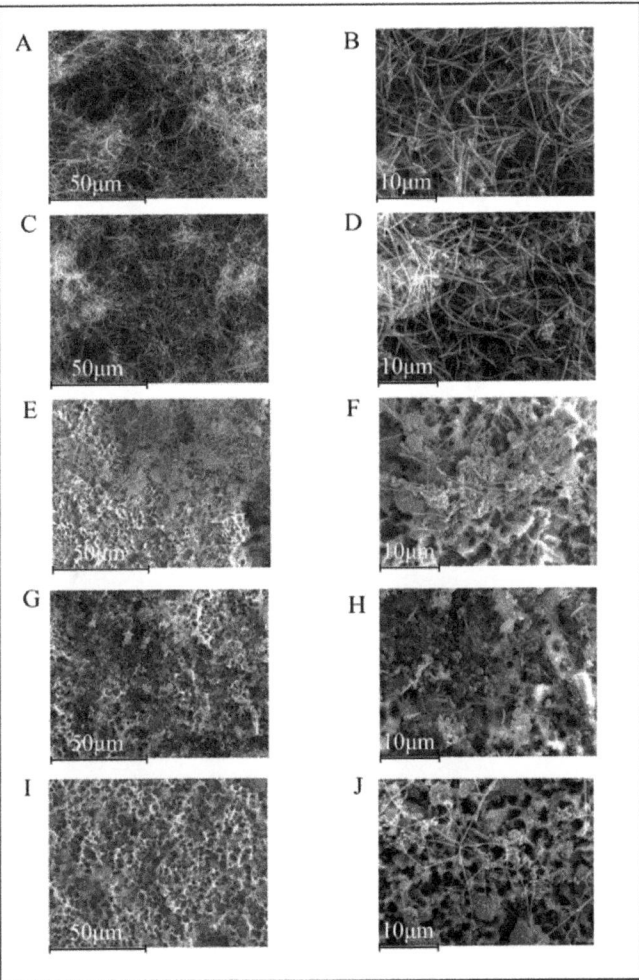

Figure 3. Scanning electron microscopy images of 48 h biofilms depicting increased area of the titanium disc covered by bacteria. In the negative control group (**A,B**) and the undoped nanoparticles (NPs) (**C,D**), the biofilm was spread throughout the disc surface, while in calcium-doped NPs (**E,F**) and zinc-doped NPs (**G,H**), some areas of titanium were still uncovered. Titanium disc covered with doxycycline-doped NPs (**I,J**) did not show biofilm formation, but rather a few scattered visible cells. Magnification: (**A,C,E,G,I**) 1000×; (**B,D,F,H,J**) 3000×.

At 72 h, a mature biofilm was present in the negative control group (Figure 4A,B). In addition, on the TiDs with Un-NPs, Ca-NPs, and Zn-NPs, the biofilms developed a similar structure, but still isolated areas of the TiDs were free from biofilm deposits (Figure 4C–H). On the TiDs with Dox-NPs, there was no biofilm formation (Figure 4I,J).

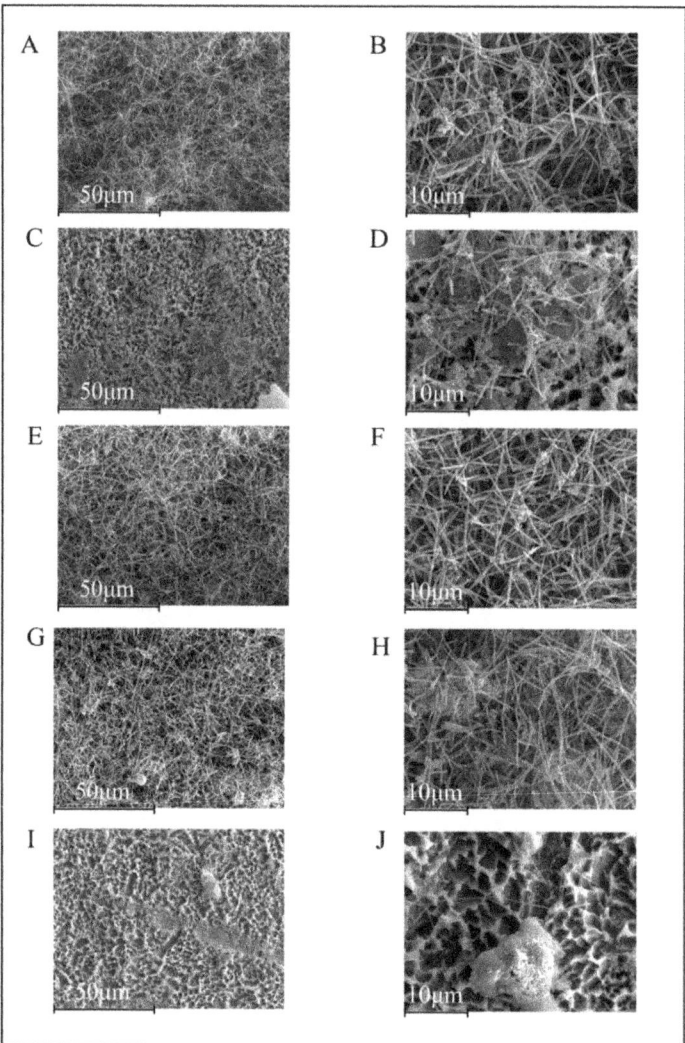

Figure 4. Scanning electron microscopy images of 72 h biofilms on titanium discs coated with: (**A,B**) phosphate buffered saline (PBS); (**C,D**) undoped nanoparticles (NPs); (**E,F**) calcium NPs; (**G,H**) zinc NPs; (**I,J**) doxycycline NPs. A mature biofilm could be seen in the negative control group (**A,B**). In unloaded NPs, calcium NPs and zinc NPs (**C–H**), there were similar biofilm structures and development, with isolated areas of the discs free from biofilm growth. Doxycycline NPs was the only group without an observable biofilm on the coated titanium surface. Magnification: (**A,C,E,G,I**) 1000×; (**B,D,F,H,J**) 3000×.

3.2. Analysis of Biofilms Vitality and Thickness by Confocal Laser Scanning Microscopy (CLSM)

After 72 h of biofilm development, biofilms were similar in thickness in all groups (ranging between 16 and 23 μm) ($p > 0.05$), although with statistically significant higher dead cell biomass in the coated TiDs as compared with the negative control. Dead cells percentages were 70.9%, 80.7%, 70.5%, and 85.9% for Un-NPs, Ca-NPs, Zn-NPs, and Dox-NPs, respectively, versus 47.2% in the negative control (Figure 5). Biofilms in TiDs doped with Dox-NPs also developed a statistically significantly lower ratio of viable/dead cells as compared with the negative control group (Control 2.16 and Dox-NPs 0.21) ($p < 0.01$) (Table 1).

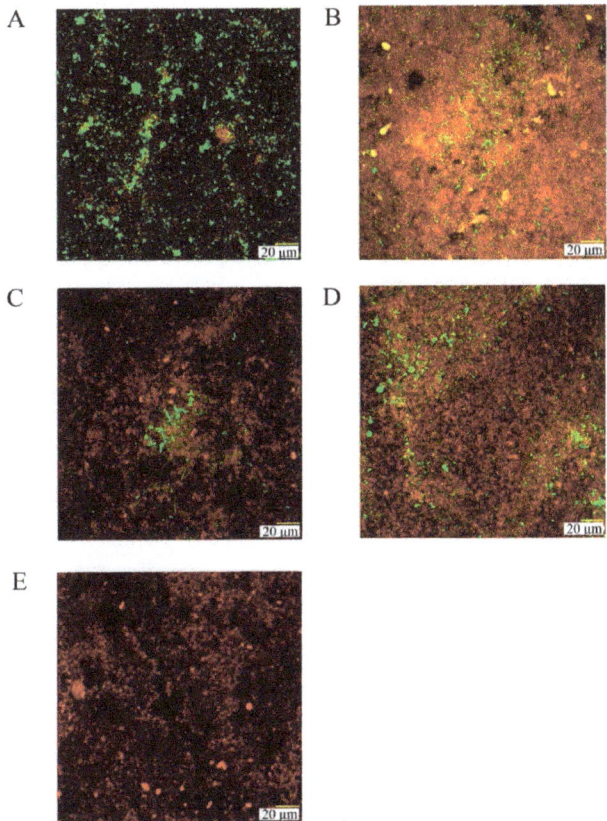

Figure 5. Confocal laser scanning micrographs after 72 h of growth on titanium discs (TiDs) coated with: (**A**) phosphate buffer saline (PBS) as control; (**B**) undoped nanoparticles (NPs); (**C**) NPs doped with calcium; (**D**) NPs doped with zinc; (**E**) NPs functionalized with doxycycline. A LIVE/DEAD® BacLight™ Bacterial Viability Kit was used to assess the vitality of cells. TiDs coated with NPs presented a reduced cell vitality in their biofilms as compared with control groups (living cells are presented in green and dead cells in red).

Table 1. Mean (standard deviation) of the viable and dead cell biomass, viable/dead cell ratios, percentages of dead and living cells, and biofilm thickness measured by confocal laser scanning microscopy ($n = 6$), for the different experimental groups. Titanium discs were treated with phosphate buffer saline (control) or covered with the four types of nanoparticles (NPs), i.e., undoped NPs (Un-NPs), NPs doped with calcium (Ca-NPs), NPs doped with zinc (Zn-NPs), and NPs doped with doxycycline (Dox-NPs).

	Control	Un-NPs	Ca-NPs	Zn-NPs	Dox-NPs
Viable cell biomass	68,750.66 (56,340.99)	96,387.27 (68,263.95)	89,579.52 (151,360)	67,989.72 (56,604.86)	30,954.77 (36,063.15)
Dead cell biomass	48,959.41 (45,304.36)	217,722.52 * (112,422.09)	214,538.38 * (102,620.69)	162,149.91 (107,249.13)	184,033.22 * (74,629.22)
Viable/dead ratio	2.16 (3.03)	0.47 (0.32)	0.34 (0.45)	0.41 (0.21)	0.21 (0.27) *
% Dead cells	47.21%	70.89%	80.70%	70.50%	85.87%
% Viable cells	52.79%	29.11%	19.30%	29.50%	14.13%
Thickness (µm)	23 (5.1)	21 (2.5)	23 (11.2)	22 (6.1)	16 (4.5)

* Statistically significant differences as compared with a negative control ($p < 0.01$). Numbers in parentheses are standard deviation values.

3.3. Bacterial Load and Presence of Specific Bacteria Analysis by DNA Isolation and Quantitative Polymerase Chain Reaction (qPCR)

The effect of the tested NPs on bacterial counts (CFUs/mL) of the six tested species grown in the biofilm model is shown in Table 2.

Table 2. Number of bacteria (colony forming units (CFU)/biofilm), expressed as mean and standard deviations (SDs) of *S. oralis*, *A. naeslundii*, *V. parvula*, *F. nucleatum*, *A. actinomycetemcomitans*, and *P. gingivalis* grown as multispecies biofilm at the different times of incubation, measured by quantitative real-time polymerase chain reaction (qPCR) ($n = 9$ for each incubation time and material). Titanium discs (TiDs) were used as a negative control as compared with TiDs coated with non-doped nanoparticles (Un-NPs), or doped with calcium (Ca-NPs), zinc (Zn-NPs), and doxycycline (Dox-NPs).

Bacterial Species	Time of Incubation	Number of Bacteria (CFU/Biofilm, Expressed as Mean (SD))				
		Control	Un-NPs	Ca-NPs	Zn-NPs	Do×-NPs
So	12 h	4.36×10^7 (1.12×10^7)	7.99×10^7 (1.89×10^7) *	8.80×10^7 (1.94×10^7) *	1.05×10^8 (3.28×10^7) *	2.03×10^4 (1.30×10^4) *
	24 h	4.92×10^7 (3.92×10^7)	1.08×10^8 (6.78×10^7)	1.09×10^8 (7.19×10^7)	7.99×10^7 (4.58×10^7)	7.75×10^5 (1.19×10^6) *
	48 h	4.02×10^7 (2.17×10^7)	4.45×10^7 (4.65×10^7)	5.45×10^7 (2.34×10^7)	5.19×10^7 (3.07×10^7)	6.63×10^4 (6.83×10^4) *
	72 h	7.42×10^7 (3.19×10^7)	1.33×10^8 (5.75×10^7)	9.88×10^7 (3.59×10^7)	1.37×10^8 (5.35×10^7) *	1.97×10^5 (2.76×10^5) *
An	12 h	3.47×10^5 (2.28×10^5)	5.11×10^5 (3.62×10^5)	4.98×10^5 (2.35×10^5)	4.68×10^5 (2.41×10^5)	3.72×10^4 (2.17×10^4) *
	24 h	9.68×10^6 (8.94×10^6)	1.10×10^7 (1.06×10^7)	1.30×10^7 (1.49×10^7)	7.21×10^6 (4.50×10^6)	4.64×10^4 (9.98×10^3) *
	48 h	4.14×10^6 (2.47×10^6)	3.90×10^6 (2.90×10^6)	4.37×10^6 (3.14×10^6)	4.34×10^6 (2.92×10^6)	5.10×10^4 (2.41×10^4) *
	72 h	7.72×10^6 (4.92×10^6)	8.87×10^6 (5.90×10^6)	7.74×10^6 (4.73×10^6)	9.71×10^6 (3.82×10^6)	5.06×10^4 (9.75×10^3) *

Table 2. Cont.

Bacterial Species	Time of Incubation	Number of Bacteria (CFU/Biofilm, Expressed as Mean (SD))				
		Control	Un-NPs	Ca-NPs	Zn-NPs	Do $\times$ -NPs
Vp	12 h	3.98×10^6 (5.02×10^6)	4.67×10^6 (5.14×10^6)	5.79×10^6 (6.80×10^6)	1.28×10^7 (2.09×10^7)	9.72×10^3 (5.71×10^3) *
	24 h	2.71×10^8 (2.49×10^8)	2.93×10^8 (2.43×10^8)	3.87×10^8 (3.30×10^8)	2.87×10^8 (3.15×10^8)	3.15×10^5 (4.41×10^5) *
	48 h	3.39×10^8 (2.88×10^8)	3.03×10^8 (4.47×10^8)	3.68×10^8 (2.19×10^8)	3.06×10^8 (2.59×10^8)	8.82×10^5 (1.17×10^6) *
	72 h	8.41×10^8 (6.59×10^8)	9.66×10^8 (8.82×10^8)	5.55×10^8 (2.75×10^8)	9.60×10^8 (6.83×10^8)	6.87×10^5 (7.52×10^5) *
Fn	12 h	2.56×10^6 (1.40×10^6)	5.93×10^6 (4.98×10^6)	4.35×10^6 (2.58×10^6)	8.63×10^6 (7.15×10^6)	4.95×10^5 (2.71×10^5) *
	24 h	6.41×10^6 (3.73×10^6)	1.34×10^7 (1.16×10^7)	1.02×10^7 (3.18×10^6)	1.23×10^7 (1.03×10^7)	2.17×10^5 (1.34×10^5) *
	48 h	2.20×10^7 (1.85×10^7)	2.45×10^7 (3.03×10^7)	2.77×10^7 (1.86×10^7)	2.65×10^7 (2.33×10^7)	2.81×10^5 (2.95×10^5) *
	72 h	3.95×10^7 (3.08×10^7)	4.30×10^7 (3.32×10^7)	3.19×10^7 (7.63×10^6)	5.20×10^7 (3.07×10^7)	2.78×10^5 (1.23×10^5) *
Aa	12 h	1.18×10^7 (1.70×10^7)	2.39×10^7 (3.41×10^7)	1.93×10^7 (2.69×10^7)	2.51×10^7 (3.75×10^7)	9.16×10^5 (1.32×10^6)
	24 h	2.48×10^6 (1.78×10^6)	5.42×10^6 (3.61×10^6)	5.04×10^6 (3.52×10^6)	5.76×10^6 (4.63×10^6)	1.25×10^5 (7.29×10^4) *
	48 h	3.64×10^5 (2.46×10^5)	5.88×10^5 (3.87×10^5)	7.16×10^5 (6.27×10^5)	5.60×10^5 (3.63×10^5)	1.10×10^5 (6.76×10^4) *
	72 h	8.70×10^6 (1.10×10^7)	8.84×10^6 (1.70×10^7)	3.97×10^6 (3.42×10^6)	5.67×10^6 (4.60×10^6)	9.82×10^4 (3.88×10^4) *
Pg	12 h	5.60×10^5 (4.07×10^5)	1.26×10^6 (4.74×10^5) *	1.17×10^6 (6.64×10^5)	1.52×10^6 (5.10×10^5) *	3.30×10^5 (1.49×10^5)
	24 h	8.22×10^5 (9.92×10^5)	2.13×10^6 (1.84×10^6)	2.21×10^6 (2.17×10^6)	2.10×10^6 (2.05×10^6)	2.47×10^5 (2.19×10^5)
	48 h	1.81×10^6 (2.42×10^6)	2.32×10^6 (3.18×10^6)	2.65×10^6 (4.17×10^6)	2.92×10^6 (5.33×10^6)	2.71×10^5 (2.77×10^5)
	72 h	2.18×10^7 (2.78×10^7)	5.30×10^7 (7.75×10^7)	5.11×10^7 (7.48×10^7)	9.03×10^7 (1.34×10^8)	5.75×10^5 (8.43×10^5) *

* Statistically significant differences as compared with negative control titanium discs ($p < 0.01$). No statistical differences were found among Un-NPs, Ca-NPs, and Zn-NPs. Numbers in parentheses are standard deviation values. So, Streptococcus oralis; An, Actinomyces naeslundii; Vp, Veillonela parvula; Fn, Fusobacterium nucleatum; Pg, Porphyromonas gingivalis; Aa, Aggregatibacter actinomycetemcomitans.

In 12-hour biofilms, the counts of specific bacterial species were significantly reduced in the Dox-NPs as compared with the negative control group for *F. nucleatum*, *A. naeslundii*, *S. oralis*, and *V. parvula* ($p < 0.001$). In the coated-NP groups, with Un-NPs, Ca-NPs, and Zn-NPs, there was a significant increase in the bacterial load of *S. oralis* ($p < 0.001$) as compared with the negative control group. No statistically significant differences were found among Un-NPs, Ca-NPs, and Zn-NPs.

In 24-hour biofilms, counts of specific bacterial species were significantly reduced in the Dox-NPs as compared with the negative control group, for all target species, except for *P. gingivalis* ($p = 0.136$). Dox-NPs reduced bacterial load in all cases ($p < 0.001$).

In 48-hour biofilms, all tested species, except *P. gingivalis* ($p < 0.04$), were significantly reduced with Dox-NPs as compared with the control group ($p < 0.001$).

In 72-hour biofilms, all tested species were significantly reduced with Dox-NPs as compared with the control group ($p < 0.01$).

4. Discussion

The present investigation, using an in vitro subgingival biofilm model, has demonstrated the antibacterial effect of coating TiDs with NPs, when NPs are loaded with doxycycline. Discs covered with Dox-NPs, as compared with negative control discs, demonstrated statistically significant reduced bacterial vitality, lower live/dead cells ratio, and significant reductions in bacterial load of all tested species, after 72 h. Furthermore, when observed with SEM, no relevant biofilm formation was identified on the Dox-NPs TiDs.

NPs have previously been found to be non-toxic and non-apoptotic after being tested with a human fibroblast cell line [19]. NPs are composed of 2-hydroxyethyl methacrylate, ethylene glycol dimethacrylate, and methacrylic acid, but the synthesis process is characterized by an efficient method, performed in the absence of harmful solvents or non-polymerized compounds, which later may interfere with cellular biological processes and cytocompatibility. Taking this into account, and assuming that dissolution of these particles is unlikely, and also considering that un-polymerized monomers are absent, all toxicity observed in bacterial cells may not be related to the polymer compounds of NPs, but to the number of ions or doxycycline loaded on NPs. It may also be that NPs are physically interrupting some biological bacterial processes.

In the Dox-NPs group, the 48- and 72-hour biofilms showed scarce biofilm formation, probably due to the presence of immature and weak biofilms, which become easily detached during SEM preparation, as previously reported in the literature [20]. In contrast, the morphology of the developed biofilms on the discs covered with Un-NPs, Ca-NPs or Zn-NPs were not significantly altered. Similarly, there were no differences in biofilm thicknesses as compared the different NPs, with and without doping, versus the negative control TiDs. This effect may be explained since the initial bacterial adhesion and growth is determined by physical forces, due to the tribological properties of the surfaces [26]. The tested titanium surfaces, covered or not by NPs, were made of grade 2 titanium Straumann SLA® surfaces, with a mean roughness (Ra) of 1.50 µm ± 0.11 and a three-dimensional topography with vertical changes (Rz) of around 20 µm [27,28]. Roughness microtopography not only facilitates initial bacterial adhesion, but also serves as a surface protection for initial cell growth and biofilm development, thus, providing anchorage to the bacterial community [26].

Moreover, observations with CLSM also resulted in no significant differences in viable cells between the different NP-coated groups as compared with the negative control, although the percentage of dead cells and the dead/viable cell ratio were significantly higher in the Dox-NPs group. Sánchez et al. 2019 [20] found similar results, when these NPs were applied onto hydroxyapatite discs. However, the bacterial mortality in the present research was about 10% lower than in the previous report. This may be due to the higher roughness of SLA titanium disc as compared with hydroxyapatite discs [28]. Similarly, the Dox-NPs demonstrated, in the 72-h biofilms, a statistically significant reduction in bacterial load as compared with the control group. These results clearly demonstrate the antibacterial potential of these Dox-NPs, probably due to the bactericidal effect of the loaded antibiotic [29]. Doxycycline may act against most bacteria by inhibiting the microbial protein synthesis. The mechanism of action is a result of binding the ribosome, to prevent ribonucleic acid synthesis by avoiding addition of more amino acid to the polypeptide [29]. Interestingly, clinical trials have shown that the combination of local doxycycline application, as an adjunct with mechanical debridement, did not always show a significant effect in peri-implantitis management [30], which may be explained by a limited diffusion of the antibiotic in biofilms forming within titanium surface irregularities. This supports the idea that the slow release of drug delivered from the titanium surface, at the bottom of the biofilm, may be more appropriate for this treatment indication. The use of polyglycolic acid (PLGA) nanospheres loaded with doxycycline has been clinically tested in periodontitis and peri-implantitis patients with promising results [31,32]. Lecio et al. (2020) [31] used 20% doxycycline-loaded PLGA nanospheres, as an adjunctive therapy for periodontitis, and found positive results in terms of reductions in bleeding on probing, probing pocket depths, and clinical attachment loss. The NPs used in the present study are non-resorbable materials with a burst release of doxycycline up to 7 days, and a maintained release of about 8 µg/mL for 21 days [21]. This concentration is high above the minimum inhibitory concentration (MIC) reported by Kulik et al. (2019), as required to inhibit the growth of 90% of organisms (1 µg/mL of doxycycline) for *P. gingivalis* and *A. actinomycetemcomitans* in a planktonic state [33]. In a single species biofilm, *P. gingivalis* showed an increased MIC of 12.5 µg/mL for doxycycline, while the MIC for *A. actinomycetemcomitans* was 20 µg/mL [34].

Although Dox-NPs have shown antibacterial effects in previous reports [19,20], these antimicrobial surface delivery systems need to be constructed considering other characteristics, such as: (1) enhancing or, at least, maintaining the chemical and physical properties of the titanium surfaces; (2) non-cytotoxic effect to host tissues; and (3) predictable long-term drug release with an effective and stable drug concentration to avoid bacterial resistance [35,36]. Toledano-Osorio et al., in 2018 [21], showed that Zn-NPs in solution (10 mg/mL) produced a sustained release of Zn^{2+} for up to 28 days, reaching a peak of 0.044 µg/mL. In the case of Ca-NPs, the release was 2.03 µg/mL at 28 days. Interestingly, Navarro-Requena et al. (2018) showed that Ca^{2+} concentrations that ranged between 100 and 150 µg/mL exerted, in dermal fibroblasts, a higher metabolic state, migration, collagen production, and, in general, an increase in gene expression related to wound healing [37].

In the present investigation, NPs were doped with calcium, zinc, and doxycycline, and were tested for their anti-biofilm effect using a multispecies in vitro biofilm model. Most of the previously introduced treatments/coatings have been tested using mainly single-species biofilms [36], which may be useful for initial screening purposes, but proper evaluations should take into account the diversity of the oral microbiota [36]. Still, the results from the present investigation should be interpreted with caution due to the limitations of the in vitro model. In this case, only six bacterial species were used for biofilm formation. In addition, dental implants with different surfaces, with different roughness and metal alloys, are available in the market and the present results may only be valid for the specific titanium surface tested. This demonstrated antibacterial and antibiofilm effect of the NPs, especially those doped with doxycycline, should be further investigated in more advanced preclinical and clinical research models.

One of the main advantages of the tested NPs is that they may be easily employed clinically, for example, NPs suspended on PBS may be spread with a micro-brush onto titanium surfaces during a surgical intervention for the treatment of peri-implantitis. During surgery, once access is gained to the affected area, and once the affected/infected soft tissue is eliminated and the implant surface is disinfected/decontaminated, NPs may be easily applied onto the titanium surface.

Doped nanoparticles may have antimicrobial properties as well as anti-inflammatory and healing-promoting activities. In fact, zinc- and calcium-doped NPs may have the ability to sequester calcium and phosphate onto their surfaces, when immersed in simulated body fluid solution [17–19] and, hence, could promote bone regeneration. Similarly, doxycycline has a broad-spectrum antibiotic effect, as well as the ability to reduce bone loss [38] and promote bone formation by reducing inflammation and osteoclastogenesis [39]. It has been recently found that doxycycline may increase up to 20 times the gene expression of OPG/RANKL ratio in cultured osteoblasts, favoring bone formation [40]. It may also act as an immunomodulatory agent [41,42], promoting bone healing. These properties should also be investigated using the appropriate research models, since peri-implantitis is a chronic inflammatory disease.

5. Conclusions

Within the limitations of the present in vitro study, non-resorbable polymeric nanoparticles doped with doxycycline were able to decrease the bacterial load in biofilms and alter their dynamics of formation. These doxycycline functionalized nanoparticles should be further investigated as a potential useful tool in the treatment of peri-implant diseases.

Author Contributions: Conceptualization, J.B., M.T.-O., M.T., R.O., M.S. and D.H.; methodology, J.B., L.V., M.T.-O., E.F., A.L.M.-C., M.S. and D.H.; validation, J.B., L.V., M.T.-O., E.F., A.L.M.-C., M.T., R.O., M.S. and D.H.; formal analysis, J.B., M.T., E.F., R.O., M.S. and D.H.; investigation, J.B., L.V., M.T.-O., E.F., A.L.M.-C., M.T., R.O., M.S. and D.H.; resources, M.T., R.O., E.F., D.H. and M.S.; data curation, J.B., L.V., E.F., R.O., M.S. and D.H.; writing—original draft preparation, J.B., L.V., M.T.-O., E.F., A.L.M.-C., M.T., R.O., M.S. and D.H.; writing—review and editing, J.B., L.V., M.T.-O., E.F., A.L.M.-C., M.T., R.O., M.S. and D.H.; visualization, J.B., L.V., M.T.-O., E.F., A.L.M.-C., M.T., R.O., M.S. and D.H.; supervision, M.S., D.H., R.O., E.F. and M.T.; project administration, M.S., D.H., R.O. and M.T.; funding acquisition, M.S., D.H., R.O. and M.T. All authors have read and agreed to the published version of the manuscript.

Funding: This research was funded by the Ministry of Economy and Competitiveness (MINECO) and the European Regional Development Fund (FEDER), grant number (PID2020-114694RB-I00 MINECO/AEI/FEDER/UE). M. Toledano-Osorio holds a FPU fellowship from the Ministry of Universities (FPU20/00450).

Institutional Review Board Statement: Not applicable.

Informed Consent Statement: Not applicable.

Data Availability Statement: The data presented in this study are available on request from the corresponding author. The data are not publicly available due to general data protection restrictions.

Conflicts of Interest: The authors declare no conflict of interest.

References

1. Zitzmann, N.U.; Hagmann, E.; Weiger, R. What is the prevalence of various types of prosthetic dental restorations in Europe? *Clin. Oral Implant. Res.* **2007**, *18* (Suppl. 3), 20–33. [CrossRef]
2. Hanif, A.; Qureshi, S.; Sheikh, Z.; Rashid, H. Complications in implant dentistry. *Eur. J. Dent.* **2017**, *11*, 135–140. [CrossRef]
3. Heitz-Mayfield, L.J.A.; Salvi, G.E. Peri-implant mucositis. *J. Periodontol.* **2018**, *89* (Suppl. 1), S257–S266. [CrossRef] [PubMed]
4. Schwarz, F.; Derks, J.; Monje, A.; Wang, H.-L. Peri-implantitis. *J. Periodontol.* **2018**, *89* (Suppl. 1), S267–S290. [CrossRef] [PubMed]
5. Berglundh, T.; Armitage, G.; Araujo, M.G.; Avila-Ortiz, G.; Blanco, J.; Camargo, P.M.; Chen, S.; Cochran, D.; Derks, J.; Figuero, E.; et al. Peri-implant diseases and conditions: Consensus report of workgroup 4 of the 2017 World Workshop on the Classification of Periodontal and Peri-Implant Diseases and Conditions. *J. Clin. Periodontol.* **2018**, *45* (Suppl. 20), S286–S291. [CrossRef] [PubMed]
6. Derks, J.; Tomasi, C. Peri-implant health and disease. A systematic review of current epidemiology. *J. Clin. Periodontol.* **2015**, *42* (Suppl. 16), S158–S171. [CrossRef]
7. Salvi, G.E.; Cosgarea, R.; Sculean, A. Prevalence of Periimplant Diseases. *Implant. Dent.* **2019**, *28*, 100–102. [CrossRef]
8. Rodrigo, D.; Sanz-Sánchez, I.; Figuero, E.; Llodrá, J.C.; Bravo, M.; Caffesse, R.G.; Vallcorba, N.; Guerrero, A.; Herrera, D. Prevalence and risk indicators of peri-implant diseases in Spain. *J. Clin. Periodontol.* **2018**, *45*, 1510–1520. [CrossRef]
9. Carcuac, O.; Abrahamsson, I.; Albouy, J.-P.; Linder, E.; Larsson, L.; Berglundh, T. Experimental periodontitis and peri-implantitis in dogs. *Clin. Oral Implant. Res.* **2013**, *24*, 363–371. [CrossRef]
10. Carcuac, O.; Berglundh, T. Composition of human peri-implantitis and periodontitis lesions. *J. Dent. Res.* **2014**, *93*, 1083–1088. [CrossRef]
11. Zitzmann, N.U.; Berglundh, T.; Marinello, C.P.; Lindhe, J. Experimental peri-implant mucositis in man. *J. Clin. Periodontol.* **2001**, *28*, 517–523. [CrossRef]
12. Salvi, G.E.; Aglietta, M.; Eick, S.; Sculean, A.; Lang, N.P.; Ramseier, C.A. Reversibility of experimental peri-implant mucositis compared with experimental gingivitis in humans. *Clin. Oral Implant. Res.* **2012**, *23*, 182–190. [CrossRef]
13. Figuero, E.; Graziani, F.; Sanz, I.; Herrera, D.; Sanz, M. Management of peri-implant mucositis and peri-implantitis. *Periodontol. 2000* **2014**, *66*, 255–273. [CrossRef] [PubMed]
14. Bhavikatti, S.K.; Bhardwaj, S.; Prabhuji, M.L.V. Current applications of nanotechnology in dentistry: A review. *Gen. Dent.* **2014**, *62*, 72–77. [PubMed]
15. Padovani, G.C.; Feitosa, V.P.; Sauro, S.; Tay, F.R.; Durán, G.; Paula, A.J.; Durán, N. Advances in Dental Materials through Nanotechnology: Facts, Perspectives and Toxicological Aspects. *Trends Biotechnol.* **2015**, *33*, 621–636. [CrossRef]
16. Ajay v, S.; Krunal K, M. Top-Down Versus Bottom-Up Nanoengineering Routes to Design Advanced Oropharmacological Products. *Curr. Pharm. Des.* **2016**, *22*, 1534–1545. [CrossRef]
17. Bueno, J.; Sánchez, M.C.; Toledano-Osorio, M.; Figuero, E.; Toledano, M.; Medina-Castillo, A.L.; Osorio, R.; Herrera, D.; Sanz, M. Antimicrobial effect of nanostructured membranes for guided tissue regeneration: An in vitro study. *Dent. Mater.* **2020**, *36*, 1566–1577. [CrossRef]

18. Fröber, K.; Bergs, C.; Pich, A.; Conrads, G. Biofunctionalized zinc peroxide nanoparticles inhibit peri-implantitis associated anaerobes and Aggregatibacter actinomycetemcomitans pH-dependent. *Anaerobe* **2020**, *62*, 102153. [CrossRef] [PubMed]
19. Osorio, R.; Alfonso-Rodríguez, C.A.; Medina-Castillo, A.L.; Alaminos, M.; Toledano, M. Bioactive Polymeric Nanoparticles for Periodontal Therapy. *PLoS ONE* **2016**, *11*, e0166217. [CrossRef]
20. Sánchez, M.C.; Toledano-Osorio, M.; Bueno, J.; Figuero, E.; Toledano, M.; Medina-Castillo, A.L.; Osorio, R.; Herrera, D.; Sanz, M. Antibacterial effects of polymeric PolymP-n Active nanoparticles. An in vitro biofilm study. *Dent. Mater.* **2019**, *35*, 156–168. [CrossRef]
21. Toledano-Osorio, M.; Babu, J.P.; Osorio, R.; Medina-Castillo, A.L.; García-Godoy, F.; Toledano, M. Modified Polymeric Nanoparticles Exert In Vitro Antimicrobial Activity Against Oral Bacteria. *Materials* **2018**, *11*, 1013. [CrossRef]
22. Medina-Castillo, A.L. Thermodynamic Principles of Precipitation Polymerization and Role of Fractal Nanostructures in the Particle Size Control. *Macromolecules* **2020**, *53*, 5687–5700. [CrossRef]
23. Medina-Castillo, A.L.; Fernandez-Sanchez, J.F.; Segura-Carretero, A.; Fernandez-Gutierrez, A. Micrometer and Submicrometer Particles Prepared by Precipitation Polymerization: Thermodynamic Model and Experimental Evidence of the Relation between Flory's Parameter and Particle Size. *Macromolecules* **2010**, *43*, 5804–5813. [CrossRef]
24. Sánchez, M.C.; Llama-Palacios, A.; Blanc, V.; León, R.; Herrera, D.; Sanz, M. Structure, viability and bacterial kinetics of an in vitro biofilm model using six bacteria from the subgingival microbiota. *J. Periodontal Res.* **2011**, *46*, 252–260. [CrossRef] [PubMed]
25. Singh, A.; Baylan, S.; Park, B.; Richter, G.; Sitti, M. Hydrophobic pinning with copper nanowhiskers leads to bactericidal properties. *PLoS ONE* **2017**, *12*, e0175428. [CrossRef]
26. Sánchez, M.C.; Llama-Palacios, A.; Fernández, E.; Figuero, E.; Marín, M.J.; León, R.; Blanc, V.; Herrera, D.; Sanz, M. An in vitro biofilm model associated to dental implants: Structural and quantitative analysis of in vitro biofilm formation on different dental implant surfaces. *Dent. Mater.* **2014**, *30*, 1161–1171. [CrossRef] [PubMed]
27. Garrett, T.R.; Bhakoo, M.; Zhang, Z. Characterisation of bacterial adhesion and removal in a flow chamber by micromanipulation measurements. *Biotechnol. Lett.* **2008**, *30*, 427–433. [CrossRef] [PubMed]
28. Rizo-Gorrita, M.; Luna-Oliva, I.; Serrera-Figallo, M.-A.; Torres-Lagares, D. Superficial Characteristics of Titanium after Treatment of Chorreated Surface, Passive Acid, and Decontamination with Argon Plasma. *J. Funct. Biomater.* **2018**, *9*, 71. [CrossRef]
29. Cha, J.-K.; Paeng, K.; Jung, U.-W.; Choi, S.-H.; Sanz, M.; Sanz-Martín, I. The effect of five mechanical instrumentation protocols on implant surface topography and roughness: A scanning electron microscope and confocal laser scanning microscope analysis. *Clin. Oral Implant. Res.* **2019**, *30*, 578–587. [CrossRef]
30. Gamal, A.Y.; Kumper, R.M. A novel approach to the use of doxycycline-loaded biodegradable membrane and EDTA root surface etching in chronic periodontitis: A randomized clinical trial. *J. Periodontol.* **2012**, *83*, 1086–1094. [CrossRef]
31. Toledano, M.; Osorio, M.T.; Vallecillo-Rivas, M.; Toledano-Osorio, M.; Rodríguez-Archilla, A.; Toledano, R.; Osorio, R. Efficacy of local antibiotic therapy in the treatment of peri-implantitis: A systematic review and meta-analysis. *J. Dent.* **2021**, *113*, 103790. [CrossRef] [PubMed]
32. Lecio, G.; Ribeiro, F.V.; Pimentel, S.P.; Reis, A.A.; da Silva, R.V.C.; Nociti-Jr, F.; Moura, L.; Duek, E.; Casati, M.; Casarin, R.C.V. Novel 20% doxycycline-loaded PLGA nanospheres as adjunctive therapy in chronic periodontitis in type-2 diabetics: Randomized clinical, immune and microbiological trial. *Clin. Oral Investig.* **2020**, *24*, 1269–1279. [CrossRef] [PubMed]
33. Moura, L.A.; Ribeiro, F.V.; Aiello, T.B.; Duek, E.A.D.R.; Sallum, E.A.; Nociti Junior, F.H.; Casati, M.Z.; Sallum, A.W. Characterization of the release profile of doxycycline by PLGA microspheres adjunct to non-surgical periodontal therapy. *J. Biomater. Sci. Polym. Ed.* **2015**, *26*, 573–584. [CrossRef]
34. Kulik, E.M.; Thurnheer, T.; Karygianni, L.; Walter, C.; Sculean, A.; Eick, S. Antibiotic Susceptibility Patterns of Aggregatibacter actinomycetemcomitans and Porphyromonas gingivalis Strains from Different Decades. *Antibiotics* **2019**, *8*, 253. [CrossRef] [PubMed]
35. Eick, S.; Seltmann, T.; Pfister, W. Efficacy of antibiotics to strains of periodontopathogenic bacteria within a single species biofilm—An in vitro study. *J. Clin. Periodontol.* **2004**, *31*, 376–383. [CrossRef] [PubMed]
36. Xing, R.; Witsø, I.L.; Jugowiec, D.; Tiainen, H.; Shabestari, M.; Lyngstadaas, S.P.; Lönn-Stensrud, J.; Haugen, H.J. Antibacterial effect of doxycycline-coated dental abutment surfaces. *Biomed. Mater.* **2015**, *10*, 055003. [CrossRef]
37. Souza, J.G.S.; Bertolini, M.M.; Costa, R.C.; Nagay, B.E.; Dongari-Bagtzoglou, A.; Barão, V.A.R. Targeting implant-associated infections: Titanium surface loaded with antimicrobial. *iScience* **2021**, *24*, 102008. [CrossRef]
38. Navarro-Requena, C.; Pérez-Amodio, S.; Castaño, O.; Engel, E. Wound healing-promoting effects stimulated by extracellular calcium and calcium-releasing nanoparticles on dermal fibroblasts. *Nanotechnology* **2018**, *29*, 395102. [CrossRef]
39. Botelho, M.A.; Martins, J.G.; Ruela, R.S.; Queiroz, D.B.; Ruela, W.S. Nanotechnology in ligature-induced periodontitis: Protective effect of a doxycycline gel with nanoparticules. *J. Appl. Oral Sci. Rev. FOB* **2010**, *18*, 335–342. [CrossRef]
40. Zhang, P.; Ding, L.; Kasugai, S. Effect of doxycycline doped bone substitute on vertical bone augmentation on rat calvaria. *Dent. Mater. J.* **2019**, *38*, 211–217. [CrossRef]

41. Toledano-Osorio, M.; Manzano-Moreno, F.J.; Toledano, M.; Osorio, R.; Medina-Castillo, A.L.; Costela-Ruiz, V.J.; Ruiz, C. Doxycycline-doped membranes induced osteogenic gene expression on osteoblastic cells. *J. Dent.* **2021**, in press. [CrossRef] [PubMed]
42. Toledano-Osorio, M.; Manzano-Moreno, F.J.; Toledano, M.; Medina-Castillo, A.L.; Costela-Ruiz, V.J.; Ruiz, C.; Osorio, R. Doxycycline-Doped Polymeric Membranes Induced Growth, Differentiation and Expression of Antigenic Phenotype Markers of Osteoblasts. *Polymers* **2021**, *13*, 1063. [CrossRef] [PubMed]

Review

Recent Advancements in Polysulfone Based Membranes for Fuel Cell (PEMFCs, DMFCs and AMFCs) Applications: A Critical Review

Rajangam Vinodh [1,†], Raji Atchudan [2,†], Hee-Je Kim [3,*] and Moonsuk Yi [1,*]

1. Department of Electronics Engineering, Pusan National University, Busan 46241, Korea; vinoth6482@gmail.com
2. Department of Chemical Engineering, Yeungnam University, Gyeongsan 38541, Korea; atchudanr@yu.ac.kr
3. Department of Electrical and Computer Engineering, Pusan National University, Busan 46241, Korea
* Correspondence: heeje@pusan.ac.kr (H.-J.K.); msyi@pusan.ac.kr (M.Y.)
† These authors contributed equally to this work.

Abstract: In recent years, ion electrolyte membranes (IEMs) preparation and properties have attracted fabulous attention in fuel cell usages owing to its high ionic conductivity and chemical resistance. Currently, perfluorinatedsulfonicacid (PFSA) membrane has been widely employed in the membrane industry in polymer electrolyte membrane fuel cells (PEMFCs); however, Nafion[TM] suffers reduced proton conductivity at a higher temperature, requiring noble metal catalyst (Pt, Ru, and Pt-Ru), and catalyst poisoning by CO. Non-fluorinated polymers are a promising substitute. Polysulfone (PSU) is an aromatic polymer with excellent characteristics that have attracted membrane scientists in recent years. The present review provides an up-to-date development of PSU based electrolyte membranes and its composites for PEMFCs, alkaline membrane fuel cells (AMFCs), and direct methanol fuel cells (DMFCs) application. Various fillers encapsulated in the PEM/AEM moiety are appraised according to their preliminary characteristics and their plausible outcome on PEMFC/DMFC/AMFC. The key issues associated with enhancing the ionic conductivity and chemical stability have been elucidated as well. Furthermore, this review addresses the current tasks, and forthcoming directions are briefly summarized of PEM/AEMs for PEMFCs, DMFCs, AMFCs.

Keywords: polysulfone; polymer electrolyte membrane; DMFCs; AMFCs; sulfonation; Nafion[TM]; fillers; inorganic/organic hybrid membranes

1. Introduction

1.1. Fuel Cells

The fuel cell is an electro-chemical energy conversion design; it alters the chemical energy of the reactants directly into electric energy along with heat and potable water. As global environmental and energy issues become more and more acute, incredible efforts are being made to explore new energy selections. As a new energy technology, fuel cells have been shown to be highly efficient and have an excellent ability to convert conventional fossil fuel energies due to low or zero-emission [1–3]. Fuel cells and batteries share multiple similarities: both are based on the anode-to-cathode electronic transfer principle and convert chemical energy into electric energy; they both require an electrolyte and external load to perform useful work and generate low DC voltages. Fuel cells are stacked similarly to batteries as well. Extensive power and voltage output is achieved by combining many cells in series. The main differences between fuel cells and batteries are the nature of their electrodes. Batteries use metallic anodes (lithium or zinc) and cathodes (generally metallic oxides). During operation, batteries consume the anode and the cathode, which will need recharge or replacement. In contrast, fuel cells operate with externally supplied reactants and do not consume any part working part of the cell. Therefore, fuel cells need no recharge and can continue operating as long as the reactant is supplied. Such repeated charging and discharging resulted in decreasing the life-time of the battery compared to that in the case of the fuel cell. In addition, fuel cells provide an inherently clean source of energy, with no adverse environmental impact during operation, as the byproducts are simply heat and water [4]. Nevertheless, the recent constraint in fuel cell commercialization stalks from the expensive nature of the raw materials (Nafion^{TM} electrolyte membrane and noble metal catalysts) and of the manufacturing method [5]. Furthermore, fuel cell electric vehicles (FCEV) are under progress by many automobile companies and are effectively verified due to their several advantages. Even though many advantages are in the fuel cells, there is a gap in the implementation of the fuel cells in on-road vehicles due to some practical issues.

1.2. Types of FCs

Fuel cells are divided into direct and indirect fuel cells according to their working temperature, the fuel cell components, and the type of electrolytes used. Fuel cells are classified according to their operational temperatures, such as low-temperature fuel cells and high-temperature fuel cells, as shown in Figure 1a [6]. The PEMFC, AMFC, and DMFC belong to low-temperature fuel cells. Molten carbonate fuel cell (MCFC), phosphoric acid fuel cell (PAFC), and solid oxide fuel cell (SOFC), fall into the high-temperature fuel cells group. In this cataloging, PEMFCs are more promising and consistent than other fuel cells owing to their versatile applications, extraordinary efficacy, and tiny emission of impurities, and can be the basis for DMFCs and AFCs. The acidic or alkaline concentrations are applied as electrolytes in fuel cells termed as mobile electrolyte systems, whereas electrolytes are immersed in a porous-based (pores enriched) material, defined as an inert/immobile electrolyte system or matrix system [7–10].

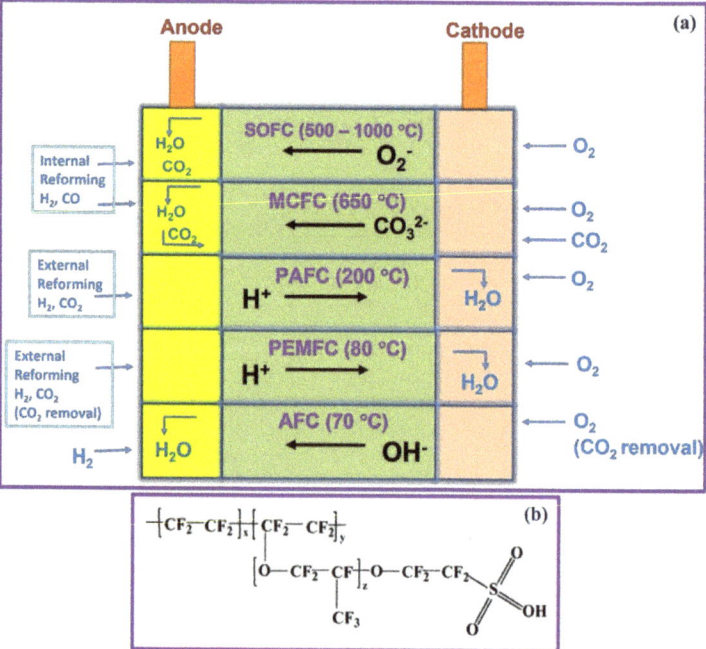

Figure 1. (a) Different types of fuel cell; (b) structure of PFSA.

1.2.1. PEMFCs

In a typical PEMFC, the cation exchange membrane (CEM or PEM) is accountable for the proton conductivity, which permits the passage of H$^+$ from anode to the cathode, establishing the essential component of the electrochemical device. In various types of fuel cells, membranes constructed with perfluorinatedsulfonicacid (PFSA) is predominantly employed because of its excellent proton conductivity and adequate chemical/mechanical characteristics; they are worked at temperatures between 120 and 180 °C in high pressure [11,12]. The structure of PFSA is illustrated in Figure 1b; the sulfonic acid group is connected to the perfluoroethereal side chains of the PFSA. Proton conductivity is due to the significant phase separation between hydrophilic and hydrophobic domains in PFSA when hydrated; and the chemical/mechanical stability is caused by the rigid structure of the polytetrafluoroethylene (PTFE) backbone and strong C-F bond even in the side chains. However, this type of membrane exhibited a severe defect at temperatures below zero degrees Celsius and above hundred degrees Celsius [13,14]. Another kind of membrane, Nafion^{TM}, developed and introduced by Dupont in the 1960s, has been extensively studied and is a commercially available proton-conducting membrane in PEMFC applications. Nafion^{TM} shows excellent characteristics, such as high electrochemical and chemical stability, low permeability to reactant species, selective and high ionic conductivity, and the ability to provide electronic insulation. However, the Nafion^{TM} membrane showed poor proton conductivity at higher temperatures due to dehydration of water, which controlled the number of water-filled channels [15–17]. To solve these problems, researchers have developed alternative ways of proposing other polymeric materials, such as SPEEK (sulfonated poly(ether ether ketone)) [18,19], polybenzimidazole (SPBI) [20,21], and polysulfone (SPSU or SPSF) [22,23]. These membranes show their strengths in different features of water uptake %, ionic conductivity, and mechanical and thermal stability. The pictorial illustration of PEMFC is depicted in Figure 2a along with cell reaction.

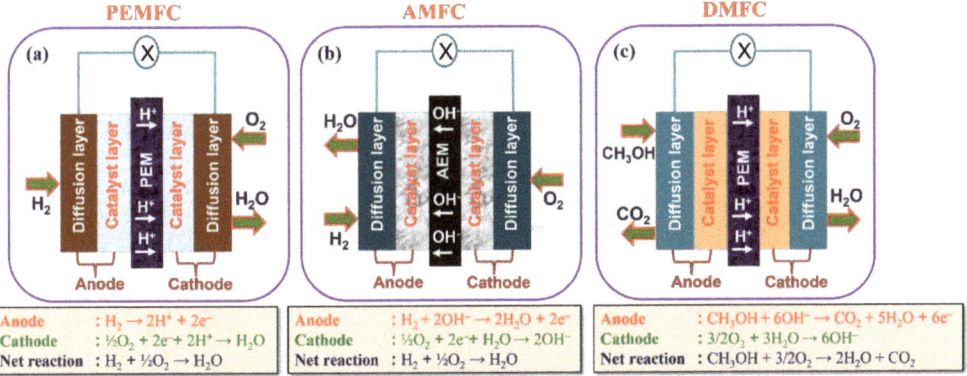

Figure 2. Schematic illustration of (**a**) PEMFC; (**b**) AMFC, and (**c**) DMFC with half-cell reaction.

1.2.2. DMFCs

The DMFCs have few merits of efficiently working at low temperature, easy strategy, and eco-friendly characteristics. The usage/handle of methanol is also easy since it exhibits liquid properties at ambient temperature. More specifically, unlike PEMFCs, aqueous methanol-based DMFCs do not require a humidification system and peculiar thermal management aids. They also have superior energy and power density as compared to indirect fuel cells and recently established lithium-ion batteries (LIBs). A plausible usage of the DMFC comprises portable electronic gadgets, military communications, transportation services, and traffic lights/signals [23–25]. The schematic of DMFC is presented in Figure 2c. The major issue with DMFCs is methanol cross-over as it permeates methanol along with water from anode to cathode direction. During DMFC operation, methanol cross-over outcomes in low power-output due to methanol oxidation at the cathode with the aid of cathode catalysts, leading to (i) electrode depolarization, (ii) mixed potential, consequently open-circuit voltage (OCV) of the DMFC less than 0.8 V, (iii) consuming of oxygen, (vi) CO poisoning, and (v) severe water accretion at the cathode, which restricts oxygen contact to cathode catalyst spots. In addition, the presence of excessive methanol cross-over lowers the overall performance of the fuel cell [26–28].

1.2.3. AMFCs

In principle, AMFC is a feasible substitute for PEMFC and is currently receiving new consideration. In AMFCs, the AEM conducts OH^- (hydroxide) or CO_3^{2-} (carbonate) anions while an electric current is flowing, which has numerous advantages, (i) in a high alkaline environment, both oxygen reduction reaction and methanol oxidation are more predominant; electro-osmotic drag by OH^- moves from cathode to anode, which reduces anode to cathode methanol cross-over, simplifies water management, and (ii) allows the use of non-noble metal catalysts. These AEMs are cheap and have improved mechanical/chemical characteristics compared to PEMs. In recent years, more research attempts have been performed to synthesize novel AEMs to enhance their ionic (OH^-) conductivity [29] along with alkaline stability [30,31]. Polymer back bones such as polystyrene (ethylene butylene) polystyrene [32], poly(2,6-dimethyl-1,4-phenylene oxide) [33], polystyrene [34], poly(ether ether ketone) [35], poly(vinyl alcohol) [36], and polyether sulfone [37,38] have been expansively explored to synthesize alkaline membranes. The afore-mentioned polymeric materials can be readily functionalized with the following cationic groups, quaternary phosphonium [39,40], guanidinium [41,42], quaternary ammonium [43], or imidazolium [44–46] which are accountable for creating the polymer backbone conductive.

1.3. Ion Exchange Membranes

For all FCs, the membrane (PEM or AEM) is the heart of the FC. It plays a prominent part in the transportation of ions within a fuel cell via the following aspects: (1) friction through the pore walls, (2) the energy of the membrane swelling process, (3) complete blockage of transport due to insufficient water absorption, (4) hydrophobic/hydrophilic contact between solvation shells and water dipoles, (5) effects of double-layer and (6) surface diffusion [47,48]. The main difference between CEMs and AEMs are tabulated in Table 1.

Table 1. Difference between cation and anion exchange membrane.

IEM	CEM	AEM
Counter ion	H^+ conductive	OH^- conductive
Ion-exchange group	$-SO_3^-$; $-PO_4^-$; $-CO_2^-$	Quaternary ammonium cation, 1-methyl pyridinium
Features	High ionic conductivity, excellent ionomer solution	Non-noble metal catalyst can be used. Oxygen reduction reaction and methanol oxidation reaction are more facile.
Issues	High-cost materials, fuel crossover, chemical, and mechanical stability, practical lifetime	Low ionic conductivity, low thermostability, influence of CO_2, durability, chemical, and mechanical stability

1.4. Preliminary Characteristic of IEM

The prepared IEMs were subjected to the following preliminary characterization studies: water uptake (WU), ion exchange capacity (IEC), ionic conductivity, permeability of methanol (p), and alkaline stability test to check the appropriateness of the IEMs in FC applications, and the pictorial protocol is illustrated in Figure 3.

Figure 3. Characteristics of IEMs for FC applications.

1.4.1. Water Uptake and Schroeder's Paradox

The WU of the IEM was measured by calculating the weights of the dry and wet membrane samples. The dry membrane weight (W_{dry}) is obtained by drying the sample at 100 °C for 12 h immediately before weighing it. The weight of the corresponding membrane in wet conditions (W_{wet}) is obtained by immersing the membrane sample in deionized water (DI water) at room temperature for about 1 day, wiping off the surface moisture with filter paper and then quickly weighing it. The water uptake (%) was determined from the subsequent equation [49]:

$$\text{WU (\%)} = \frac{W_{wet} - W_{dry}}{W_{dry}} \times 100\% \tag{1}$$

In addition, the sorption may be measured by bringing a membrane to equilibrium with a liquid by either immersion of the membrane into the liquid (directly) or by contact with the vapor phase (isopiestically). Since the solution, the vapor, and the sample are all in equilibrium, it is believed that there is no difference between the two methods. The uptake of water by PFSA from a liquid reservoir and a saturated vapor reservoir differs under the same conditions. This phenomenon is called Schroeder's paradox, and more recently, attempts have been made to explain this phenomenon theoretically.

1.4.2. Ion Exchange Capacity

IEC is a quantity of the capacity of an insoluble substance to endure ions displacement with formerly attached and lightly encapsulated into its architecture by oppositely charged ions existing in the adjacent solution. IEC was calculated by a back titration method with the following formula [50]:

$$\text{IEC}\left(\text{meq.g}^{-1}\right) = \frac{\text{Titre value} \times \text{Normality of tirant}}{\text{Membrane weight (dry)}} \tag{2}$$

1.4.3. Ionic Conductivity

The ionic conductivity of the IEM was measured by AC impedance spectroscopy. Prior to the testing, the membranes (IEMs) of various forms were fully hydrated overnight in DI water. The measuring device with IEM was positioned in DI water to maintain the relative humidity (RH) at 100% throughout the experiment. Membrane resistance was measured from the difference in the resistance between the blank cell and the one with IEM separates the counter electrode and working electrode compartment and is converted into ionic conductivity values using the below formula [51]:

$$\text{Ionic conductivity } (\text{S cm}^{-1}) = \frac{L}{R \times A} \quad (3)$$

where R is resistance of IEM (ohm); L is width of IEM (cm); A is area of IEM (cm^2).

1.4.4. Methanol Permeability

The methanol permeability (p) is studied at ambient temperature using a two-portion diffusion cell comprising of a collector (C) and a reservoir (R). C and R were separated by the investigated IEM, occupied with DI water, methanol, respectively. Both the portions are stirred continuously during the permeability test. The methanol permeability is calculated from the time versus concentration curve of the methanol collector slope values according to the following equation [52].

$$p \left(\text{cm}^2 \text{ s}^{-1}\right) = \frac{m \times V_C \times d}{A \times C_R} \quad (4)$$

where m represents the linear plot slope; V_C signifies the methanol solution volume in the C; A and d illustrate the area and thickness of the IEM; C_R is the methanol concentration in the tank.

1.4.5. Selectivity Ratio

Especially for DMFCs, the IEM must have two significant characteristics. The proton/hydroxide ionic conductivity should be maximal and have minimal methanol diffusion. Therefore, the higher the ratio of ionic conductivity to methanol permeability (termed as selectivity ratio), the better the IEM performance of the DMFC. This selectivity ratio indicates the performance of the IEM [53].

1.4.6. Oxidative Stability

Oxidative resistance is studied by Fenton's test in terms of weight loss over a period. In Fenton's reagent, degradation of the polymer is caused by free radicals attacking the electrophilic sites, leading to weight loss.

2. Polysulfone

Polysulfone is a commercially existing aromatic polymer. The relentless attention of the membrane researchers for PSU is because of its outstanding properties [54], such as soluble tendency in a wide range of solvents (dimethylformamide, dimethyl sulfoxide, halogen derivative, dimethyl acetamide, halogen derivatives), excellent film forming capacity, withstanding in high temperatures, wide range of operating pH, outstanding mechanical strength, and reasonable reactivity in aromatic electrophilic substitution reactions (acylation, chloromethylation, nitration, sulfonation, etc.) [55]. The chemical structure of polysulfone is shown in Figure 4. In this present review, we have seen the recent developments of the polysulfone-based membrane and its composites for PEMFCs, DMFCs, and AMFCs applications.

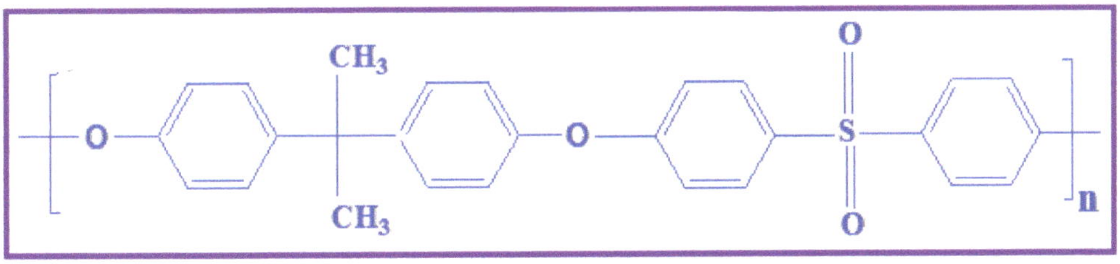

Figure 4. Chemical structure of PSU.

2.1. Membranes Derived from Polysulfone and Its Composites for PEMFCs Application

PEMFC technology has evolved quickly over the past 2 decades, with many advantages over traditional energy storage devices, such as batteries and internal combustion engines. PEMFCs are more energy-efficient related with diesel/gas engines. They also produce no hazardous by-products [56–58]. However, the practical feasibility of this technology is highly dependent on the PEM and its characteristics [59,60]. Hence, PEM is a crucial component in PEMFCs devices. The outcome of PEM depends not only on excellent mechanical and thermal resistance but also on the other characteristics, such as film-forming capacity, excellent proton conductivity, and reduced methanol cross-over [61,62]. Recently, nano fillers have been widely explored to adjust polymeric membranes to enhance the outcome of PEMs. These enhancements are reached by introducing a nonstop proton transfer path in the polymer environment and enhanced mechanical/ thermal characteristics of the polymer [63,64].

Metal-organic frameworks (MOFs) as carriers for proton-conducting material have received remarkable attraction from many experimental scientists owing to their high surface area compared to usual filler materials that permits encapsulation of proton transfer material [65,66]. For example, Leila Ahmadian-Alam and Hossein Mahdavi reported a ternary composite membrane composed of MOF and sulfonic acid functionalized silica (MOF/SO$_3$H-f-Si) nanoparticles with polysulfone for PEMFCs [67]. The implanting of MOF/SO$_3$H-f-Si nanoparticles on sulfonated PSU ensued in substantial enhancement of the thermal and mechanical properties of the composite membrane. The ion conductivity and transport properties of the composite membrane were increased to 0.017 S cm^{-1} by adding only 5% of MOF/SO$_3$H-f-Si nanoparticles. Furthermore, the nanocomposite exhibited a supreme power density (PD) of 40.80 mW cm^{-2}. Nor Azureen Mohamad et al. described cross-linked highly sulfonated polyphenylene sulfone (SPPSU) membranes comprised of carbon nanodots (CNDs) as a PEM for PEMFCs application [68]. The cross-linked membrane was prepared by pyrolysis at 453 K, where cross-linking occurs between SPPSU and CNDs. The prepared cross-linked composite membrane showed the maximum ionic conductivity of 56.3 mS cm^{-1}. Further, the authors demonstrated that the CNDs encapsulation into SPPSU membrane by pyrolysis treatment displayed a high ionic conductivity with superior dimensional stability. Recently, Balappa B. Munavalli and Mahadevappa Y. Kariduraganavar have prepared PEM based composite membrane by two step methods. First, sulfanilic acid (H$_2$N-C$_6$H$_4$-SO$_3$H) functionalized poly(1,4-phenylene ether ether sulfone) (SPEESSA) was synthesized. Then, different weight percentages of -SO$_3$H functionalized zeolites have been incorporated into the prepared composite membrane [69]. The composite membranes, Na-ZSM-5 zeolite, Na-β zeolite, and Na-Mordenite zeolite, exhibited the ionic conductivities of 102, 112, and 124 mS cm^{-1}, respectively. Furthermore, the composite membrane with 8 weight% Na-ZSM-5 zeolite, Na-Beta zeolite, and Na-Mordenite zeolite exhibited outstanding PD of 0.37, 2.042, and 0.45 W cm^{-2}, respectively, in H$_2$/O$_2$ fuel cells. In addition, the obtained PEMFCs results were much better than the commercially existing Nafion® 117 membranes. Jinzhao Li et al. reported graphene oxide-based nanoscale ionic materials (NIMs-GO) by sulfonation with 3-(trihydroxysilyl)-1-propanesulfonic acid (SIT) and consequent neutralization with amino-terminated polyoxypropylene (PO)-polyoxyethylene (EO) block co-polymer [70]. The schematic illustration of the NIMs-GO synthesis is depicted in Figure 5(A1). Transmission electron microscopy (TEM) was employed to analyze the morphology of the prepared GO, SIT-GO, and NIMs-GO (Figure 5a–f). Despite sulfonation by SIT, GO nanosheets exhibit a wrinkled and folded configuration, an intrinsic property of GO due to their large surface area and intramolecular attraction. Remarkably, the NIMs-GO exhibited greatly stretched features (Figure 5e,f), after being ion-exchanged with M2070. The authors stated that the change in morphology confirms that the M2070 has been ionically bonded to the GO surface via -SO_3H/-NH_2 interactions. The resulting NIMs-GO with acid-base pairs and hygroscopic EO units were incorporated into sulfonated polysulfone (SPSF) to fabricate nanocomposite membranes. The water uptake and retention ability of the SPSF/NIMs-GO nanocomposite membranes were enhanced due to the hydrophilic EO units of NIMs-GO. Furthermore, the maximum PD of 167.6 mW cm^{-2} was attained for SPSF/NIMs-GO-3 at 60 °C/100% RH, which is higher than that of Nafion® 117 and the pristine SPSF membrane (Figure 5(B1)). When the relative humidity drops to 50% (Figure 5(B2)), the maximum PD of 33.3, 17, and 23.2% decreases by SPSF, SPSF/NIMs-GO-3, and Nafion® 117, respectively. All these results are due to the increased H$^+$ conductivity of the fuel cell in both hydrated and low relative humidity conditions.

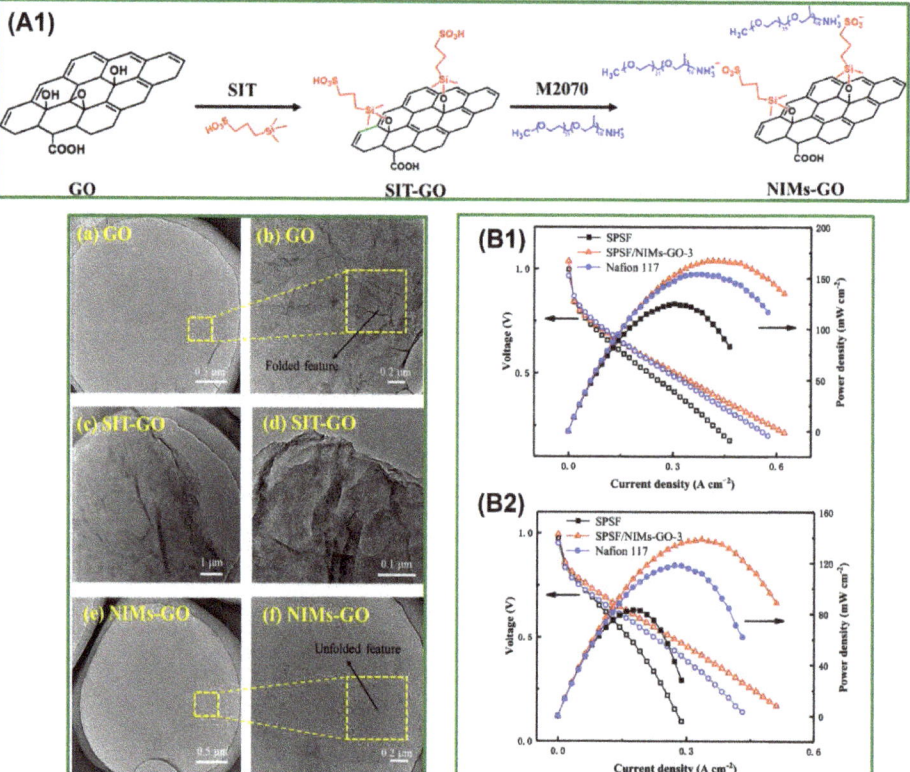

Figure 5. (**A1**) Reaction protocol of SIT-GO and NIMs-GO; (**a–f**) represents the TEM images of the prepared GOs; H_2/O_2 fuel cell performances at (**B1**) 60 °C/100% RH and (**B2**) 60 °C/50% RH. Reproduced with permission from [70]. Copyright 2019 American Chemical Society.

Recently, Cataldo Simari et al. synthesized sulfonated polysulfone (SPSF)/layered double hydroxide (LDH) nanocomposite membranes with various weight percentage filler content by an easiest solution intercalation method to replace Nafion® electrolyte in PEMFCs applications [71]. The comprehensive exfoliation and nano dispersion of the LDH platelets into the polymer improve the thermomechanical resistance, water retention capability, and dimensional stability of the electrolyte membranes. The photographic images of the prepared PEMs were depicted in Figure 6a. All membranes except the sPSU-LDH$_4$ membrane are transparent. In addition, no inorganic particles are noticed in both the sPSU-LDH$_2$ and sPSU-LDH$_3$ membranes illustrating no agglomeration. The power density and polarization curves are shown in Figure 6b,c. The maximum PD of 204.5 mW cm^{-2} at 110 °C/25% RH was achieved for the sPSU-LDH$_3$ composite membrane, which is double the value achieved by the Nafion® membrane. Such a superficial performance was attributed by the establishment of extremely interconnected ion pathways encouraging an efficient Ghrotthus-type mechanism for the H$^+$ passage even in dehydrated environments.

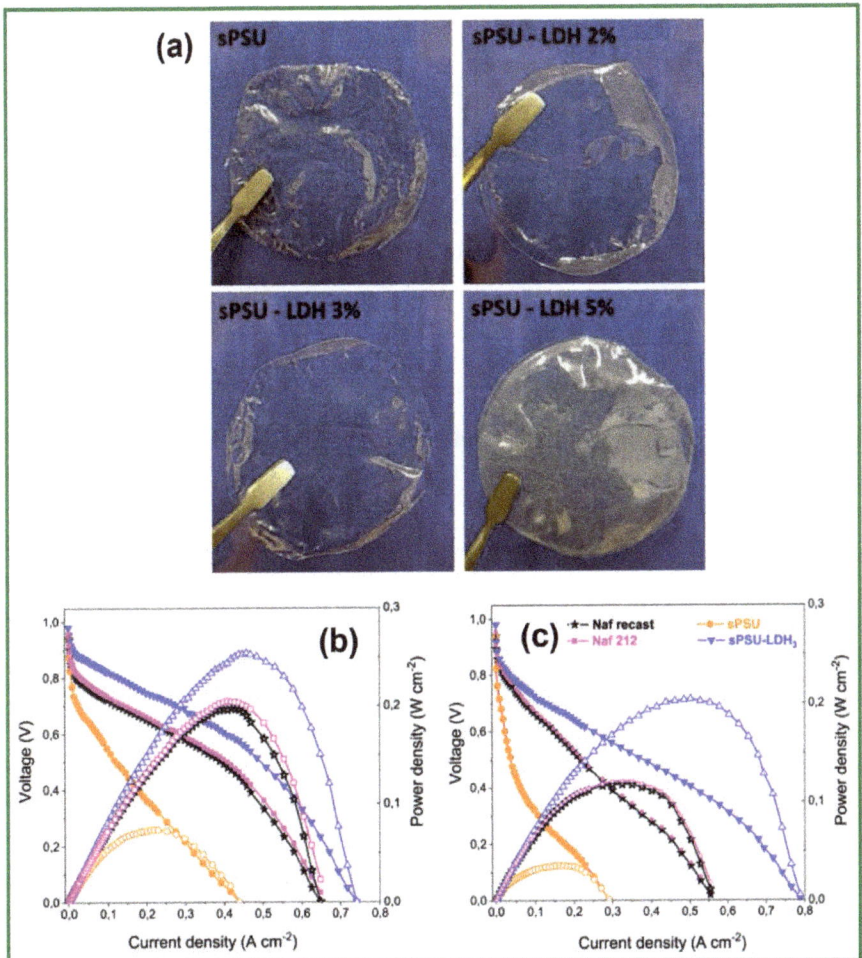

Figure 6. (a) Pictorial representation of the prepared membranes; cell voltage and PD plots of H$_2$/O$_2$ fuel cell at (b) 80 °C/30% RH and (c) 110 °C/25% RH. Reproduced with permission from [71]. Copyright 2020 Elsevier.

Ting Pan et al. described novel composite membrane from functionalized PSU with high sulfonic acid groups, N,N-bis (sulfopropyl)aminyl-4-phenyl polysulfone (PSF-N-$C_3H_6SO_3H$) and O,O'-bis(sulfopropyl)resorcinol-5-yl-4-phenyl polysulfone (PSF-O-$C_3H_6SO_3H$) [72]. The above polymers prepared by grafting amino phenyl group and dimethoxy phenyl groups to the polymer backbone through bromination of PSU followed by Suzuki cross-coupling reaction, and the introduction of the sulfopropyl groups through sulfone ring-opening reaction. Furthermore, the prepared composite membrane exhibited the highest proton conductivity of 46.66 mS cm^{-1} at 95 °C/90% RH. In addition, the prepared membrane exhibited an adequate swelling ratio and water uptake and reduced methanol cross-over. The outstanding presentation of the composite membrane is due to the phase separation between the hydrophobic and hydrophilic subphases and the establishment of the hydrogen-bonding network in the hydrophilic subphase. Very recently, Berlina Maria Mahimai et al. prepared a series of nanocomposites from PSF, SPANI (sulfonated polyaniline), and Nb_2O_5 (niobium pentoxide) by the solution casting method [73]. The composite membrane with 10 wt% Nb_2O_5/PSF/SPANI displayed the highest ionic conductivity of 0.0674 S cm^{-1}. Furthermore, the authors demonstrated that incorporating Nb_2O_5 into virgin PSF enhanced the proton conductivity and improved the thermal and oxidative stability.

In order to find different polymer electrolyte materials other than Nafion^{TM}, polymeric membranes functionalized with H_3PO_3 (phosphonic acid) groups have encouraged much research consideration owing to their enhanced ionic conductivity at high temperature under dehydration environment ascribing to the self-ionization of H_3PO_3 groups within an infused hydrogen-bonding network [74,75]. When compared with SO_3H and COOH, the H_3PO_3 group has moderate acidity and low water solubility and swelling ability, so it has a high ability for hydrogen bonding [76]. Furthermore, the bond that exists in phosphonic acid (-C-P-) is more thermally and electrochemically stable than the sulfonic acid (-C-S-) bond and carboxylic acid (-C-C-) bond, and therefore, more appropriate for PEMFCs application [77,78]. For example, Lesi Yu et al. reported proton-conducting composite membrane from SPSF and polysulfone grafted (phosphonated polystyrene) (SPSF/PPSF) through controlled atom transfer radical polymerization (ATRP) for PEMFCs application [79]. The supreme ionic conductivity of 0.01723 S cm^{-1} at 95 °C/90% RH was achieved. Furthermore, the SPSF/PPSF membrane exhibited promising thermal stability, adequate swelling ratio, and water uptake, notably enhanced mechanical stability. In addition, the permeability of methanol decreased from 5.74×10^{-8} cm^2 s^{-1} for PPSF to 0.96×10^{-8} cm^2 s^{-1} for the composite membrane.

PEMFCs operating at high temperatures (HT-PEMFCs) have received considerable attraction owing to their improved electrode reaction kinetics and simplified humidification and thermal management [80,81]. In the HT-PEMFC devices, the PEM is a vital element for carrying H$^+$ (protons) and allocating fuel and oxygen. Hence, HT-PEMs necessitate both good ionic conductivity and adequate mechanical stability. There have been incredible efforts to progress HT-PEMs with high proton transport capacity at higher temperatures (120–300 °C). Recently, Hongying Tang et al. have prepared phosphate poly(phenylene sulfone) (P-PPSU) by post-phosphonylation of brominated poly(phenylene sulfone) (Br-PPSU), followed by acidification [82]. In addition, the prepared P-PPSU material can act as a binder material in the catalyst layer to decrease the decay of operating performance of HT-PEMFC operations. The ionic conductivity of P-PPSU membrane at a high temperature without extra humidification is only 0.30 mS cm^{-1} at 160 °C, the PD of 242 mW cm^{-2} is attained in fuel cell operation at 160 °C. The obtained values are low when compared with Nafion binder material; however, the excellent stability of 200 h is noticed in FCs worked at 160 °C with P-PPSU polymer binder with no noteworthy decrease in the fuel cell evaluation. Hence, the authors demonstrated that the prepared P-PSSU is a viable candidate as a binder material in the catalyst layer for extremely robust HT-PEMFCs. Jujia Zhang et al. have also prepared 2,4,6-tri(dimethylaminomethyl)-phenol (TDAP) with three tertiary amine groups that were grafted to PSF (TDAP-PSF) to attain higher phosphoric acid uptake at lower grafting degree from HT-PEMFCs [83]. Furthermore, the single cell reaches the PD of 453 mW cm^{-2} and has excellent stability without exterior humidification. Huijuan Bai et al. also described a new strategy for grafting poly(1-vinylimidazole) with phosphoric acid doping sites on the PSF backbone via ATRP [84]. The authors demonstrated that the high H$^+$ conductivity is attained due to the establishment of micro-phase separated structures, and mechanical properties are maintained due to the decreased plasticizing effect produced by the separation of phosphoric acid adsorption sites and the polymer backbone. The obtained phosphoric acid incorporated membranes have outstanding ionic conductivity of 127 mS cm^{-1} at 160 °C and excellent tensile strength of 7.94 MPa. On the other hand, the single H$_2$/O$_2$ fuel cell performance with the optimized membrane is inspiring, achieving a peak PD of 559 mW cm^{-2} at 160 °C. Table 2 summarized the preliminary characteristics of various proton-conducting polysulfone-based composite membranes along with their fuel cell evaluations.

Table 2. Preliminary characteristics of various proton-conducting polysulfone based composite membranes along with their fuel cell evaluation.

Membrane	Membrane Characteristics						Fuel Cell Performance	Ref.
	WA (%)	IEC (meq. g^{-1})	Ionic Conductivity (S cm^{-1})	Methanol Permeability	Selectivity Ratio	Oxidative Stability		
PSF/MOF/Si nanocomposite	16.50	0.86	0.017 @ 70 °C	-	-	-	OCV: 0.90 V; PD: 40.80 mW cm^{-2} @ 160 °C	[67]
Crosslinked CNDs-SPPSU	134	1.67	0.0563 @ 80 °C	-	-	-	OCV: 1.0224 V @ 100% RH	[68]
SPEESSA/sulfonic acid zeolite composite	29.12	3.189	0.124	-	-	-	OCV: 0.91 V; PD: 0.45 W cm^{-2} @ 1.1 A cm^{-2}	[69]
SPSU/NIMs-GO composite	34.1	1.49	0.23 @ 75 °C	-	-	-	OCV: 1.038 V; PD: 167.6 mW cm^{-2} @ 60 °C	[70]
SPSU-LDH composite	31	1.49	0.0137 @ 120 °C	-	-	-	PD: 204.5 mW cm^{-2} @ 110 °C	[71]
PSF-N-C$_3$H$_6$SO$_3$H/ PSF-O-C$_3$H$_6$SO$_3$H	60	2.03	0.04666	2.65 × 10^{-8} cm^2 s^{-1}	-	94.12% residual mass remains at 80 °C for 1 h in Fenton's solution	-	[72]
PSU/SPANI/Nb$_2$O$_5$ nanocomposite	17.6	1.50	0.0674	-	-	98.6% residual mass remains in Fenton's solution	-	[73]
PSU-g-phosphonated polystyrene/SPSU composite	23.07	-	0.0172 @ 95 °C	0.96 × 10^{-8} cm^2 s^{-1}	-	>95% residual mass remains at 25 °C for 120 h in Fenton's solution	-	[79]
Phosphonated PSU	6.6	2.75	0.0003 @ 160 °C	-	-	87.7% residual mass remains for 70 h in Fenton's solution	-	[82]
PA doped TDAP-g-PSU	-	-	0.056 @ 160 °C	-	-	-	OCV: 0.92 V; PD: 453 mW cm^{-2} @ 150 °C	[83]
Poly(1-vinylimidazole)-g-PSU	220.3	-	0.127 @ 160 °C	-	-	-	OCV: 0.98 V; PD: 559 mW cm^{-2} @ 160 °C	[84]

In summary, the potential of sulfonated polysulfone and its composites, phosphonated polysulfone, and several grafted polymers of sulfonated polysulfone for low and high-temperature polymer electrolyte membranes for PEMFC has been discussed. In general, the incorporation of nano sized inorganic filler or metal organic frameworks or zeolites has enhanced a phenomenal result in both the mechanical characteristics and ion conducting properties.

2.2. Polysulfone and Its Composites for DMFCs

DMFCs provide numerous distinct advantages associated with reasonable working temperatures, easy handling and storage of liquid fuel (methanol), offering power in the utmost effective way. Furthermore, there is no need to recharge the DMFC because liquid fuel can be delivered directly to the anode, and electricity can be produced immediately. Significantly, it might be the major energy basis for portable electronic instruments and automobiles with no toxic gases associated with combustion engines [85–87]. Sulfonated polysulfone (SPSU) exhibited exceptional mechanical strength and extraordinary methanol resistance (even at 100% sulfonation), illustrating its tremendous potential for the fabrication of novel polymeric membranes used in DMFC technology. Nevertheless, the very low ionic conductivity of SPSU still remains one of the most serious drawbacks. A favorable and cost-effective method to report this issue is to: (i) blend SPSU with other polymers. In practice, this approach is commonly applied in order to modify the characteristics of a virgin macromolecule, attaining superior properties in the resulting blended materials [88–91]. (ii) The preparation of composite membranes by dispersion of inorganic fillers including silica (SiO_2) [92], titania (TiO_2) [93], zeolites [94], and heteropoly acids inside the polymer matrix have been demonstrated to satisfactorily enhance the ionic conductivity of the resulting electrolyte without sacrificing its mechanical resistance [95,96]; and (iii) the introduction of functionalized 2D-layered materials (example, graphene oxide, smectite clay, layered double hydroxides (LDHs), and siliceous layered materials) effectively lowers the methanol permeability in Nafion-based membranes and simultaneously improves their proton conductivity, water retention capacity, and thermo-mechanical resistance [97–100]. Among these inorganic fillers, LDHs have recently gained more attention, a class of nanostructured materials belonging to the anionic clay family, with unique physicochemical properties [101–104]. For instance, E. Lufrano et al. described the incorporation of hygroscopic LDH particles into SPSU for DMFCs [105]. The substantial enhancement in the water and methanol absorption and dimensional stability of the SPSU/LDH composite membrane was observed when compared with both pristine SPSU and Nafion® 212 membranes. Furthermore, the fabricated single DMFC achieved the remarkable PD of 150 mW cm^{-2} at 80 °C at higher methanol concentration (5 M methanol) solution. Xianlin Xu et al. reported bio-inspired amino acid-functionalized cellulose whiskers impregnated SPSU as PEM for DMFCs [106]. The maximum ionic conductivity of 0.234 S cm^{-1} at 80 °C achieved for 10 wt% L-Serine-functionalized cellulose whiskers. In addition, enhanced water uptake and reduced methanol cross-over were observed. Therefore, the composition of filler and mixed matrix display outstanding characteristics, and H^+ conducting mixed-matrix membranes are promising materials in DMFCs. Adnan Ozden et al. prepared SPSU/zirconium hydrogen phosphate (ZrP) composite membranes with different degrees of sulfonation (20, 35, and 42%) and a uniform weight percentage of ZrP (2.5%) to alleviate the practical tasks related to the usage of traditional Nafion® membranes in DMFCs [107]. The SPSU/ZrP-42 composite membrane exhibited a maximum OCV of 0.75 V and PD of 119 mW cm^{-2} at 80 °C. Nattinee Krathumkhet et al. synthesized composite membrane from sulfonated ZSM-5 zeolite and SPSU by solution casting method [108]. First, sulfonated ZSM-5 zeolite was synthesized by an organo-functionalization method using poly(2-acrylamido-2-methylpropanesulfonic acid). Then, SPSU was prepared by the conventional method. The composite membrane, ZSM-5/SPSU, significantly enhanced the ionic conductivity, water uptake, methanol cross-over, and IEC relative to the pristine SPSU membrane. Recently, C. Simari et al. reported blended electrolyte membranes comprised of SPSU and SPEEK (SPSU/SPEEK) with two different ratios, 50/50 and 25/75, through a facile and modest solution casting method for DMFC applications [109]. The fabricated blend membrane showed enhancement of the proton transport along with the reduced methanol cross-over which is one of the essential criteria for DMFC operation. Furthermore, the DMFC performance with 25/75 blend membrane showed a PD of 130 mW cm^{-2} at 353 K in 4 M methanol. Faizah Altaf et al. also prepared sulfonated polysulfone (SPSU) based composite PEM filled with polydopamine (PD) anchored carbon nanotubes (PD-CNTs) by phase inversion methodology with varying the filler (PCSPSU) [110] and the detailed reaction protocol was given in Figure 7. The composite membrane, 0.5 weight% PD-CNTs, displayed a 43% rise in ionic conductivity compared to the original SPSU membrane, increasing from 0.085 S cm^{-1} for pristine to 0.1216 S cm^{-1} for the composite membrane at 80 °C. The prepared composite membrane also exhibited a remarkable 75% reduction in methanol permeability (5.68 × 10^{-7} cm^2 s^{-1}) compared to recast Nafion® 117 membranes (23.00 × 10^{-7} cm^2 s^{-1}). The obtained outcomes suggested that the PD functionalized CNTs based PEMs as a potential candidate for DMFCs.

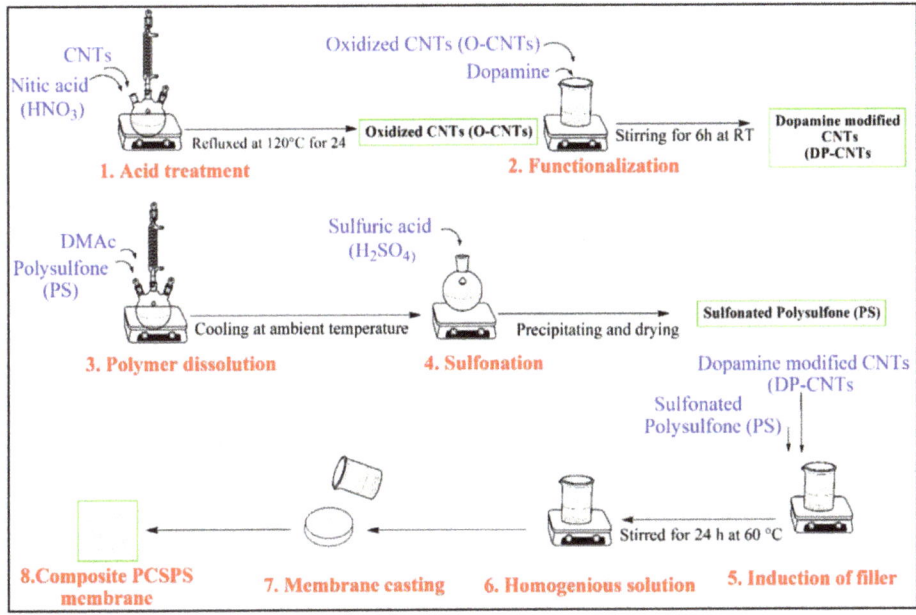

Figure 7. Reaction protocols entailed in the preparation of composite membrane, PCSPSU. Reproduced with permission from [110]. Copyright 2020 Elsevier.

In summary, SPSU membranes-based composite membranes were widely used as PEM for DMFCs to enhance its ionic conductivity, methanol cross-over, and single cell performance. Nevertheless, as previously discussed, in the performances of SPSU composites, blends, and LDH based SPSUs, many inconsistencies with the experimental results in relation to ionic conductivity, water uptake, and so on are perceptible. Each method used to improve the performance of composite and or blend SPSU based membranes offers benefits and drawbacks. Table 3 consists of SPSU and its composites for DMFCs application.

Table 3. Sulfonate polysulfone and its composites for DMFCs.

Membrane	Membrane Characteristics						Fuel Cell Performance	Ref.
	WA (%)	IEC (meq.g^{-1})	Ionic Conductivity (S cm^{-1})	Methanol Permeability (cm^2 s^{-1})	Selectivity Ratio (sS cm^{-3})	Oxidative Stability		
SPSU/LDH nanocomposite	29	1.49	0.102 @ 120 °C	116 mA cm^{-2}	-	-	OCV: 0.82 V; PD: 150 mW cm^{-2} @ 80 °C in 5 M CH$_3$OH	[105]
Amino-acid functionalized cellulose whiskers/SPSU	68	-	0.234 @ 80 °C	7.6 × 10^{-7}	-	-	OCV: 0.73 V; PD: 73.757 mW cm^{-2} @ 60 °C in 2 M CH$_3$OH	[106]
SPSU/ZrP	38	-	0.156 @ 80 °C	-	-	96.66% of weight retention after Fenton test	OCV: 0.75 V; PD: 119 mW cm^{-2} @ 80 °C	[107]
Sulfonated ZSM-5/SPSU	45.41	1.03	0.00965 @ RT	2.24 × 10^{-6}	4309.03	-	-	[108]
SPSU/SPEEK	34	-	0.073 @ 120 °C	-	-	-	OCV: 0.81 V; PD: 130 mW cm^{-2} @ 80 °C in 4 M CH$_3$OH	[109]
PD-CNT/SPSU composite	32	-	0.1216 @ 80 °C	5.68 × 10^{-7}	-	-	-	[110]

2.3. Alkaline Based Polysulfone and Its Composites for AMFCs

Recently, the progress of AMFCs has improved significantly, primarily due to the advantages of the existence of these systems over the widely known PEMFCs. The alkaline medium produced by AEM in the fuel cell favors electrode kinetics [111] and subsequently avoids the usage of expensive and noble metal catalysts. Hence, it is possible to use non-precious metals (cobalt, nickel and aluminium) [112], thereby reducing the cost of the system [113]. Nieves Urena et al. reported on amphiphilic semi-interpenetrating polymer networks for AEMFC applications with three dissimilar ionic groups, namely, tetramethyl ammonium, 1-methylimidazolium, and 1,2-dimethylimidazolium and cross-linked with N,N,N',N'-tetramethylethylenediamine (TMEDA) [114]. The resulting membrane exhibits the following characteristic: (i) at low temperatures (lower than 100 °C) has high thermal stability, (ii) lower water uptake at ambient temperature, (iii) acceptable hydroxyl ion conductivity, (iv) outstanding chemical stability, (v) excellent dimensional stability because of the inferior water uptake. Furthermore, the membrane showed excellent alkaline stability. Recently, Yang Bai et al. prepared quaternized polysulfone-based AEMS cross-linked with rGO (CQPSU-X-rGO) functionalized with different chain length small molecules [115]. Especially, the functionalized CQPSU-X-rGO showed improved ionic conductivity and chemical stability. The maximum ionic conductivity of 0.140 S cm^{-1} at 80 °C was achieved for rGO cross-linked AEMS. Tiantian Li et al. synthesized PSU based anion exchange membrane via Friedel-Crafts alkylation method contains pendant imidazolium functionalized side chain to avoid conventional carcinogenic chloromethylation. It does not require any special functional groups on the polymeric materials, which is the main advantage compared with other mentioned chloromethylation-free routes in the literature [116]. Furthermore, the membranes synthesized in this methodology displayed excellent ionic conductivity and swelling ratio along with good mechanical, thermal, and alkaline stabilities. Very recently, Lingling Ma et al. synthesized a series of AEMs modified with bulky rigid -cyclodextrin (CD) and long flexible multiple quaternary ammonium (MQ) membrane for AMFC applications [117]. The resulting AEM with a relatively low IEC of 1.50 meq. g^{-1} exhibits a good ionic conductivity of 112.4 mS cm^{-1} at 80 °C, whereas its counterpart without CD modification shows 83.0 mS cm^{-1} despite a similar ion exchange capacity (1.60 meq. g^{-1}). This is because large CD units can impart a high free volume to the membrane, dropping the ion transfer resistance, while the hydrophilicity of the external surface of the CD can promote the formation of ion transport channels across the long flexible MQ cross-links. The fabricated H_2/O_2 FC provides a maximum PD of 288 mW cm^{-1} at 60 °C. Mona Iravaninia et al. prepared AEM from polysulfone membrane by a conventional three-step method, chloromethylation, amination, alkalization with functionalized trimethylamine and N,N,N',N'-tetramethyl-1.6-hexanediamine [118]. The prepared membrane exhibited ionic conductivity of 2–42 mS cm^{-1} at 25–80 °C in different RH. The IECs, anion transport numbers, and hydration numbers were within the range of 1.6-2.1 meq. g^{-1}, 0.95–0.98 and 9–16, respectively. Furthermore, the single H_2/O_2 fuel cell showed a OCV of 1.05 V and a maximum PD of 110 mW cm^{-2} at 60 °C. Yang Bai et al. proposed a facile strategy to construct rGO stable cross-linked PSU-based AEMs with enhanced properties [119]. The cross-linked AEMS can constrict the internal packing structure and improve alkaline stability, ion conductivity, and oxidative stability. The rGO cross-linked AEM showed higher ionic conductivity of 117.7 mS cm^{-1} at 80 °C. Wan Liu et al. derived AEM from QPSU and exfoliated LDH for fuel cell applications [120]. The composite membrane comprising 5% LDH sheets showed good performance, displaying an ionic conductivity of 0.0235 S cm^{-1} at 60 °C. Yuliang Jiang et al. reported a series of PSU-based AEMs with cross-linker, 4, 4'-trimethyenedipiperidine (TMDP) [121]. The cross-linked aminated polysulfone (CAPSF) displayed supreme alkaline stability compared with non-crosslinked aminated polysulfone (APSF) in 1 M KOH for 15 days at 333 K. Furthermore, the CAPSF exhibits better dimensional stability as compared with the non-cross-linked APSF membrane owing to the compact interconnected architecture formation. From the above results, the authors concluded that the prepared crosslinked AEM is a potential candidate for AMFCs. Maria Teresa Perez-Prior et al. prepared crosslinked polysulfone AEMs using 1,4-diazabicyclo [2,2,2] octane (DABCO) as cross-liner [122]. The obtained results revealed that the cross-linked membranes displayed exceptional thermal stability, improved water uptake and dimensional stability as compared with non-cross-linked AEM. Prerana Sharma et al. described a novel strategy to synthesize alkaline membrane of chloromethylated polysulfone using cross-linker, 4,4'(3,3'-bis(chloromethyl)-[1,1'-biphenyl]-4,4-diyl)bis(oxy))dianiline) (BCBD) [123]. The detailed reaction pathway of cross-linked quaternary polysulfone (CR-QPS) membrane is shown in Figure 8. The cross-linked membrane performed well in AMFCs and exhibited maximum OCV of 0.813 V and PD of 103.6 mW cm^{-2} at 260 mA cm^{-2}.

P. F. Msomi et al. reported a sequence of AEM comprised of poly(2,6-dimethyl-1,4-phenylene) (PPO) and PSF blended with titania (QPPO/PSF/TiO_2) [124]. The swelling ratio, ionic conductivity, water uptake, and IEC of the composite were enhanced by multiplying the titania filler content. Furthermore, the QPSU/PSF/2% TiO_2 displayed a supreme PD of 118 mW cm^{-2} at 60 °C with excellent membrane stability over 60 h. K. Rambabu et al. described imidazolium functionalized PSF membranes modified with zirconia (Im-PSF/ZrO_2) by solution casting method for AMFC applications [125]. The enhanced water absorption, IEC (2.84 meq. g^{-1}), hydroxyl ion conductivity (80.2 mS cm^{-1} at 50 °C), and thermal resistance achieved for Im-PSF/ZrO_2 composite membrane as compared with pristine Im-PSF, which confirms the strong adhesion and property enhancement caused by zirconia. Furthermore, the composite membrane with Im-PSF/10% ZrO_2 showed a maximum PD of 270 mW cm^{-2} with OCV of 1.04 C in H_2/O_2 fueled AMFCs.

In summary, prominent developments have been made for the use of quaternized polysulfone with AEM in alkaline membrane fuel cells with respect to thermal, electrochemical, mechanical stability, and hydroxyl ion conductivity. Furthermore, virtuous advancement has been achieved regarding the impregnation of various inorganic filler or ionic liquids or polymer blend into various polymeric assemblies where the resultant AEMs accomplished rational performance when tested in AMFCs.

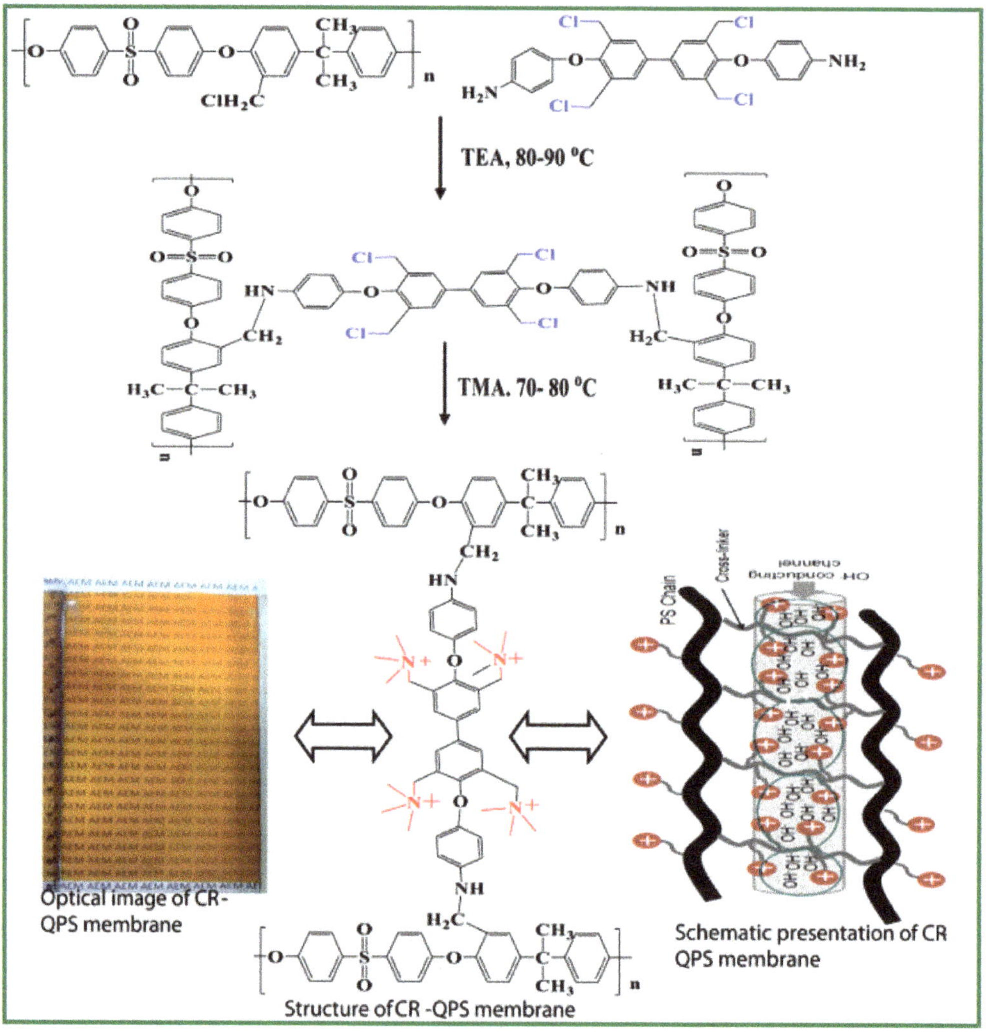

Figure 8. Schematic illustration for synthesis of CR-QPS AEM. Reproduced with the permission from [123]. Copyright 2020 Elsevier.

3. Conclusions and Future Perspectives

The emerging fuel cell market is a strong driving force for the scientific community to achieve new, affordable, and high-performance membrane materials. The present review deals with the recent advancements of polysulfone-based proton exchange membrane/anion exchange membrane for PEMFCs, DMFCs, and AMFCs application. Polysulfone-derived PEM/AEM and its composites are exploited a crucial role in the fuel cell applications as evidenced by the ample literature that is available. For PEMFCs/DMFCs, sulfonated polysulfone and its composites with inorganic fillers, layered double hydroxides, metal-organic frameworks have been investigated in this present review. Specifically, water uptake, ionic (H^+) conductivity, methanol permeability, alkaline stability, and the performance of fuel cell substantially enhanced as compared with pristine sulfonated polysulfone. Furthermore, many polymer electrolyte membranes reported in this review showed a better fuel cell performance and reduced methanol crossover compared with commercially available Nafion membranes in both PEMFCs and DMFCs operation. However, still Nafion membranes were used in the industrial sector and transport vehicles. Therefore, the commercialization of the PEMs is the utmost priority to every researcher in the membrane study to overcome Nafion membrane for PEMFCs/DMFCs applications.

As deliberated, the use of AEMs in electrochemical systems could potentially eliminate the common issues such as fuel crossover, confronted in PEMFCs. Additionally, the use of AEMs has several advantages, such as being used in alkaline environments, which enables the use of non-precious metal catalysts. Nevertheless, numerous problems need to be fixed such as poor ionic conductivity (which is accountable for poor voltage efficiency and ohmic losses), insufficient membrane stability in alkaline and oxidative atmospheres, and a lack of suitable alkaline ionomers, especially for AMFCs. Several conventional methods have been extensively studied to improve the ionic conductivity of AEMs. Recently, interpenetrating polymer network (IPN) and pore-enriched composite AEMs have efficiently imitated the Nafion-like morphology, where the hydrophobic polyolefin and the hydrophilic quaternized polymer moiety are well disconnected. As a result, a fabulous enhancement in the ionic conductivity could be attained. Inclusive data regarding the oxidative stability of AEMs can inspire further work towards the modification of existing materials or the development of new materials for AEMs. The development of AEMs based on PEEK, polybenzimidazole, and functional group chemistries based on imidazolium and guanidinium are still in the early stages. Therefore, the chemical stability of these AEMs can be studied in detail and their performance in electrochemical systems can be explored extensively.

Furthermore, the aminated/quaternized polysulfone blended with other polymers or the incorporation of inorganic fillers, such as silica, titania, zirconia, zeolites, metal-organic frameworks, etc., hinder the ionic conductivity and may reduce the chemical stability of the AEM. Despite their low alkaline stability, AEMs is still an important research field with a great outlook due to their outstanding advantages over PEMFCs. Therefore, there is an urgent need to progress novel AEMs that attain a high ionic conductivity and selectivity and exhibit outstanding chemical stability in alkaline conditions and high temperatures.

Author Contributions: Investigation and writing-original draft, R.V.; Investigation and visualization, R.A.; Conceptualization and validation, H.-J.K.; Project administration and supervision, M.Y. All authors have read and agreed to the published version of the manuscript.

Funding: This research was funded by BK21 Plus Creative Human Resource Education and Research Programs for ICT Convergence in the 4th Industrial Revolution, Pusan National University, Busan, South Korea.

Institutional Review Board Statement: Not applicable.

Informed Consent Statement: Not applicable.

Data Availability Statement: The data presented in this study are available on request from the corresponding author.

Acknowledgments: The authors gratefully acknowledge financial support from BK21 Plus Creative Human Resource Education and Research Programs for ICT Convergence in the 4th Industrial Revolution, Pusan National University, Busan, South Korea.

Conflicts of Interest: The authors declare no conflict of interest.

References

1. Kordesch, K.V.; Simader, G.R. Environmental impact of fuel cell technology. *Chem. Rev.* **1995**, *95*, 191–207. [CrossRef]
2. Yan, Q.; Toghiani, H.; Causey, H. Steady state and dynamic performance of proton exchange membrane fuel cells (PEMFCs) under various operating conditions and load changes. *J. Power Sources* **2006**, *161*, 492–502. [CrossRef]
3. Barbir, F.; Gómez, T. Efficiency and economics of proton exchange membrane (PEM) fuel cells. *Int. J. Hydrogen Energy* **1997**, *22*, 1027–1037. [CrossRef]
4. Kacprzak, A. Hydroxide electrolyte direct carbon fuel cells-technology review. *Int. J. Energy Res.* **2018**, *43*, 65–85. [CrossRef]

5. Zakaria, Z.; Kamarudin, S.K.; Timmiati, S. Membranes for direct ethanol fuel cells: An overview. *Appl. Energy* **2016**, *163*, 334–342. [CrossRef]
6. Steele, B.; Heinzel, A. Materials for fuel-cell technologies. *Nature* **2001**, *414*, 345–352. [CrossRef] [PubMed]
7. Carrette, L.; Friedrich, K.A.; Stimming, U. Fuel cells-fundamentals and applications. *Fuel Cells* **2001**, *1*, 5–39. [CrossRef]
8. Araya, S.S.; Andreasen, S.J.; Nielsen, H.V.; Kaer, S.K. Investigating the effects of methanol-water vapor mixture on a PBI-based high temperature PEM fuel cell. *Int. J. Hydrogen Energy* **2012**, *37*, 18231–18242. [CrossRef]
9. Haile, S.M. Fuel cell materials and components. *Acta Mater.* **2003**, *51*, 5981–6000. [CrossRef]
10. Kordesch, K.; Simader, G. *Fuel Cells and Their Applications*; VCH: Weinheim, Germany, 1996.
11. Pineri, M.; Eisenberg, A. *Structure and Properties of Ionomers*; Springer: Dordrecht, The Netherlands, 1987.
12. Samms, S.R.; Wasmus, S.; Savinell, R.F. Thermal stability of nafion® in simulated fuel cell environments. *J. Electrochem. Soc.* **1996**, *143*, 1498. [CrossRef]
13. Uosaki, K.; Okazaki, K.; Kita, H. Conductivity of Nation membranes at low temperatures. *J. Electroanal. Chem. Interfacial. Electrochem.* **1990**, *287*, 163–169. [CrossRef]
14. Cappadonia, M.; Erning, J.W.; Niaki, S.M.S.; Stimming, U. Conductance of Nafion 117 membranes as a function of temperature and water content. *Solid State Ion.* **1995**, *77*, 65–69. [CrossRef]
15. Adjemian, K.T.; Srinivasan, S.; Benziger, J.; Bocarsly, A.B. Investigation of PEMFC operation above 100 °C employing perfluorosulfonic acid silicon oxide composite membranes. *J. Power Sources* **2002**, *109*, 356–364. [CrossRef]
16. Kim, Y.M.; Choi, S.H.; Lee, H.C.; Hong, M.Z.; Kim, K.; Lee, H.-I. Organic-inorganic composite membranes as addition of SiO_2 for high temperature-operation in polymer electrolyte membrane fuel cells (PEMFCs). *Electrochim. Acta* **2004**, *49*, 4787–4796. [CrossRef]
17. Antonucci, P.L.; Arico, A.S.; Creti, P.; Ramunni, E.; Antonucci, V. Investigation of a direct methanol fuel cell based on a composite Nafion®-silica electrolyte for high temperature operation. *Solid State Ion.* **1999**, *125*, 431–437. [CrossRef]
18. Seyed, H.-S.; Gholamreza, B.; Mohammad, S.L. Performance of the sulfonated poly ether ether ketone proton exchange membrane modified with sulfonated polystyrene and phosphotungstic acid for microbial fuel cell applications. *J. Appl. Polym. Sci.* **2021**, *138*, 50430.
19. Martina, P.; Gayathri, R.; Raja Pugalenthi, M.; Cao, G.; Liu, C.; Ramesh Prabhu, M. Nanosulfonated silica incorporated SPEEK/SPVdF-HFP polymer blend membrane for PEM fuel cell application. *Ionics* **2020**, *26*, 3447–3458. [CrossRef]
20. Mader, J.A.; Benicewicz, B.C. Sulfonated Polybenzimidazoles for High Temperature PEM Fuel Cells. *Macromolecules* **2010**, *43*, 6706–6715. [CrossRef]
21. Singha, S.; Jana, T.; Modestra, J.A.; Kumar, A.N.; Mohan, S.V. Highly efficient sulfonated polybenzimidazole as a proton exchange membrane for microbial fuel cells. *J. Power Sources* **2016**, *317*, 143–152. [CrossRef]
22. Lufrano, F.; Squadrito, G.; Patti, A.; Passalacqua, E. Sulfonated polysulfone as promising membranes for polymer electrolyte fuel cells. *J. Appl. Polym. Sci.* **2000**, *77*, 1250–1256. [CrossRef]
23. Zhao, T.S.; Kreuer, K.D.; Nguyen, T.V. *Advances in Fuel Cells*; Elsevier Ltd.: Amsterdam, The Netherlands, 2007.
24. Srinivasan, S. *Fuel Cells: From Fundamentals to Applications*; Springer: Berlin, Germany, 2006.
25. Kamarudin, S.K.; Daud, W.R.W.; Ho, S.L.; Hasran, U.A. Overview on the challenges and developments of micro-direct methanol fuel cells (DMFC). *J. Power Sources* **2007**, *163*, 743–754. [CrossRef]
26. Zaidi, S.M.; Matsuura, T. *Polymer Membranes for Fuel Cells*; Springer Science: Berlin, Germany, 2009.
27. Li, N.; Fane, A.; Ho, W.S.; Matsuura, T. *Advanced Membrane Technology and Applications*; John Wiley & Sons, Inc.: Hoboken, NJ, USA, 2008.
28. Zhao, T.S.; Xu, C.; Chen, R.; Yang, W.W. Mass transport phenomena in direct methanol fuel cells. *Prog. Energy Combust. Sci.* **2009**, *35*, 275–292. [CrossRef]
29. Dai, J.; He, G.; Ruan, X.; Zheng, W.; Pan, Y.; Yan, X. Constructing a rigid crosslinked structure for enhanced conductivity of imidazolium functionalized polysulfone hydroxide exchange membrane. *Int. J. Hydrogen Energy* **2016**, *41*, 10923–10934. [CrossRef]
30. Mohanty, A.D.; Tignor, S.E.; Krause, J.A.; Choe, Y.K.; Bae, C. Systematic alkaline study of polymer backbones for anion exchange membrane applications. *Macromolecules* **2016**, *49*, 3361–3372. [CrossRef]
31. Pan, Y.; Wang, T.Y.; Yan, X.M.; Xu, X.W.; Zhang, Q.D.; Zhao, B.L.; El Hamouti, I.; Hao, C.; He, G.H. Novel benzimidazolium functionalized polysulfone-based anion exchange membranes with improved alkaline stability. *Chin. J. Polym. Sci.* **2018**, *36*, 129–138. [CrossRef]
32. Vinodh, R.; Ilakkiya, A.; Elamathi, S.; Sangeetha, D. A novel anion exchange membrane from polystyrene (ethylene butylene) polystyrene: Synthesis and characterization. *Mater. Sci. Eng. B* **2010**, *167*, 43–50. [CrossRef]
33. Liu, L.; Chu, X.; Liao, J.; Huang, Y.; Li, Y.; Ge, Z.; Hickner, M.A.; Li, N. Tuning the properties of Poly(2,6-dimethyl-1,4-phenylene oxide) anion exchange membranes and their performance in H_2/O_2 fuel cells. *Energy Environ. Sci.* **2018**, *11*, 435–446. [CrossRef]
34. Zhang, W.; Liu, Y.; Horan, J.L.; Jin, Y.; Ren, X.; Ertem, S.P.; Seifert, S.; Liberatore, M.W.; Herring, A.M.; Coughlin, E.B. Crosslinked anion exchange membranes with connected cations. *J. Polym. Sci. Part A Polym. Chem.* **2017**, *56*, 618–625. [CrossRef]
35. Shukla, G.; Shahi, V.K. Poly(arylene ether ketone) copolymer grafted with amine groups containing long alkyl chain by chloroacetylation for improved alkaline stability and conductivity of anion exchange membrane. *ACS Appl. Energy Mater.* **2018**, *1*, 1175–1182. [CrossRef]

36. Zuo, D.; Gong, Y.; Yan, Q.; Zhang, H. Preparation and characterization of hydroxyl ion-conducting interpenetrating polymer network based on PVA and PEI. *J. Polym. Res.* **2016**, *23*, 126–132. [CrossRef]
37. Guo, D.; Lai, A.N.; Lin, C.X.; Zhang, Q.G.; Zhu, A.M.; Liu, Q.L. Imidazolium-functionalized poly(arylene ether sulfone) anion-exchange membranes densely grafted with flexible side chains for fuel cells. *ACS Appl. Mater. Interfaces* **2016**, *8*, 25279–25288. [CrossRef] [PubMed]
38. Pérez-Prior, M.T.; Várez, A.; Levenfeld, B. Synthesis and characterization of benzimidazolium-functionalized polysulfones as anion-exchange membranes. *J. Polym. Sci. Part A Polym. Chem.* **2015**, *53*, 2363–2373. [CrossRef]
39. Yan, X.; Gu, S.; He, G.; Wu, X.; Zheng, W.; Ruan, X. Quaternary phosphonium-functionalized poly(ether ether ketone) as highly conductive and alkali-stable hydroxide exchange membrane for fuel cells. *J. Membr. Sci.* **2014**, *466*, 220–228. [CrossRef]
40. Jangu, C.; Long, T.E. Phosphonium cation-containing polymers: From ionic liquids to polyelectrolytes. *Polymer* **2014**, *55*, 3298–3304. [CrossRef]
41. Liu, L.; Li, Q.; Dai, J.; Wang, H.; Jin, B.; Bai, R. A facile strategy for the synthesis of guanidinium-funcionalized polymer as alkaline anion exchange membrane with improved alkaline stability. *J. Membr. Sci.* **2014**, *453*, 52–60. [CrossRef]
42. Kim, D.S.; Fujimoto, C.H.; Hibbs, M.R.; Labouriau, A.; Choe, Y.-K.; Kim, Y.S. Resonance stabilized perfluorinated ionomers for alkaline membrane fuel cells. *Macromolecules* **2013**, *46*, 7826–7833. [CrossRef]
43. Cha, M.S.; Lee, J.Y.; Kim, T.-H.; Jeong, H.Y.; Shin, H.Y.; Oh, S.-G.; Hong, Y.T. Preparation and characterization of crosslinked anion exchange membrane (AEM) materials with poly(phenylene ether)-based short hydrophilic block for use in electrochemical applications. *J. Membr. Sci.* **2017**, *530*, 73–83. [CrossRef]
44. Huang, X.L.; Lin, C.X.; Hu, E.N.; Soyekwo, F.; Zhang, Q.G.; Zhu, A.M.; Liu, Q.L. Imidazolium-functionalized anion exchange membranes using poly(ether sulfone)s as macrocrosslinkers for fuel cells. *RCS Adv.* **2017**, *7*, 27342–27353. [CrossRef]
45. Gong, X.; Yan, X.; Li, T.; Wu, X.; Chen, W.; Huang, S.; Wu, Y.; Zhen, D.; He, G. Design of pendent imidazolium side with flexible ether-containing spacer for alkaline anion exchange membrane. *J. Membr. Sci.* **2017**, *523*, 216–224. [CrossRef]
46. Hugar, K.M.; Kostalik, H.A., IV; Coates, G.W. Imidazolium cations with exceptional alkaline stability: A systematic study of structure-stability relationships. *J. Am. Chem. Soc.* **2015**, *137*, 8730–8737. [CrossRef] [PubMed]
47. Grew, K.N.; Chiu, W.K.S. A dusty fluid model for predicting hydroxyl anion conductivity in alkaline anion exchange membranes. *J. Electrochem. Soc.* **2010**, *157*, B327. [CrossRef]
48. Hren, M.; Božič, M.; Fakin, D.; Kleinschek, K.S.; Gorgieva, S. Alkaline membrane fuel cells: Anion exchange membranes and fuels. *Sustain. Energy Fuels* **2021**, *5*, 604–637. [CrossRef]
49. Sun, C.; Zlotorowicz, A.; Nawn, G.; Negro, E.; Bertasi, F.; Pagot, G.; Vezzù, K.; Pace, G.; Guarnieri, M.; Noto, V.D. [Nafion/(WO$_3$)$_x$] hybrid membranes for vanadium redox flow batteries. *Solid State Ion.* **2018**, *319*, 110–116. [CrossRef]
50. Vinodh, R.; Sangeetha, D. Quaternized Poly(Styrene Ethylene Butylene Poly Styrene)/Multiwalled Carbon Nanotube Composites for Alkaline Fuel Cell Applications. *J. Nanosci. Nanotechnol.* **2013**, *13*, 5522–5533. [CrossRef]
51. Vinodh, R.; Sangeetha, D. Efficient utilization of anion exchange composites using silica filler for low temperature alkaline membrane fuel cells. *Int. J. Plast. Technol.* **2013**, *17*, 35–50. [CrossRef]
52. Di, S.; Yan, L.; Han, S.; Yue, B.; Feng, Q.; Xie, L.; Chen, J.; Zhang, D.; Sun, C. Enhancing the high-temperature proton conductivity of phosphoric acid doped poly(2,5-benzimidazole) by preblending boron phosphate nanoparticles to the raw materials. *J. Power Sources* **2012**, *211*, 161–168. [CrossRef]
53. Vinodh, R.; Sangeetha, D. Comparative study of composite membranes from nano-metal-oxide-incorporated polymer electrolytes for direct methanol alkaline membrane fuel cells. *J. Appl. Polym. Sci.* **2013**, *128*, 1930–1938. [CrossRef]
54. Santoyo, A.B.; Carraso, J.L.G.; Gomez, E.G.; Martin, F.M.; Montesinos, A.M.H. Application of reverse osmosis to reduce pollutants present in industrial waste water. *Desalination* **2003**, *155*, 101–108. [CrossRef]
55. Nechifor, G.; Voicu, S.I.; Nechifor, A.C.; Garea, S. Nanostructured hybrid membrane polysulfone-carbon nanotubes for hemodialysis. *Desalination* **2009**, *241*, 342–348. [CrossRef]
56. Arico, A.S.; Srinivasan, S.; Antonucci, V. DMFCs from fundamental aspects to technology development. *Fuel Cells* **2001**, *1*, 133–161. [CrossRef]
57. Dillon, R.; Srinivasan, S.; Arico, A.S.; Antonucci, V. International activities in DMFC R&D: Status of technologies and potential applications. *J. Power Sources* **2004**, *127*, 112–126.
58. Kariduraganavar, M.Y.; Munavalli, B.B.; Torvi, A.I. Proton conducting polymer electrolytes for fuel cells via electrospinning technique. In *Organic-Inorganic Composite Polymer Electrolyte Membranes: Preparation, Properties and Fuel Cell Applications*; Inamuddin, Mohammad, A., Asiri, A.M., Eds.; Springer International Publishing: New York, NY, USA, 2017; pp. 421–458.
59. Kraytsberg, A.; Eli, Y.E. A review of advanced, materials for proton exchange membrane fuel cells. *Energy Fuels* **2014**, *28*, 7303–7330. [CrossRef]
60. Muller, F.; Ferreira, C.A.; Azambuja, D.S.; Aleman, C.; Armelin, E. Measuring the proton conductivity of ion-exchange membranes using electrochemical impedance spectroscopy and through-plane cell. *J. Phys. Chem. B* **2014**, *118*, 1102–1112. [CrossRef] [PubMed]
61. Peighambardoust, S.J.; Rowshanzamir, S.; Amjadi, M. Review of the proton exchange membranes for fuel cell applications. *Int. J. Hydrogen Energy* **2010**, *35*, 9349–9384. [CrossRef]
62. Munavalli, B.; Torvi, A.; Kariduraganavar, M. A facile route for the preparation of proton exchange membranes using sulfonated side chain graphite oxides and crosslinked sodium alginate for fuel cell. *Polymer* **2018**, *142*, 293–309. [CrossRef]

63. Li, Z.; He, G.; Zhao, Y.; Cao, Y.; Wu, H.; Li, Y.; Jiang, Z. Enhanced proton conductivity of proton exchange membranes by incorporating sulfonated metal-organic frameworks. *J. Power Sources* **2014**, *262*, 372–379. [CrossRef]
64. Liang, X.; Zhang, F.; Feng, W.; Zou, X.; Zhao, C.; Na, H.; Liu, C.; Sun, F.; Zhu, G. From metal-organic framework (MOF) to MOF-polymer composite membrane: Enhancement of low-humidity proton conductivity. *Chem. Sci.* **2013**, *4*, 983–992. [CrossRef]
65. Ren, Y.; Chia, G.H.; Gao, Z. Metal-organic frameworks in fuel cell technologies. *Nano Today* **2013**, *8*, 577–597. [CrossRef]
66. Horike, S.; Umeyama, D.; Kitagawa, S. Ion conductivity and transport by porous coordination polymers and metal-organic frameworks. *Acc. Chem. Res.* **2013**, *46*, 2376–2384. [CrossRef] [PubMed]
67. Ahmadian-Alam, L.; Mahdavi, H. A novel polysulfone-based ternary nanocomposite membrane consisting of metal-organic framework and silica nanoparticles: As proton exchange membrane for polymer electrolyte fuel cells. *Renew. Energy* **2018**, *126*, 630–639. [CrossRef]
68. Nor, N.A.M.; Nakao, H.; Jaafar, J.; Kim, J.-D. Crosslinked carbon nanodots with highly sulfonated polyphenylsulfone as proton exchange membrane for fuel cell applications. *Int. J. Hydrogen Energy* **2020**, *45*, 9979–9988.
69. Munavalli, B.B.; Kariduraganavar, M.Y. Development of novel sulfonic acid functionalized zeolites incorporated composite proton exchange membranes for fuel cell application. *Electrochim. Acta* **2017**, *296*, 294–307. [CrossRef]
70. Li, J.; Wu, H.; Cao, L.; He, X.; Shi, B.; Li, Y.; Xu, M.; Jiang, Z. Enhanced proton conductivity of sulfonated polysulfone membranes under low humidity via the incorporation of multifunctional graphene oxide. *ACS Appl. Nano Mater.* **2019**, *2*, 4734–4743. [CrossRef]
71. Simari, C.; Lufrano, E.; Brunetti, A.; Barbieri, G.; Nicotera, I. Highly-performing and low-cost nanostructured membranes based on Polysulfone and layered doubled hydroxide for high-temperature proton exchange membrane fuel cells. *J. Power Sources* **2020**, *471*, 228440. [CrossRef]
72. Pan, T.; Yue, B.; Yan, L.; Zeng, G.; Hu, Y.; He, S.; Lu, W.; Zhao, H.; Zhang, J. N,N-bis(sulfopropyl)aminyl-4-phenyl polysulfone and O,O'-bis(sulfopropyl)resorcinol-5-yl-4-phenyl polysulfone composite membrane for proton exchange membrane fuel cells. *Int. J. Hydrogen Energy* **2020**, *45*, 23490–23503. [CrossRef]
73. Mahimai, B.M.; Kulasekaran, P.; Deivanayagam, P. Novel polysulfone/sulfonated polyaniline/niobium pentoxide polymer blend nanocomposite membranes for fuel cell applications. *J. Appl. Polym. Sci.* **2021**, *138*, 51207. [CrossRef]
74. Han, S.; Yue, B.; Yan, L. Research progress in the development of high-temperature proton exchange membranes based on phosphonic acid group. *Acta Phys. Chim. Sin.* **2014**, *30*, 8–21.
75. Stone, C.; Daynard, T.; Hu, L.; Mah, C.; Steck, A. Phosphonic acid functionalized proton exchange membranes for PEM fuel cells. *J. New Mat. Electr. Sys.* **2000**, *3*, 43–50.
76. Bock, T.; Möhwald, H.; Mülhaupt, R. Arylphosphonic acid-functionalized polyelectrolytes as fuel cell membrane material. *Macromol. Chem. Phys.* **2007**, *208*, 1324–1340. [CrossRef]
77. Herath, M.B.; Creager, S.E.; Kitaygorodskiy, A.; DesMarteau, D.D. Perfluoroalkyl phosphonic and phosphinic acids as proton conductors for anhydrous proton-exchange membranes. *Chem. Phys. Chem.* **2010**, *11*, 2871–2878. [CrossRef]
78. Paddison, S.J.; Kreuer, K.-D.; Maier, J. About the choice of the protogenic group in polymer electrolyte membranes: Ab initio modelling of sulfonic acid, phosphonic acid, and imidazole functionalized alkanes. *Phys. Chem. Chem. Phys.* **2006**, *8*, 4530–4542. [CrossRef] [PubMed]
79. Yu, L.; Yue, B.; Yan, L.; Zhao, H.; Zhang, J. Proton conducting composite membranes based on sulfonated polysulfone and polysulfone-g-(phosphonated polystyrene) via controlled atom-transfer radical polymerization for fuel cell applications. *Solid State Ion.* **2019**, *338*, 103–112. [CrossRef]
80. Yang, J.; Li, Q.; Cleemann, L.N.; Jensen, J.O.; Pan, C.; Bjerrum, N.J.; He, R. Crosslinked hexafluoropropylidene polybenzimidazole membranes with chloromethyl polysulfone for fuel cell applications. *Adv. Energy Mater.* **2013**, *3*, 622–630. [CrossRef]
81. Bai, H.; Wang, H.; Zhang, J.; Wu, C.; Zhang, J.; Xiang, Y.; Lu, S. Simultaneously enhancing ionic conduction and mechanical strength of poly(ether sulfones)-poly(vinyl pyrrolidone) membrane by introducing graphitic carbon nitride nanosheets for high temperature proton exchange membrane fuel cell application. *J. Membr. Sci.* **2018**, *558*, 26–33. [CrossRef]
82. Tang, H.; Geng, K.; Hu, Y.; Li, N. Synthesis and properties of phosphonated polysulfones for durable high-temperature proton exchange membranes fuel cell. *J. Membr. Sci.* **2020**, *605*, 118107. [CrossRef]
83. Zhang, J.; Zhang, J.; Bai, H.; Tan, Q.; Wang, H.; He, B.; Xiang, Y.; Lu, S. A new high temperature polymer electrolyte membrane based on trifunctional group grafted polysulfone for fuel cell application. *J. Membr. Sci.* **2019**, *572*, 496–503. [CrossRef]
84. Bai, H.; Wang, H.; Zhang, J.; Zhang, J.; Lu, S.; Xiang, Y. High temperature polymer electrolyte membrane achieved by grafting poly (1-vinylimidazole) on polysulfone for fuel cells application. *J. Membr. Sci.* **2019**, *592*, 117395. [CrossRef]
85. Devi, A.U.; Muthumeenal, A.; Sabarathinam, R.; Nagendran, A. Fabrication and electrochemical properties of SPVdF-co-HFP/SPES blend proton exchange membranes for direct methanol fuel cells. *Renew. Energy* **2017**, *102*, 258–265. [CrossRef]
86. Azimi, M.; Peighambardoust, S. Methanol crossover and selectivity of nafion/heteropolyacid/. montmorillonite nanocomposite proton exchange membranes for DMFC Applications. *Iran. J. Chem. Eng.* **2017**, *14*, 65–81.
87. Yılmaz, E.; Can, E. Cross-linked poly (aryl ether sulfone) membranes for direct methanol fuel cell applications. *J. Polym. Sci. B Polym. Phys.* **2018**, *56*, 558–575. [CrossRef]
88. Bhavani, P.; Sangeetha, D. Blend membranes for direct methanol and proton exchange membrane fuel cells. *Chin. J. Polym. Sci.* **2012**, *30*, 548–560. [CrossRef]

89. Muthumeenal, A.; Neelakandan, S.; Kanagaraj, P.; Nagendran, A. Synthesis and properties of novel proton exchange membranes based on sulfonated polyethersulfone and N-phthaloyl chitosan blends for DMFC applications. *Renew. Energy* **2016**, *86*, 922–929. [CrossRef]
90. Clarizia, G.; Tasselli, F.; Simari, C.; Nicotera, I.; Bernardo, P. Solution casting blending: An effective way for tailoring gas transport and mechanical properties of poly(vinyl butyral) and Pebax2533. *J. Phys. Chem. C* **2019**, *123*, 11264–11272. [CrossRef]
91. Simari, C.; Lufrano, E.; Coppola, L.; Nicotera, I. Composite gel polymer electrolytes based on organo-modified nanoclays: Investigation on lithium-ion transport and mechanical properties. *Membranes* **2018**, *8*, 69. [CrossRef]
92. Nohara, T.; Koseki, K.; Tabata, K.; Shimada, R.; Suzuki, Y.; Umemoto, K.; Takeda, M.; Sato, R.; Rodbuntum, S.; Arita, T.; et al. Core size-dependent proton conductivity of silica filler-functionalized polymer electrolyte membrane. *ACS Sustain. Chem. Eng.* **2020**, *8*, 14674–14678. [CrossRef]
93. Devrim, Y.; Erkan, S.; Bac, N.; Eroğlu, I. Preparation and characterization of sulfonated polysulfone/titanium dioxide composite membranes for proton exchange membrane fuel cells. *Int. J. Hydrogen Energy* **2009**, *34*, 3467–3475. [CrossRef]
94. Borduin, R.; Li, W. Fabrication of foamed polyethersulfone–zeolite mixed matrix membranes for polymer electrolyte membrane fuel cell humidification. *J. Manuf. Sci. Eng.* **2017**, *139*, 021004. [CrossRef]
95. Sakamoto, M.; Nohara, S.; Miyatake, K.; Uchida, M.; Watanabe, M.; Uchida, H. Effects of incorporation of SiO2 nanoparticles into sulfonated polyimide electrolyte membranes on fuel cell. performance under low humidity conditions. *Electrochim. Acta* **2014**, *137*, 213–218. [CrossRef]
96. de Bonis, C.; Simari, C.; Kosma, V.; Mecheri, B.; D'Epifanio, A.; Allodi, V.; Mariotto, G.; Brutti, S.; Suarez, S.; Pilar, K.; et al. Enhancement of proton mobility and mitigation of methanol crossover in sPEEK fuel cells by an organically modified titania nanofiller. *J. Solid State Electrochem.* **2016**, *20*, 1585–1598. [CrossRef]
97. Simari, C.; Stallworth, P.; Peng, J.; Coppola, L.; Greenbaum, S.; Nicotera, I. Graphene oxide and sulfonated-derivative: Proton transport properties and electrochemical behavior of Nafion based nanocomposites. *Electrochim. Acta* **2019**, *297*, 240–249. [CrossRef]
98. Simari, C.; Potsi, G.; Policicchio, A.; Perrotta, I.; Nicotera, I. Clay carbon nanotubes hybrid materials for nanocomposite membranes: Advantages of branched structure for proton transport under low humidity conditions in PEMFCs. *J. Phys. Chem. C* **2016**, *120*, 2574–2584. [CrossRef]
99. Nicotera, I.; Simari, C.; Boutsika, L.G.; Coppola, L.; Spyrou, K.; Enotiadis, A. NMR investigation on nanocomposite membranes based on organosilica layered materials bearing different functional groups for PEMFCs. *Int. J. Hydrogen Energy* **2017**, *42*, 7940–27949. [CrossRef]
100. Nicotera, I.; Kosma, V.; Simari, C.; Angioni, S.; Mustarelli, P.; Quartarone, E. Ion dynamics and mechanical properties of sulfonated polybenzimidazole membranes for high temperature proton exchange membrane fuel cells. *J. Phys. Chem. C* **2015**, *119*, 9745–9753. [CrossRef]
101. Aramendía, M.A.; Borau, V.; Jiménez, C.; Marinas, J.M.; Ruiz, J.R.; Urbano, F.J. Catalytic transfer hydrogenation of citral on calcined layered double hydroxides. *Appl. Catal. Gen.* **2001**, *206*, 95–101. [CrossRef]
102. Vaccari, A. Preparation and catalytic properties of cationic and anionic clays. *Catal. Today* **1998**, *41*, 53–71. [CrossRef]
103. Sels, B.F.; De Vos, D.E.; Jacobs, P.A. Hydrotalcite-like anionic clays in catalytic organic reactions. *Catal. Rev. Sci. Eng.* **2001**, *43*, 443–488. [CrossRef]
104. Climent, M.J.; Corma, A.; Iborra, S.; Primo, J. Base catalysis for fine chemical production: Claisen-schmidt condensation on zeolites and hydrocalcites for the production of chalcones and flavanones of pharmaceutical interest. *J. Catal.* **1995**, *151*, 60–66. [CrossRef]
105. Lufrano, E.; Simari, C.; Lo Vecchio, C.; Arico, A.S.; Baglio, V.; Nicotera, I. Barrier properties of sulfonated polysulfone/layered double hydroxides nanocomposite membrane for direct methanol fuel cell operating at high methanol concentrations. *Int. J. Hydrogen Energy* **2020**, *45*, 20647–20658. [CrossRef]
106. Xua, X.; Zhaoa, G.; Wang, H.; Li, X.; Feng, X.; Cheng, B.; Shi, L.; Kang, W.; Zhuang, X.; Yin, Y. Bio-inspired amino-acid-functionalized cellulose whiskers incorporated into sulfonated polysulfone for proton exchange membrane. *J. Power Sources* **2019**, *409*, 123–131. [CrossRef]
107. Ozden, A.; Ercelik, M.; Devrim, Y.; Colpan, C.O.; Hamdullahpur, F. Evaluation of sulfonated polysulfone/zirconium hydrogen phosphate composite membranes for direct methanol fuel cells. *Electrochim. Acta* **2017**, *256*, 196–210. [CrossRef]
108. Krathumkhet, N.; Vongjitpimol, K.; Chuesutham, T.; Changkhamchom, S.; Phasuksom, K.; Sirivat, A.; Wattanakul, K. Preparation of sulfonated zeolite ZSM-5/sulfonated polysulfone composite membranes as PEM for direct methanol fuel cell application. *Solid State Ion.* **2018**, *319*, 278–284. [CrossRef]
109. Simari, C.; Lo Vecchio, C.; Baglio, V.; Nicotera, I. Sulfonated polyethersulfone/polyetheretherketone blend as high performing and cost-effective electrolyte membrane for direct methanol fuel cells. *Renew. Energy* **2020**, *159*, 336–345. [CrossRef]
110. Altaf, F.; Gill, R.; Batool, R.; Rehman, Z.-U.; Majeed, H.; Abbas, G.; Jacob, K. Synthesis and applicability study of novel poly(dopamine)-modified carbon nanotubes based polymer electrolyte membranes for direct methanol fuel cell. *J. Environ. Chem. Eng.* **2020**, *8*, 104118. [CrossRef]
111. McLean, G.F.; Niet, T.; Prince-Richard, S.; Djilali, N. An assessment of alkaline fuel cell technology. *Int. J. Hydrogen Energy* **2002**, *27*, 507–526. [CrossRef]

112. Lu, S.; Pan, J.; Huang, A.; Zhuang, L.; Lu, J. Alkaline polymer electrolyte fuel cells completely free from noble metal catalysts. *Proc. Natl. Acad. Sci. USA* **2008**, *105*, 20611–20614. [CrossRef]
113. Hu, Q.; Li, G.; Pan, J.; Tan, L.; Lu, J.; Zhuang, L. Alkaline polymer electrolyte fuel cell with Ni-based anode and Co-based cathode. *Int. J. Hydrogen Energy* **2013**, *38*, 16264–16268. [CrossRef]
114. Ureña, N.; Pérez-Prior, M.T.; del Rio, C.; Várez, A.; Levenfeld, B. New amphiphilic semi-interpenetrating networks based on polysulfone for anion-exchange membrane fuel cells with improved alkaline and mechanical stabilities. *Polymer* **2021**, *226*, 123824. [CrossRef]
115. Bai, Y.; Yuan, Y.; Miao, L.; Lü, C. Functionalized rGO as covalent crosslinkers for constructing chemically stable polysulfone-based anion exchange membranes with enhanced ion conductivity. *J. Membr. Sci.* **2019**, *570–571*, 481–493. [CrossRef]
116. Li, T.; Yan, X.; Liu, J.; Wu, X.; Gong, X.; Zhen, D.; Sun, S.; Chen, W.; He, G. Friedel-Crafts alkylation route for preparation of pendent side chain imidazolium-functionalized polysulfone anion exchange membranes for fuel cells. *J. Membr. Sci.* **2019**, *573*, 157–166. [CrossRef]
117. Ma, L.; Qaisrani, N.A.; Hussain, M.; Li, L.; Jia, Y.; Ma, S.; Zhou, R.; Bai, L.; He, G.; Zhang, F. Cyclodextrin modified, multication cross-linked high performance anion exchange membranes for fuel cell application. *J. Membr. Sci.* **2020**, *607*, 118190. [CrossRef]
118. Iravaninia, M.; Azizi, S.; Rowshanzamir, S. A comprehensive study on the stability and ion transport in cross-linked anion exchange membranes based on polysulfone for solid alkaline fuel cells. *Int. J. Hydrogen Energy* **2017**, *42*, 17229–17241. [CrossRef]
119. Bai, Y.; Yuan, Y.; Yang, Y.; Lu, C. A facile fabrication of functionalized rGO crosslinked chemically stable polysulfone-based anion exchange membranes with enhanced performance. *Int. J. Hydrogen Energy* **2019**, *44*, 6618–6630. [CrossRef]
120. Liu, W.; Liang, N.; Peng, P.; Qu, R.; Chen, D.; Zhang, H. Anion-exchange membranes derived from quaternized polysulfone and exfoliated layered double hydroxide for fuel cells. *J. Solid State Chem.* **2017**, *246*, 324–328. [CrossRef]
121. Jiang, Y.; Wang, C.; Pan, J.; Sotto, A.; Shen, J. Constructing an internally cross-linked structure for polysulfone to improve dimensional stability and alkaline stability of high performance anion exchange membranes. *Int. J. Hydrogen Energy* **2019**, *44*, 8279–8289. [CrossRef]
122. Teresa Perez-Prior, M.; Urena, N.; Tannenberg, M.; del Rio, C.; Levenfeld, B. DABCO-Functionalized Polysulfones as Anion-Exchange Membranes for Fuel Cell Applications: Effect of Crosslinking. *J. Polym. Sci. Part B Polym. Phys.* **2017**, *55*, 1326–1336. [CrossRef]
123. Sharma, P.; Manohar, M.; Kumar, S.; Shahi, V.K. Highly charged and stable cross-linked polysulfone alkaline membrane for fuel cell applications: 4,4,-((3,3'-bis(chloromethyl)-(1,1'-biphenyl)-4,4-diyl) bis(oxy))dianiline (BCBD) a novel cross-linker. *Int. J. Hydrogen Energy* **2020**, *45*, 18693–18703. [CrossRef]
124. Msomi, P.F.; Nonjola, P.T.; Ndungu, P.G.; Ramontja, J. Poly (2, 6-dimethyl-1, 4-phenylene)/polysulfone anion exchange membrane blended with TiO2 with improved water uptake for alkaline fuel cell application. *Int. J. Hydrogen Energy* **2020**, *45*, 29465–29476. [CrossRef]
125. Rambabu, K.; Bharath, G.; Arangadi, A.F.; Velu, S.; Banat, F.; Show, P.L. ZrO_2 incorporated polysulfone anion exchange membranes for fuel cell applications. *Int. J. Hydrogen Energy* **2020**, *45*, 29668–29680. [CrossRef]

Article

Grape Pomace Extracted Tannin for Green Synthesis of Silver Nanoparticles: Assessment of Their Antidiabetic, Antioxidant Potential and Antimicrobial Activity

Rijuta Ganesh Saratale [1], Ganesh Dattatraya Saratale [2], Somin Ahn [2] and Han-Seung Shin [2,*]

1 Research Institute of Biotechnology and Medical Converged Science, Dongguk University-Seoul, Ilsandong-gu, Goyang-si 10326, Gyeonggi-do, Korea; rijutaganesh@gmail.com
2 Department of Food Science and Biotechnology, Dongguk University-Seoul, Ilsandong-gu, Goyang-si 10326, Gyeonggi-do, Korea; gdsaratale@dongguk.edu (G.D.S.); griju22@gmail.com (S.A.)
* Correspondence: spartan@dongguk.edu

Citation: Saratale, R.G.; Saratale, G.D.; Ahn, S.; Shin, H.-S. Grape Pomace Extracted Tannin for Green Synthesis of Silver Nanoparticles: Assessment of Their Antidiabetic, Antioxidant Potential and Antimicrobial Activity. *Polymers* **2021**, *13*, 4355. https://doi.org/10.3390/polym13244355

Academic Editor: Suguna Perumal

Received: 23 November 2021
Accepted: 10 December 2021
Published: 13 December 2021

Publisher's Note: MDPI stays neutral with regard to jurisdictional claims in published maps and institutional affiliations.

Copyright: © 2021 by the authors. Licensee MDPI, Basel, Switzerland. This article is an open access article distributed under the terms and conditions of the Creative Commons Attribution (CC BY) license (https://creativecommons.org/licenses/by/4.0/).

Abstract: In nanoscience, the "green" synthesis approach has received great interest as an eco-friendly and sustainable method for the fabrication of a wide array of nanoparticles. The present study accounts for an expeditious technique for the synthesis of silver nanoparticles (AgNPs) utilizing fruit waste grape pomace extracted tannin. Grape pomace tannin (Ta) involved in the reduction and capping of AgNPs and leads to the formation of stable Ta-AgNPs. Various conditions were attempted to optimize the particle size and morphology of Ta-AgNPs which was further analyzed using various analytical tools for different characteristic motives. UV-visible spectroscopy showed a characteristic peak at 420 nm, indicating successful synthesis of AgNPs. Energy disperses spectroscopy (EDS) analysis proved the purity of the produced Ta-AgNPs and manifested a strong signal at −2.98 keV, while Fourier-transform infrared spectrophotometer (FTIR) spectra of the Ta-AgNPs displayed the existence of functional groups of tannin. Zeta potential measurements (−28.48 mV) showed that the Ta-AgNPs have reasonably good stability. High resolution transmission electron microscopy (HR-TEM) analysis confirmed the average dimension of the synthesized NPs was estimated about 15–20 nm. Ta-AgNPs potentials were confirmed by in vitro antidiabetic activity to constrain carbohydrate digesting enzymes, mainly α-amylase and α-glucosidase, with a definite concentration of sample displaying 50% inhibition (IC_{50}), which is about 43.94 and 48.5 µg/mL, respectively. Synthesized Ta-AgNPs exhibited significant antioxidant potential with respect to its 2,2′-azino-bis(3-ethylbenzothi-azoline-6-sulfonic acid) (ABTS) (IC_{50} of 40.98 µg/mL) and 2,2-diphenyl-1-picrylhydrazyl (DPPH) (IC_{50} of 53.98 µg/mL) free radical scavenging activities. Ta-AgNPs exhibited extraordinary antibacterial activity against selected pathogenic strains and showed comparable antimicrobial index against ampicillin as a positive control.

Keywords: grape pomace; silver nanoparticles (AgNPs); in vitro antidiabetic activity; DPPH; antibacterial activity

1. Introduction

Owing to distinctive physical, chemical, optical, catalytic, and magnetic properties, nanomaterials have gained considerable attention for various biological, pharmaceutical, and electronic applications [1]. In recent times, research interest towards nanotechnology has improved which leads to the augmented growth in the production of nanomaterial and its market. Based on their size, nanomaterials are differently grouped, such as nanoparticles, dendrimers, nanotubes and nanofilms [2]. Further, this upsurges the diversity of nanoscale materials. A plethora of advancements in the methodologies for the synthesis of nanoparticles with different characteristics put them together as the most applicable and widely used in materials science. Two conventional production processes, mainly (a) electrochemical and chemical reduction and (b) photochemical and physical vapor condensation, are used for industrial scale nanomaterial production to achieve perfect shapes and higher purity [3]. However, these processes are energy demanding and require hazardous reagents (stabilizing and reducing agents), thus are not eco-friendly and cost-effective. Hence, there is an extensive demand for the definition of less demanding production technologies that would expand the affordability of the whole nanotechnology industry [4,5].

Green synthesis is accomplished by combining metal salts with natural reducing agents (such as plant extracts, fruit extracts, and their secondary metabolites), microbial extracts and their by-products (such as vitamins, sugars, and biodegradable polymers) to create nanomaterials [2,6]. The green synthesis of NPs by employing green chemistry principles (Figure 1) is gaining abundant attraction for the development of these future nanosized materials. Plant extract-based nanoparticle synthesis is a non-toxic, eco-friendly, sustainable, and economical way and can perform under aqueous conditions, with low energy requirements and does not require toxic chemicals. Moreover, nanoparticle synthesis by plant extracts is comparatively much faster than the microbial route and easily scalable to produce NPs in huge quantities [7,8]. The fruit and fruit peel extracts contain various pharmacological compounds which function as reducing and capping compounds in the fabrication of different kinds of nanoparticles [9–11].

Figure 1. Overview of green chemistry principles applied in green nanotechnology.

Silver nanoparticles (AgNPs) have drawn more attention from various entrepreneurs due to their wide range of scope in numerous industries, such as agriculture, pharmacy, pigments, catalysis, electronics, and cosmetics. Other incredible properties of AgNPs include higher conductivity nature, chemical stability, which increases its potential in pharmaceutical applications mainly, cancer treatment, medical imaging, and drug delivery with reduced undesired toxicity [6,12].

Grape (*Vitas vinifera*) can be considered as one of the largest fruit crops; about >67 million tons of grapes are produced per annum globally. Grapes are mainly used for wine production. All through the manufacture of grapes-wine, a major extent of solid organic by-product as a grape pomace is produced (about 40%). Grape pomace signifies a vital source of phenolic antioxidants and can be utilized as an animal feed supplement with health promoting factors [13,14]. However, utilization of the huge amount of grape pomace is still scarce, thus it accumulates near wine industries as a waste product and causes environmental and disposal complications. High levels of condensed tannins are held back as residue, based on the low extraction during winemaking. Tannin is a polyphenolic compound, having human health benefits due to higher antioxidant potential [11,15]. The present study was intended for silver nanoparticles green synthesis using *Vitis vinifera* (extracted grape pomace tannin), which has not been studied well in nano research. Further, the influence of several operational factors in the synthesis of Ta-AgNPs were critically examined and optimized. Characterization of Ta-AgNPs with regard to the size and size distribution morphology and structure of particle size was accomplished using various standard analytical techniques. Ta-AgNPs were assessed for multi biogenic potentials in terms of in vitro antidiabetic and antioxidant activities by employing standard enzyme assays. Finally, the antimicrobial efficacy was investigated against pathogenic bacteria cultures to raise their potential applications in biomedical sectors.

2. Materials and Methods

2.1. Grape Tannin Extraction and Reagents

Grape pomace (GP) was procured from the local wine industry, washed and oven dried, until a persistent weight was attained. Further, the dried GP was cut into small pieces and finely ground. Tannin extraction from grape pomace was performed using the methodology reported elsewhere [16]. The finely ground grape pomace was added in water comprising of Na_2CO_3 (2.5%) and Na_2SO_3 (2.5%) aqueous base solution at 80 °C for 4 h followed by the filtration, and the resultant portion was spray-dehydrated and the resulting powder was utilized for the synthesis of silver nanoparticles and for subsequent procedures. Silver nitrate ($AgNO_3$), ascorbic acid, 2,2-diphenyl-1-picryhydrazyl (DPPH), 2,2′-Azino-bis(3-ethylbenzothiazoline)-6 sulfonic acid (ABTS), sodium potassium tartrate, 3,5-dinitrosalicylic acid (DNS), acarbose, α-glucosidase, and α-amylase, were procured from Sigma-Aldrich, St. Louis, MO, USA. All other reagents and chemicals used for the study were of analytical grade quality and of higher pureness. Double distilled water was used throughout in all the experiments for solution preparations (Millipore Corporate, Billerica, MA, USA).

2.2. Green Mode Synthesis of Ta-AgNPs, Optimization of Conditions and Stability Studies

The synthesis of Ta-AgNPs was performed in the aqueous grape pomace extracted tannin, which is a reducing agent and silver nitrate ($AgNO_3$), as the precursor compound. Grape tannin (1000 ppm) and the $AgNO_3$ solution (1 mM) were individually prepared. Appropriate volumes of tannin and $AgNO_3$ solution (ratio of 1:10) in a flask were gradually mixed at 30 °C on a magnetic stirrer. At regular time intervals, the samples were collected from the reaction mixture and studied for their absorption spectrum by employing UV-visible absorption spectroscopy. The change in color (light brown and became darker) was also noted. In order to improve the properties and analytical merit of Ta-AgNPs, the reaction conditions, such as pH value (2, 3, 4, 5, 6, 7, and 8), reaction time (0, 5, 10, 20 and 30 min), concentration of $AgNO_3$ (0.5, 1.0, 2.0, and 2.5 mM), and tannin concentration by means of varied mixing ratios of Ta:$AgNO_3$ (1:1, 1:5, 1:8, 1:10, 1:15, and 1:20), were inspected in a detailed manner. The optimization of NP synthesis factors was diversified one at a time by upholding the other variable stable factors. Under optimized conditions the produced Ta-AgNPs were concentrated and separated from the reaction mixture by setting a centrifuge at 12,000 rpm for 20 min (Labogene, 1736R, Lillerød, Denmark). The resulting Ta-AgNPs pellet was washed with distilled water to exclude the impurities and further dehydrated in an oven (60 °C) for analytical studies and biogenic potentials. The synthesized Ta-AgNPs were observed for up to 3 months for their stability, by keeping them at room temperature conditions and by applying the procedure reported earlier [9]. All experiments were conducted in triplicate sets.

2.3. Characterization of Ta-AgNPs

The optical property of Ta-AgNPs was determined between 300 nm and 700 nm at regular time intervals by using a UV-visible spectrophotometer (Optizen, Model-2120, Daejeon, Korea). The participation of several functional groups of extracted tannin during the bio-reduction and synthesis of Ta-AgNPs was analyzed using a Fourier-transform infrared spectrophotometer (Perkin-Elmer, Norwalk, CT, USA). Meanwhile, the spectrum of energy disperses spectroscopy (EDS; JEOL-64000, Tokyo, Japan) and element distribution of Ta-AgNPs was also measured. Zeta potential of Ta-AgNPs was assessed by a zeta potential analyzer (ELS-8000, Tokyo, Japan). A high resolution transmission electron microscopy (HR-TEM, Tecnai G2 20 S-TWIN, FEI Company, Loughborough, UK) was used to analyze the size, shape, and exterior morphology characteristics of Ta-AgNPs. The particle size of Ta-AgNPs was measured using the standard procedure [9].

2.4. In Vitro Antidiabetic Potential of Synthesized Ta-AgNPs

Antidiabetic potential of synthesized Ta-AgNPs was evaluated by measuring the inhibition capability against two types of carbohydrate hydrolyzing enzymes (α-amylase and α-glucosidase). For the α-amylase enzyme assay, a diverse quantity of synthesized Ta-AgNPs (20, 40, 60, 80, and 100 µg/mL; about 1 mL) was added to 1 mL starch solution and kept at room temperature (30 °C) for 10 min. Through adding 1 mL of dinitrosalicylic acid color reagent, the reaction was stopped and then kept in a boiling water bath for 10 min and further cooled. Finally, the absorbance was checked for the mixture at 540 nm in a colorimeter. The α-glucosidase assay was performed according to the standard procedure and determined the inhibition of the enzyme activity by the Ta-AgNPs [17]. For the determination of both the enzyme assays, acarbose was considered as standard. Three replicated determinations were performed, and the averaged results were recognized to determine the antidiabetic potential of Ta-AgNPs. The enzyme activity was specified as of IC_{50} value (articulated as the definite concentration of a sample displaying 50% inhibition).

$$\text{Free radical scavenging (\%)} = [(AC - AT)/AC] \times 100$$

AC = absorbance of control; AT = absorbance after exposure to Ta-AgNPs.

2.5. In Vitro Antioxidant Potential of Synthesized Ta-AgNPs

Antioxidant potentials of ascorbic acid (as standard), extracted tannin, and synthesized Ta-AgNPs were investigated by quantifying the free radical scavenging activity against 2,2-diphenyl-1-picrylhydrazyl (DPPH) and 2,2-Azino-bis(3-ethylbenzothiazoline)-6 sulfonic acid (ABTS). The scavenging activity enzyme assays were conducted by employing the formerly described standard protocol [18]. The antioxidant activity of all samples was measured by taking the average mean values and standard deviation values and their scavenging potential was specified as of IC_{50} value using the previously described procedures.

2.6. Antimicrobial Activity

In vitro estimation for antibacterial efficacy of bio fabricated Ta-AgNPs was performed using Gram-negative and Gram-positive bacteria cultures (*Escherichia coli* and *Staphylococcus aureus*) through the standard Kirby–Bauer disc diffusion procedure [19]. First, in the nutrient broth the cultures were revived at 37 °C overnight to attain optimum O.D. (0.4) at 600 nm. Freshly grown overnight cultures were inoculated (100 µL) and swabbed using sterilized cotton swabs on agar plates. Further, using sterile filter paper discs, extracted tannin, Ta-AgNPs, and Ampicillin were kept on the inoculated agar medium. For the uniform perfusion of the samples initially, the petri plates were left to stand for 1 h, then incubated overnight at 37 °C for 24 h for bacterial culture growth. Zone of inhibition was calculated (mm) using a uniform scale round the disc infused with extracted tannin, Ta-AgNPs, and Ampicillin. Extracted tannin aqueous form was reflected as the negative control and ampicillin was applied as a positive control. The antimicrobial index of Ta-AgNPs against each pathogenic bacteria was measured and interpreted using the mentioned formula [20].

Antimicrobial index = (inhibition zone by Ta-AgNPs/inhibition zone by ampicillin) × 100

2.7. Statistical Analysis

All the experiments were conducted in three sets and the data of all the results calculated values were deliberated as mean ± standard error mean (SEM). The data obtained were inferred by using the one-way analysis of a variance (ANOVA) test convoyed by a Tukey–Kramer multiple comparisons test.

3. Results and Discussion

3.1. Ta-AgNPs Synthesis

Tannin as a naturally occurring predominant phytomolecule in grape pomace possesses higher antioxidant activity [13]. Tannins were extracted from grape pomace and utilized for the synthesis and fabrication of Ta-AgNPs. Synthesis of Ta-AgNPs was visually marked by checking the color alteration of the reaction solution, from slight brown to dark brownish red, and also subjected to surface plasmon resonance (SPR) analysis by using UV-visible spectroscopy at various time intervals. The spectral analysis manifested a distinct SPR peak at 420 nm after 30 min of incubation (Figure 2). The absorption spectral peaks in the range of 410–450 were used for the characterization of the Ag nanoparticles [21]. In line with the Mie theory, spherical nanoparticles show only a single surface plasmon resonance (SPR) band, which supports our results [12]. The phenolic compounds are mainly responsible for this chelating ability due to the nucleophilic nature of their aromatic rings. During the NP synthesis process, Ag^+ ions are captured and chelated by extracted tannins, which subsequently undergo a reduction, nucleation, and capping process, resulting in the development of stable Ta-AgNPs [22]. The detailed schematic representation of the grape pomace tannin mediated synthesis of Ta-AgNPs has been presented in Figure 3.

3.2. Optimization of Ta-AgNPs Synthesis Process Parameters

To regulate the size and morphology of nanoparticles, the process parameters can be either optimized or modified. The solution pH, temperature, reaction time, and phytochemical quantity are vital factors affecting the size, shape, and efficiency of the NP synthesis process [22,23]. Temperature is another significant process parameter, with an increase in temperature, the development of nucleation centers increases, which eventually upsurges the rate of nanoparticle synthesis. In this study, the SPR peak of produced Ta-AgNPs showed an increase in the absorption intensity from 30 to 40 °C, while no difference at 40 °C and 50 °C was noted (Figure 4a). At a higher temperature (60 °C) a sharp reduction in SPR peak was discerned. Similarly, in other studies of AgNPs synthesis using plant extract, low reaction temperatures for stable nanoparticles synthesis relative to high temperatures were required [24].

For instance, alteration in the pH leads to change in the overall charge of bioactive phytomolecules, which in turn facilitates their binding affinity and hence bioreduction of metal ions into nanoparticles. During the synthesis of Ta-AgNPs at different pH of the reaction media, maximum SPR distinct peak was recorded at pH 7.0 while other pH values found were not significant in the synthesis of Ta-AgNPs (Figure 4b). Nindawat and Agrawal [25] showed that *Arnebia hispidissima* extract promoted synthesis of small size AgNPs at alkaline pH levels of 7.0, 9.0 and 11.0, whereas flat UV spectrum was observed at pH 3.0, which support our results. The effect of the initial concentrations of silver nitrate from 0.5 mM to 2.5 mM on Ta-AgNPs synthesis was studied. The results suggested a distinct upsurge peak up to 1.0 mM (Figure 4c). However, as the increase in silver nitrate concentration accelerates to decrease the SPR value, it might be due to the agglomeration of synthesized NPs.

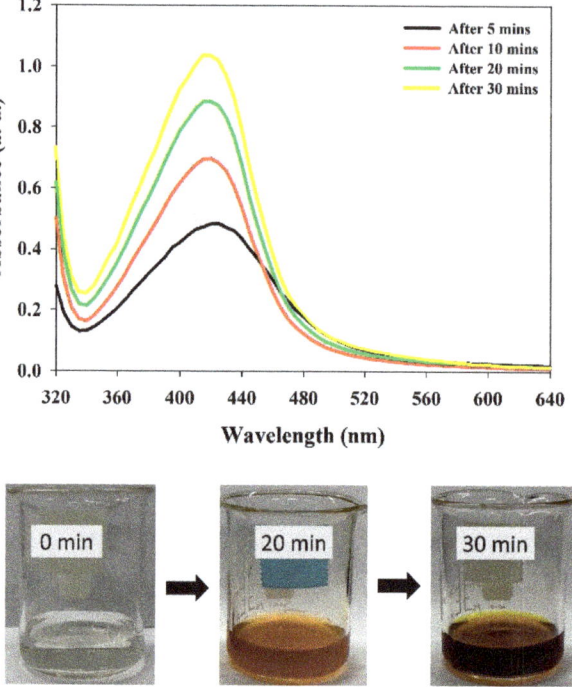

Figure 2. UV-visible absorption spectrum of synthesized Ta-AgNPs and the color changes in the reaction mixture at different time intervals.

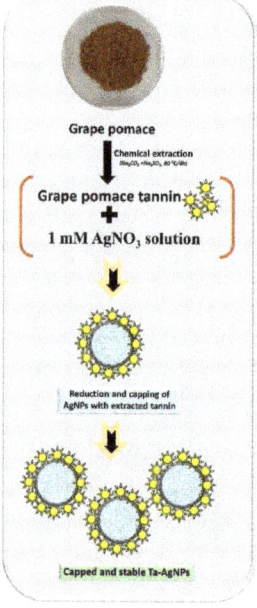

Figure 3. Schematic representation of the grape pomace tannin mediated synthesis of Ta-AgNPs.

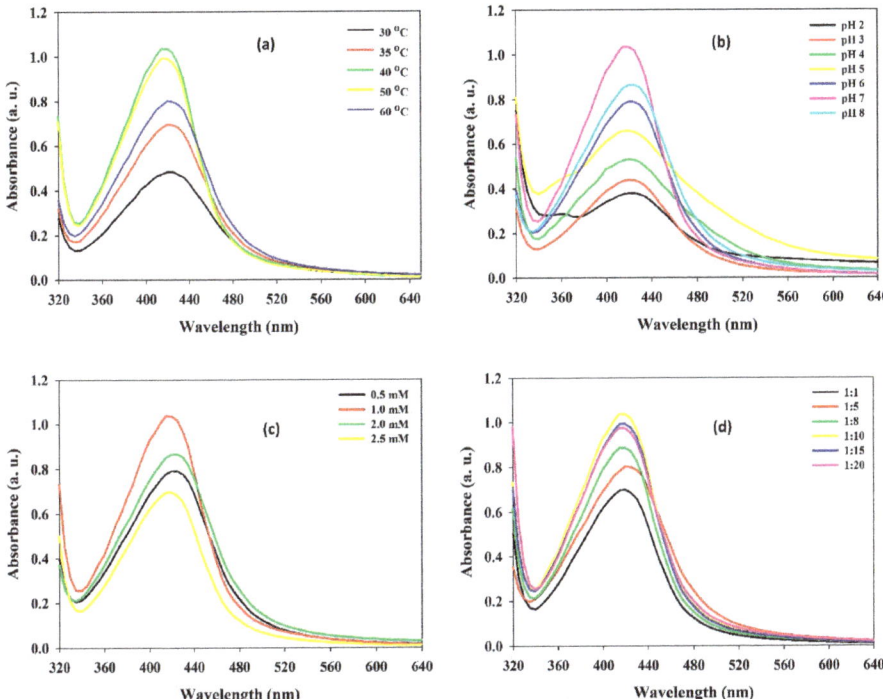

Figure 4. The effect of various influencing operational parameters (**a**) incubation temperature; (**b**) initial pH; (**c**) AgNO$_3$ concentration; and (**d**) Ta:AgNO$_3$ ratio on the synthesis of Ta-AgNPs and their UV-visible spectra.

High concentrations of polyphenols avoid coalescence and aggregation of nanoparticles; thus, determination of proper tannin concentration is essential. The effect of varying concentrations of tannins on Ta-AgNPs synthesis was investigated by making different mixing ratios of tannin and AgNO$_3$. The results suggest that lower concentrations of tannin were found to be effective and the maximum SPR was noticed at a tannin and AgNO$_3$ 1:10 ratio (Figure 4d). However, further decreasing the tannin concentration was found to be not effective in Ta-AgNPs synthesis [26]. The obtained results can be considered as noteworthy, which can help make the process worthwhile and commercially applicable. In accordance with the comprehensive results, the optimized parameters for Ta-AgNPs synthesis were 40 °C temperature, pH: 7.0, 1.0 mM AgNO$_3$, Tannin/AgNO$_3$ ratio: 1:10, and 30 min incubation time, and were selected for further experimentation.

3.2.1. Analytical Studies of Synthesized Ta-AgNPs

Generally, nanoparticles are characterized on the basis of their morphology, size, surface area, zeta potential and dispersity index. A homogenous and monodispersed solution of these nanoparticles is extremely important for numerous applications.

XRD is generally advantageous to analyze the purity and monocrystalline nature of the nanoparticles [9,27]. X-rays penetrate deep into nanoparticles that generate a diffraction pattern, which is further compared with the standard for the collection of structural details. Measurements for the XRD of the Ta-AgNPs showed four distinct peaks at 2θ angles of 38.12, 46.15, 64.75, and 76.54 attributes to (111), (200), (220), and (311) (Figure 5). The XRD spectrum of the synthesized AgNPs showed 2θ peak corresponding to 111 (at 38.1°) Bragg reflections of silver and also confirms the presence of face centered cubic (FCC) crystal structure. The results are in accordance with the silver nanoparticle synthesized by leaves of *Panax ginseng* confirmed the crystalline nature with FCC structure [28].

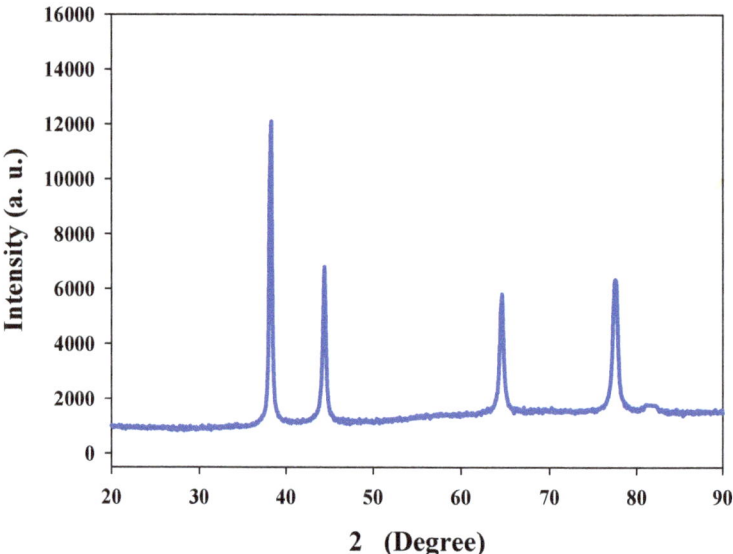

Figure 5. XRD pattern of Ta-AgNPs synthesized under optimized conditions.

3.2.2. FTIR Analysis

FTIR spectroscopy is used for the portrayal of the surface chemistry of nanoparticles and to identify the active functional groups [18]. FTIR analysis was carried out to discover the functional group deviations amongst grape pomace extracted tannins before and after it is fabricated on the surface of the AgNPs. FTIR spectra of extracted tannin and Ta-AgNPs are represented in Figure 6. The presence of broad absorption bands around 3254 cm^{-1} in both extracted tannin and synthesized Ta-AgNPs corresponds to the hydroxyl group (O-H) of polyphenol constituent [29]. Two peaks at 1582 and 1402 cm^{-1} are representative of the aromatic ring structures, whereas a small peak at 2922 cm^{-1} relates to the stretching of C-H [11]. Moreover, the peaks in the region 1000 to 1300 cm^{-1} exhibited for aromatic ring vibration [16]. In a study, analogous outcomes have been observed in the silver nanoparticle synthesis using different forms of tannins (condensed and hydrolysable) extracted from chestnut, mangrove and quebracho [30]. FTIR results suggest the presence of polyphenolic and aromatic constituents of extracted tannins on the surface of Ta-AgNPs, which are involved in the reduction, capping and stabilizing the synthesized Ta-AgNPs.

3.2.3. EDS and Zeta Potential Analysis

Energy dispersive spectroscopy (EDS) is commonly used to calculate elemental composition and the purity investigation of synthesized AgNPs. The EDS measurement of the synthesized Ta-AgNPs showed the strongest absorption peak at 2.98 keV corresponds to metallic silver due to surface plasmon resonance by silver atoms (Figure 7). The other minor peaks of C and S are related to tannin molecules, suggesting its involvement in the synthesis and fabrication of Ta-AgNPs. Few other studies confirmed that the adsorption of silver has been observed around 3 keV, which corresponds to the binding energy of elemental silver [11,28,31].

Further, the synthesized Ta-AgNps were studied by using zeta potential for the determination of surface charge. The analysis results showed the presence of higher negative surface charge (−28.48 mV), which can prevent the NPs from agglomerating. Some researchers suggest that nanoparticle zeta potential values > 30 or <−30 are comprised of high levels of stability [32]. This property is useful for the stability of synthesized NPs due to which SPR spectrum of Ta-AgNPs remain stable for about 3 months, and thus can be used in a continued way. Analogous zeta potential value −28.4 mV was perceived in the silver nanoparticles synthesized using *Vitis vinifera* skin extract [11,30].

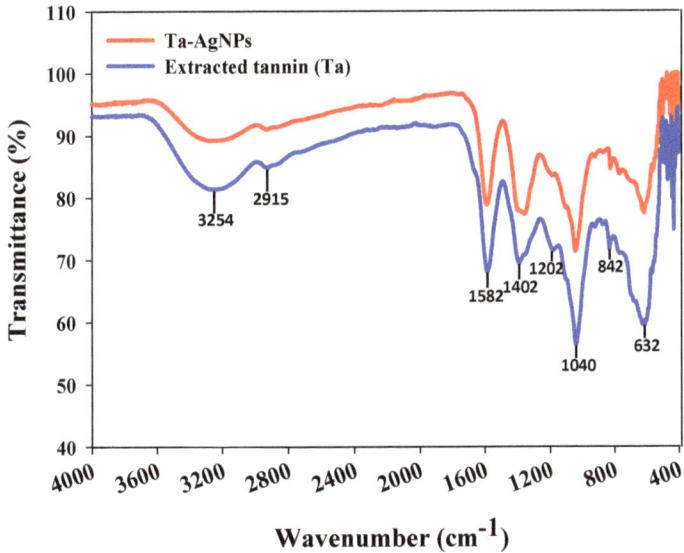

Figure 6. FTIR spectra of extracted tannin and Ta-AgNPs synthesized under optimized conditions.

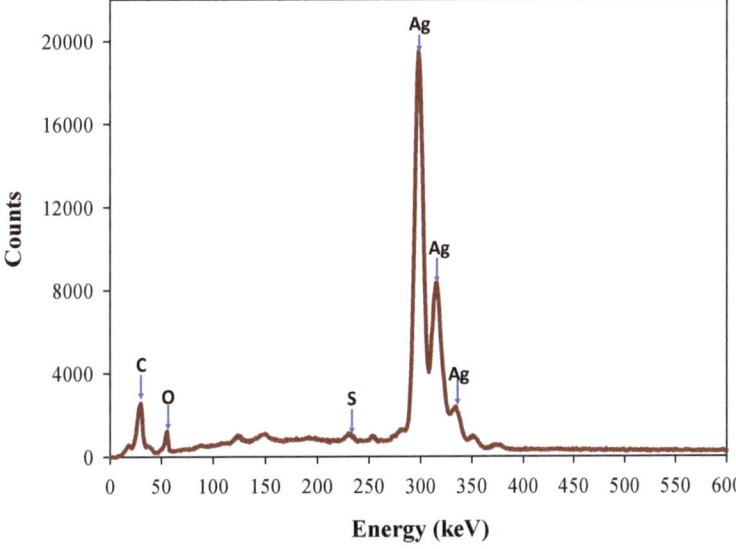

Figure 7. Energy dispersive spectroscopy (EDS) analysis of Ta-AgNPs synthesized under optimized conditions.

3.2.4. HR-TEM Analysis

Electron microscopy is another universally used technique for the analysis of the size and morphological characterization of nanoparticles. The TEM micrographs at different 50 nm and 20 nm magnifications revealed that the synthesized Ta-AgNPs are spherical in shape and uniformly distributed in the sample (Figure 8a,b), which aligns well with XRD and UV-visible spectroscopy results. TEM images also showed the presence of dark caps on the outer layer of nanoparticles which was due to the occurrence of tannin biomolecules on the surface of synthesized Ta-AgNPs. The particle histogram suggested the maximum NP size is in the range of 15 to 20 nm (Figure 8c), which increases its potential applicability in various sectors. Similar types of observations were recorded in the silver nanoparticles synthesized using *Acacia nilotica* leaf extract, jasmine flower extract, and aqueous extract of *Dracocephalum kotschyi* Boiss, respectively [31,33–35].

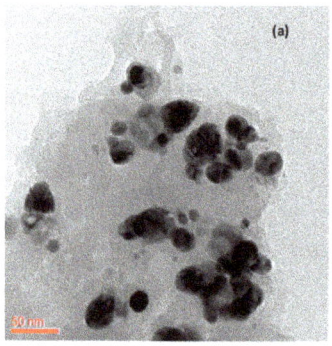
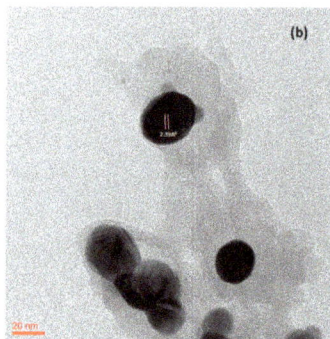
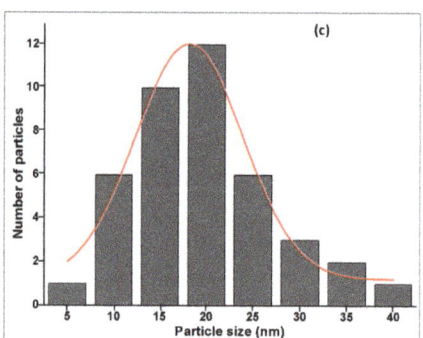

Figure 8. HR-TEM images of Ta-AgNPs: (**a**) at 50 nm; (**b**) at 20 nm amplification; and (**c**) average particle size histogram of the Ta-AgNPs produced under optimized conditions.

3.3. Biogenic Potential of Synthesized Ta-AgNPs

The biogenic potential of the green synthesized Ta-AgNPs was assessed by investigating their antidiabetic, antioxidant, and antibacterial activities.

3.3.1. Antidiabetic Potential

The green synthesized silver nanoparticles have been established as highly stable and useful candidates for drug carriage because of their ultra-small size and unique physicochemical properties [7]. The capability to fine-tune the surface charge of the nanoparticle helps in targeting specific locations and the controlled release of drugs. α-amylase and α-glucosidase are accountable for hydrolyzing oligosaccharides or polysaccharides into α-D-glucose which are adsorbed by intestinal cells, leading to postprandial hyperglycemia [36]. This unusual higher sugar level is responsible for the occurrence of type 2 diabetes (T2DM) which is troublesome and not easy to control. Acarbose, voglibose, and miglitol found clinically effective drugs to restrain or to treat T2DM by inhibiting the carbohydrate degrading enzymes [37]. However, these drugs are costly and also show adverse effects. To overcome this, there is a crucial requirement to establish effective NPs coated with natural products to control T2DM and sequential disorders. The synthesized Ta-AgNPs showed effective inhibition for both α-amylase and α-glucosidase enzyme activities in a dose dependent mode. The half-inhibitory concentration (IC_{50}) of Ta-AgNPs, extracted tannin and standard drug acarbose were determined and shown in Figure 9. The IC_{50} value of Ta-AgNPs and acarbose for α-amylase and for α-glucosidase were 43.94 and 40.2 μg/mL and 48.5 and 40.0 μg/mL, respectively (Figure 9). *Holoptelea integrifolia* leaves mediated AgNPs showed antidiabetic potential against α-amylase with significant 86.66% inhibition in enzyme activity [38]. In this study, the authors have proposed that synthesized AgNPs inhibit ATP-sensitive K^+ channel mechanism in beta cells of the pancreas.

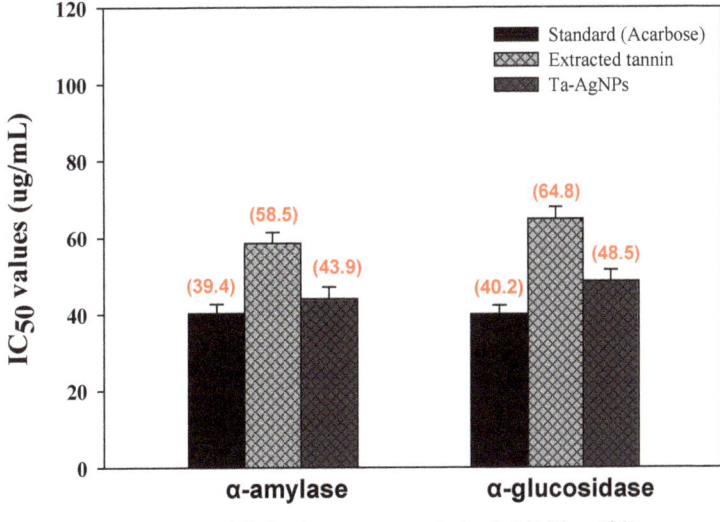

Figure 9. Antidiabetic potential of synthesized Ta-AgNPs, extracted tannin, and standard (acarbose) against α-amylase and α-glucosidase and their IC_{50} values.

3.3.2. Antioxidant Potential

Free radical generation is responsible for the existence of several pathological diseases, for instance cancer, heart disease, diabetes, Alzheimer's, hypertension, atherosclerosis, and aging [8]. The bio-synthesized Ta-AgNPs demonstrated significant antioxidant perspectives in terms of radical scavenging activities against stable free radical DPPH and ABTS and are presented in Figure 9. Ta-AgNPs displayed promising DPPH and ABTS free radical-scavenging activities in a concentration dependent mode. The standard catechol and synthesized Ta-AgNPs for DPPH and ABTS showed IC_{50} values (44.4 and 43.8 μg/mL) and (53.9 and 40.9 μg/mL), respectively (Figure 10). In both enzymes only extracted tannin was found less effective and documented higher IC_{50} value (70.8 and 65.2 μg/mL) relative to standard and Ta-AgNPs (Figure 10). In the case of grape seed and apple tannins, grape seed tannins displayed significantly more antioxidant activity than apple tannins [39]. The free-radical scavenging potential of biosynthesized nanoparticles and their application for the cure of different pathological conditions have been studied in vitro by several researchers [9,25,40]. The significant antidiabetic and antioxidant potential of Ta-AgNPs due to the tannin molecules are involved during the synthesis and fabrication of NPs. These molecules enhance the surface area of NPs, and their proper interaction leads to significant antidiabetic and antioxidant activities.

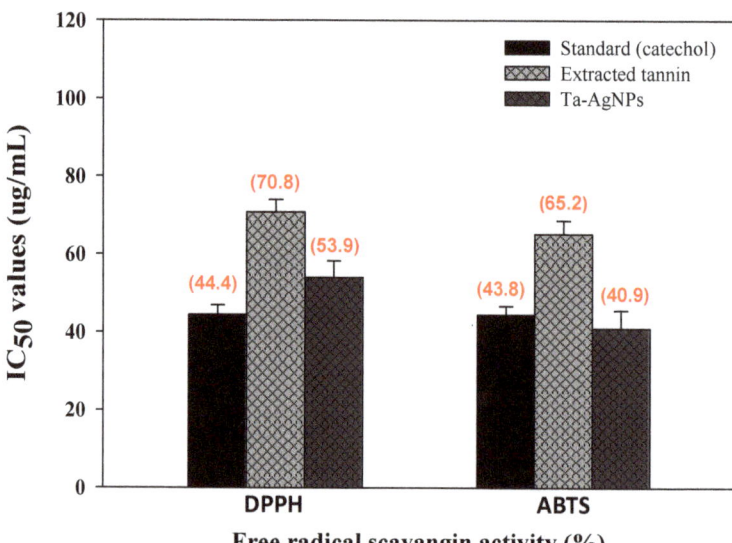

Figure 10. Antioxidant potential in terms of scavenging activity of synthesized Ta-AgNPs, extracted tannin, standard (catechol) against highly stable DPPH and ABTS and their IC_{50} values.

3.3.3. Antibacterial Potential

There has been emerging attention paid to exploring substitute approaches to developing novel antimicrobial agents since there is a continuous rise in multidrug resistant bacteria due to excess antibiotics use, mutation, and environmental circumstances [6,41]. The smaller sized nanoparticles have a superior binding surface area relative to larger NPs and show more potent antimicrobial activity. AgNPs found a potent antimicrobial agent. The synthesized Ta-AgNPs exhibited potential antibacterial activity towards the designated strains. The results are expressed as zone of the inhibition (ZOI) to define the comparative antibacterial potential of Ta-AgNPs and the results have been presented in Table 1. The obtained results revealed that the synthesized Ta-AgNPs executing significant antibacterial activity and can be used in the development of antibacterial drugs. It was supposed that Ta-AgNPs exhibited significant antibacterial effect, owing to its ability to penetrate the membrane and interact with cellular components, mainly destruction of respiratory enzymes, destruction of electron transport process, and DNA function, which leads to growth inhibition. Still, additional research is essential to understand the exact mechanisms of antibacterial activity by NPs. Disruption of membrane potential leads to cytoplasmic leakage, which results in the release of proteins and lipopolysaccharide molecules, and finally lysis of bacterial cells was observed [42]. AgNPs synthesized using the leaf extract of *Neurada procumbens* showed noteworthy antimicrobial activity against multidrug resistant Gram-negative pathogens *Klebsiella pneumoniae, Acinetobacter baumannii* and *Escherichia coli* [43]. Moreover, Hashim et al. [11], Escárcega-González et al. [41], and Kim et al. [44] also conveyed antibacterial activity of plant extract mediated AgNPs against *S. epidermidis, S. aureus, Listeria monocytogenes*; *E. coli, P. aeruginosa, P. aeruginosa, B. subtilis*; and *E. coli*, and *S. aureus*, respectively.

Table 1. Assessment of antimicrobial activity of Ta-AgNPs against pathogenic microorganisms.

Pathogen	Zone of Inhibition (mm)			
	Ta-AgNPs concentration (20 µg/mL)	Positive control (20 µg/mL)	Negative control (20 µg/mL)	Antimicrobial index (%)
Escherichia coli	14.2 ± 0.48	15.2 ± 0.52	4.85 ± 0.98	93.4 ± 2.05
Staphylococcus aureus	11.1 ± 0.82	13.7 ± 0.65	4.15 ± 0.68	81.0 ± 2.35

Positive control—Ampicillin; negative control—extracted tannin. NA—no activity. Values are mean ± standard error of three replicates.

4. Conclusions

The fruit waste grape pomace extracted tannin was exploited for the synthesis of Ta-AgNPs for the potent approach as more cost effective and non-toxic, as well as useful in lessening the burden of grape pomace waste. Optimization of synthesis parameters and their characterization using various standard analytical techniques was performed. The results suggest the synthesized NPs are spherical, monodispersed and highly stable. Tannin fabricated AgNPs showed significant antidiabetic potential by inhibiting the marker carbohydrate hydrolyzing enzymes, namely α-amylase and α-glucosidase. Additionally, Ta-AgNPs showed promising antioxidant potential and antibacterial activity against pathogenic microorganisms. In consonance with all-inclusive results, Ta-AgNPs displayed a wide array of biological solicitations and can be recognized as attractive, eco-friendly material for its possible use in drug delivery, diabetes treatment, antibacterial activity, and cancer therapy, without negative effects.

Author Contributions: Conceptualization, R.G.S. and G.D.S.; methodology, R.G.S.; validation, R.G.S., H.-S.S. and G.D.S.; formal analysis, R.G.S., H.-S.S. and G.D.S.; investigation, R.G.S.; resources, H.-S.S.; data curation, R.G.S., H.-S.S. and G.D.S.; writing—original draft preparation, R.G.S. and G.D.S.; writing—review and editing, R.G.S., H.-S.S., S.A. and G.D.S.; supervision, H.-S.S. and G.D.S.; project administration, H.-S.S.; funding acquisition, H.-S.S. All authors have read and agreed to the published version of the manuscript.

Funding: This work was supported by the Korea Institute of Planning and Evaluation for Technology in Food, Agriculture and Forestry (IPET) through the High Value-added Food Technology Development Program funded by the Ministry of Agriculture, Food and Rural Affairs (MAFRA) (121017-03).

Acknowledgments: This work was supported by the Korea Institute of Planning and Evaluation for Technology in Food, Agriculture and Forestry (IPET) through the High Value-added Food Technology Development Program funded by the Ministry of Agriculture, Food and Rural Affairs (MAFRA) (121017-03).

Conflicts of Interest: There is no conflict of interest to declare.

References

1. Bouafia, A.; Laouini, S.E.; Ahmed, A.S.A.; Soldatov, A.V.; Algarni, H.; Chong, K.F.; Ali, G.A.M. The Recent Progress on Silver Nanoparticles: Synthesis and Electronic Applications. *Nanomaterials* **2021**, *11*, 2318. [CrossRef]
2. Saratale, R.G.; Karuppusamy, I.; Saratale, G.D.; Pugazhendhi, A.; Kumar, G.; Park, Y.; Ghodake, G.S.; Bharagava, R.N.; Banu, R.; Shin, H.S. A comprehensive review on green nanomaterials using biological systems: Recent perception and their future applications. *Colloids Surf. B Biointerfaces* **2018**, *170*, 20–35. [CrossRef] [PubMed]
3. Iravani, S.; Korbekandi, H.; Mirmohammadi, S.V.; Zolfaghari, B. Synthesis of silver nanoparticles: Chemical, physical and biological methods. *Res. Pharm. Sci.* **2014**, *9*, 385–406. [PubMed]
4. Kumar, L.H.; Kazi, S.N.; Masjuki, H.H.; Zubir, M.N.M. A review of recent advances in green nanofluids and their application in thermal systems. *Chem. Eng. J.* **2021**, *429*, 132321. [CrossRef]
5. Muo, I.; Azeez, A. Green Entrepreneurship: Literature Review and Agenda for Future Research. *Int. J. Entrep. Knowl.* **2019**, *7*, 17–29. [CrossRef]
6. Ahmad, S.A.; Das, S.S.; Khatoon, A.; Ansari, M.T.; Afzal, M.; Hasnain, S.; Nayak, A.K. Bactericidal activity of silver nanoparticles: A mechanistic review. *Mater. Sci. Energy Technol.* **2020**, *3*, 756–769. [CrossRef]

7. Xue, H.; Tan, J.; Zhu, X.; Li, Q.; Cai, X. Counter-current fractionation-assisted and bioassay-guided separation of active compounds from cranberry and their interaction with α-glucosidase. *LWT Food Sci. Technol.* **2021**, *145*, 111374. [CrossRef]
8. Dauthal, P.; Mukhopadhyay, M. Noble Metal Nanoparticles: Plant-Mediated Synthesis, Mechanistic Aspects of Synthesis, and Applications. *Ind. Eng. Chem. Res.* **2016**, *55*, 9557–9577. [CrossRef]
9. Saratale, R.G.; Shin, H.-S.; Kumar, G.; Benelli, G.; Ghodake, G.S.; Jiang, Y.Y.; Kim, D.S.; Saratale, G.D. Exploiting fruit byproducts for eco-friendly nanosynthesis: Citrus × clementina peel extract mediated fabrication of silver nanoparticles with high efficacy against microbial pathogens and rat glial tumor C6 cells. *Environ. Sci. Pollut. Res.* **2018**, *25*, 10250–10263. [CrossRef] [PubMed]
10. Hamelian, M.; Zangeneh, M.M.; Shahmohammadi, A.; Varmira, K.; Veisi, H. Pistacia atlantica leaf extract mediated synthesis of silver nanoparticles and their antioxidant, cytotoxicity, and antibacterial effects under in vitro condition. *Appl. Organomet. Chem.* **2019**, *33*, e5278. [CrossRef]
11. Hashim, N.; Paramasivam, M.; Tan, J.S.; Kernain, D.; Hussin, M.H.; Brosse, N.; Gambier, F.; Raja, P.B. Green mode synthesis of silver nanoparticles using Vitis vinifera's tannin and screening its antimicrobial activity/apoptotic potential versus cancer cells. *Mater. Today Commun.* **2020**, *25*, 101511. [CrossRef]
12. Vijayaraghavan, K.; Ashokkumar, T. Plant-mediated biosynthesis of metallic nanoparticles: A review of literature, factors affecting synthesis, characterization techniques and applications. *J. Environ. Chem. Eng.* **2017**, *5*, 4866–4883. [CrossRef]
13. Fontana, A.R.; Antoniolli, A.; Bottini, R. Grape Pomace as a Sustainable Source of Bioactive Compounds: Extraction, Characterization, and Biotechnological Applications of Phenolics. *J. Agric. Food Chem.* **2013**, *61*, 8987–9003. [CrossRef] [PubMed]
14. Friedman, M. Antibacterial, Antiviral, and Antifungal Properties of Wines and Winery Byproducts in Relation to Their Flavonoid Content. *J. Agric. Food Chem.* **2014**, *62*, 6025–6042. [CrossRef] [PubMed]
15. Mangan, J.L. Nutritional Effects of Tannins in Animal Feeds. *Nutr. Res. Rev.* **1988**, *1*, 209–231. [CrossRef] [PubMed]
16. Ping, L.; Pizzi, A.; Guo, Z.D.; Brosse, N. Condensed tannins from grape pomace: Characterization by FTIR and MALDI TOF and production of environment friendly wood adhesive. *Ind. Crop. Prod.* **2012**, *40*, 13–20. [CrossRef]
17. Balan, K.; Qing, W.; Wang, Y.; Liu, X.; Palvannan, T.; Wang, Y.; Ma, F.; Zhang, Y. Antidiabetic activity of silver nanoparticles from green synthesis using Lonicera japonica leaf extract. *RSC Adv.* **2016**, *6*, 40162–40168. [CrossRef]
18. Saratale, G.D.; Saratale, R.G.; Kim, D.-S.; Kim, D.-Y.; Shin, H.S. Exploiting Fruit Waste Grape Pomace for Silver Nanoparticles Synthesis, Assessing Their Antioxidant, Antidiabetic Potential and Antibacterial Activity Against Human Pathogens: A Novel Approach. *Nanomaterials* **2020**, *10*, 1457. [CrossRef]
19. Bauer, A.W.; Kirby, M.W.M.; Sherris, J.C.; Turck, M. Antibiotic Susceptibility Testing by a Standardized Single Disk Method. *Am. J. Clin. Pathol.* **1966**, *45*, 493–496. [CrossRef]
20. Duygu, D.Y.; Erkaya, I.A.; Erdem, B.; Yalçin, D. Characterization of silver nanoparticle produced by Pseudopediastrum boryanum (Turpin) E. Hegewald and its antimicrobial effects on some pathogens. *Int. J. Environ. Sci. Technol.* **2019**, *16*, 7093–7102. [CrossRef]
21. Chung, I.-M.; Park, I.; Seung-Hyun, K.; Thiruvengadam, M.; Rajakumar, G. Plant-Mediated Synthesis of Silver Nanoparticles: Their Characteristic Properties and Therapeutic Applications. *Nanoscale Res. Lett.* **2016**, *11*, 1–14. [CrossRef] [PubMed]
22. Velusamy, P.; Kumar, G.V.; Jeyanthi, V.; Das, J.; Pachaiappan, R. Bio-Inspired Green Nanoparticles: Synthesis, Mechanism, and Antibacterial Application. *Toxicol. Res.* **2016**, *32*, 95–102. [CrossRef]
23. Pinto, R.J.B.; Lucas, J.M.F.; Morais, M.P.; Santos, S.A.; Silvestre, A.J.; Marques, P.A.; Freire, C.S. Demystifying the morphology and size control on the biosynthesis of gold nanoparticles using Eucalyptus globulus bark extract. *Ind. Crop. Prod.* **2017**, *105*, 83–92. [CrossRef]
24. Azizi, S.; Mohamad, R.; Bahadoran, A.; Bayat, S.; Rahim, R.A.; Ariff, A.; Saad, W.Z. Effect of annealing temperature on antimicrobial and structural properties of bio-synthesized zinc oxide nanoparticles using flower extract of Anchusa italica. *J. Photochem. Photobiol. B Biol.* **2016**, *161*, 441–449. [CrossRef]
25. Nindawat, S.; Agrawal, V. Fabrication of silver nanoparticles using Arnebia hispidissima (Lehm.) A. DC. root extract and unravelling their potential biomedical applications. *Artif. Cells Nanomed. Biotechnol.* **2019**, *47*, 166–180. [CrossRef] [PubMed]
26. Gangwar, C.; Yaseen, B.; Kumar, I.; Singh, N.K.; Naik, R.M. Growth Kinetic Study of Tannic Acid Mediated Monodispersed Silver Nanoparticles Synthesized by Chemical Reduction Method and Its Characterization. *ACS Omega* **2021**, *6*, 22344–22356. [CrossRef]
27. Pirtarighat, S.; Ghannadnia, M.; Baghshahi, S. Green synthesis of silver nanoparticles using the plant extract of Salvia spinosa grown in vitro and their antibacterial activity assessment. *J. Nanostruct. Chem.* **2019**, *9*, 1–9. [CrossRef]
28. Singh, P.; Kim, Y.J.; Wang, C.; Mathiyalagan, R.; Yang, D.C. The development of a green approach for the biosynthesis of silver and gold nanoparticles by using Panax ginseng root extract, and their biological applications. *Artif. Cells Nanomed. Biotechnol.* **2015**, *44*, 1–8. [CrossRef]
29. Xu, H.; Wang, L.; Su, H.; Gu, L.; Han, T.; Meng, F.; Liu, C. Making Good Use of Food Wastes: Green Synthesis of Highly Stabilized Silver Nanoparticles from Grape Seed Extract and Their Antimicrobial Activity. *Food Biophys.* **2015**, *10*, 12–18. [CrossRef]
30. Raja, P.B.; Rahim, A.A.; Qureshi, A.K.; Awang, K. Green synthesis of silver nanoparticles using tannins. *Mater. Sci.-Poland* **2014**, *32*, 408–413. [CrossRef]
31. Saratale, R.G.; Saratale, G.D.; Cho, S.-K.; Ghodake, G.; Kadam, A.; Kumar, S.; Mulla, S.I.; Kim, D.-S.; Jeon, B.-H.; Chang, J.S.; et al. Phyto-fabrication of silver nanoparticles by Acacia nilotica leaves: Investigating their antineoplastic, free radical scavenging potential and application in H2O2 sensing. *J. Taiwan Inst. Chem. Eng.* **2019**, *99*, 239–249. [CrossRef]
32. Lee, J.; Kim, T.; Choi, I.-G.; Choi, J. Phenolic Hydroxyl Groups in the Lignin Polymer Affect the Formation of Lignin Nanoparticles. *Nanomaterials* **2021**, *11*, 1790. [CrossRef]

33. Orlowski, P.; Zmigrodzka, M.; Tomaszewska, E.; Soliwoda, K.; Czupryn, M.; Antos-Bielska, M.; Szemraj, J.; Celichowski, G.; Grobelny, J.; Krzyzowska, M. Tannic acid-modified silver nanoparticles for wound healing: The importance of size. *Int. J. Nanomed.* **2018**, *13*, 991–1007. [CrossRef] [PubMed]
34. Aravind, M.; Ahmad, A.; Ahmad, I.; Amalanathan, M.; Naseem, K.; Mary, S.M.M.; Parvathiraja, C.; Hussain, S.; Algarni, T.S.; Pervaiz, M.; et al. Critical green routing synthesis of silver NPs using jasmine flower extract for biological activities and photocatalytical degradation of methylene blue. *J. Environ. Chem. Eng.* **2021**, *9*, 104877. [CrossRef]
35. Chahardoli, A.; Qalekhani, F.; Shokoohinia, Y.; Fattahi, A. Biological and catalytic activities of green synthesized silverna-noparticles from the leaf infusion of Dracocephalum kotschyi Boiss. *Glob. Chall.* **2021**, *5*, 2000018. [CrossRef] [PubMed]
36. Zhang, J.; Sun, L.; Dong, Y.; Fang, Z.; Nisar, T.; Zhao, T.; Wang, Z.-C.; Guo, Y. Chemical compositions and α-glucosidase inhibitory effects of anthocyanidins from blueberry, blackcurrant and blue honeysuckle fruits. *Food Chem.* **2019**, *299*, 125102. [CrossRef]
37. Imran, S.; Taha, M.; Ismail, N.H.; Kashif, S.M.; Rahim, F.; Jamil, W.; Hariono, M.; Yusuf, M.; Wahab, H. Synthesis of novel flavone hydrazones: In-vitro evaluation of α-glucosidase inhibition, QSAR analysis and docking studies. *Eur. J. Med. Chem.* **2015**, *105*, 156–170. [CrossRef] [PubMed]
38. Kumar, V.; Singh, S.; Srivastava, B.; Bhadouria, R.; Singh, R. Green synthesis of silver nanoparticles using leaf extract of Holoptelea integrifolia and preliminary investigation of its antioxidant, anti-inflammatory, antidiabetic and antibacterial activities. *J. Environ. Chem. Eng.* **2019**, *7*, 103094. [CrossRef]
39. Figueroa Espinoza, M.C.; Zafimahova, A.; Alvarado, P.G.; Dubreucq, E.; Legrand, C.P. Grape seed and apple tannins: Emulsifying and antioxidant properties. *Food Chem.* **2015**, *178*, 38–44. [CrossRef]
40. Rajaram, K.; Aiswarya, D.C.; Sureshkumar, P. Green synthesis of silver nanoparticle using Tephrosia tinctoria and its anti-diabetic activity. *Mater. Lett.* **2015**, *138*, 251–254. [CrossRef]
41. Escárcega-González, C.E.; Garza-Cervantes, J.A.; Vazquez-Rodríguez, A.; Montelongo-Peralta, L.Z.; Treviño-González, M.; Castro, E.D.B.; Saucedo-Salazar, E.; Morales, R.C.; Soto, D.R.; González, F.T. In vivo antimicrobial activity of silver nanoparticles produced via a green chemistry synthesis using *Acacia rigidula* as a reducing and capping agent. *Int. J. Nanomed.* **2018**, *13*, 2349–2363. [CrossRef] [PubMed]
42. Gahlawat, G.; Shikha, S.; Chaddha, B.S.; Chaudhuri, S.R.; Mayilraj, S.; Choudhury, A.R. Microbial glycolipoprotein-capped silver nanoparticles as emerging antibacterial agents against cholera. *Microb. Cell Fact.* **2016**, *15*, 1–14. [CrossRef] [PubMed]
43. Alharbi, F.A.; Alarfaj, A.A. Green synthesis of silver nanoparticles from Neurada procumbens and its antibacterial activity against multi-drug resistant microbial pathogens. *J. King Saud Univ. Sci.* **2020**, *32*, 1346–1352. [CrossRef]
44. Kim, M.; Jee, S.-C.; Shinde, S.K.; Mistry, B.M.; Saratale, R.G.; Saratale, G.D.; Ghodake, G.S.; Kim, D.-Y.; Sung, J.-S.; Kadam, A.A. Green-Synthesis of Anisotropic Peptone-Silver Nanoparticles and Its Potential Application as Anti-Bacterial Agent. *Polymers* **2019**, *11*, 271. [CrossRef] [PubMed]

Interaction Insight of Pullulan-Mediated Gamma-Irradiated Silver Nanoparticle Synthesis and Its Antibacterial Activity

Mohd Shahrul Nizam Salleh [1,2,*], Roshafima Rasit Ali [2,*], Kamyar Shameli [2], Mohd Yusof Hamzah [3], Rafiziana Md Kasmani [4] and Mohamed Mahmoud Nasef [2]

1. School of Chemical Engineering, College of Engineering, Universiti Teknologi MARA Cawangan Terengganu, Bukit Besi Campus, Dungun 23200, Terengganu, Malaysia
2. Chemical Process Engineering Department, Malaysia-Japan International Institute of Technology, Universiti Teknologi Malaysia, Jalan Sultan Yahya Ahmad Petra, Kuala Lumpur 54100, Johor, Malaysia; kamyar@utm.my (K.S.); mahmoudeithar@cheme.utm.my (M.M.N.)
3. Radiation Processing Technology Division, Malaysia Nuclear Agency, Kajang 43000, Selangor, Malaysia; m_yusof@nuclearmalaysia.gov.my
4. Department of Renewable Energy Engineering, Faculty of Chemical & Energy Engineering, Universiti Teknologi Malaysia, Johor Bahru 81310, Johor, Malaysia; rafiziana@utm.my
* Correspondence: shahrulnizam@uitm.edu.my (M.S.N.S.); roshafima@utm.my (R.R.A.)

Abstract: The production of pure silver nanoparticles (Ag-NPs) with unique properties remains a challenge even today. In the present study, the synthesis of silver nanoparticles (Ag-NPs) from natural pullulan (PL) was carried out using a radiation-induced method. It is known that pullulan is regarded as a microbial polysaccharide, which renders it suitable to act as a reducing and stabilizing agent during the production of Ag-NPs. Pullulan-assisted synthesis under gamma irradiation was successfully developed to obtain Ag-NPs, which was characterized by UV-Vis, XRD, TEM, and Zeta potential analysis. Pullulan was used as a stabilizer and template for the growth of silver nanoparticles, while gamma radiation was modified to be selective to reduce silver ions. The formation of Ag-NPs was confirmed using UV–Vis spectra by showing a surface plasmon resonance (SPR) band in the region of 410–420 nm. As observed by TEM images, it can be said that by increasing the radiation dose, the particle size decreases, resulting in a mean diameter of Ag-NPs ranging from 40.97 to 3.98 nm. The XRD analysis confirmed that silver metal structures with a face-centered cubic (FCC) crystal were present, while TEM images showed a spherical shape with smooth edges. XRD also demonstrated that increasing the dose of gamma radiation increases the crystallinity at a high purity of Ag-NPs. As examined by zeta potential, the synthesized Ag-NP/PL was negatively charged with high stability. Ag-NP/PL was then analysed for antimicrobial activity against *Staphylococcus aureus*, and it was found that it had high antibacterial activity. It is found that the adoption of radiation doses results in a stable and green reduction process for silver nanoparticles.

Keywords: silver nanoparticles; pullulan; gamma irradiation; green synthesis; antibacterial

1. Introduction

A multiphase material composed of at least one dimension below 100 nm is regarded as a nanomaterial. Polymer/silver nanomaterials that combine the advantages of the metal particles and the polymer's processability open a new gateway in developing new nanocomposite systems with improved performances [1]. Investigation of metallic nanoparticles is a continuing concern within other nanomaterials. They are considered to be very promising as they contain remarkable antibacterial properties due to their large surface area-to-volume ratio, which is of interest for researchers due to the growing microbial resistance against metal ions and antibiotics, and the development of resistant strains [2,3]. Metallic nanoparticles, particularly silver, have a better property than that because of their bulk structure. Thanks to their flexible structure, their unique properties contribute to better thermal, electronic, antimicrobial, and sensing functionalities [4]. A previous researcher also reported that incorporating an organic material into inorganic particles will prevent the agglomeration of colloidal dispersions [5]. The aggregations of particles in colloidal dispersions are expected to reduce their diffusivity, thus limiting the contact with the bacteria.

Over the past century, there has been a dramatic increase in the exploration of the potential of metal nanoparticles (NPs). NPs are reported as promising materials due to their significant composition of high-energy surface atoms [6]. Factors that influence the higher content of high-energy surface atoms in NPs are believed to be based on their exceptional physical and chemical properties, which correspond to their bulk solid counterparts [7]. More recent attention has focused on the provision of producing silver NPs (Ag-NPs) because they have several superior properties and are widely used in different fields. Ag-NPs are used in diverse fields, including the medical, catalysis, electronic, optic, environmental, biotechnology, and packaging industries [8]. The antibacterial properties are the most encouraging characteristic in Ag-NPs as they provide an antimicrobial nature in the final product embedded with Ag-NPs.

Green synthesis of Ag-NPs always focuses on using starch, fungus, yeasts, plant extracts, and other biological and natural materials as a reactant [9–11]. The use of these materials in producing Ag-NPs has been stated as a safe and eco-friendly process due to the absence of hazardous reducing agents [12]. Pullulan, in this case, which is also regarded as a biomaterial, is derived from the polymorphic fungus *Aureobasidium pullulans* and has the structure of a linear homopolysaccharide of glucose [13,14]. Pullulan is composed primarily of maltotriose units, which are units of three α-1,4-linked glucose molecules that are polymerized linearly via α-1,6-linkages. Figure 1 illustrates the proposed structure of pullulan. The fabrication of Ag-NPs using pullulan is a better approach because the pullulan will act as a capping agent. According to previous researchers, these green-synthesized Ag-NPs, in comparison to chemically synthesized synthetic Ag-NPs, were found to be less toxic [15]. Several studies have reported that the toxicity of silver nanoparticles on normal cells varies depending on the particle size, capping agent and reducing agent used, and the techniques in synthesizing the nanoparticles [16]. It is known that toxicity of Ag-NPs towards normal cells is likely but minimal if synthesizing using a biomaterial. The reactivity of Ag-NPs to generate ROS was shown to induce the toxicity of Ag-NPs [15].

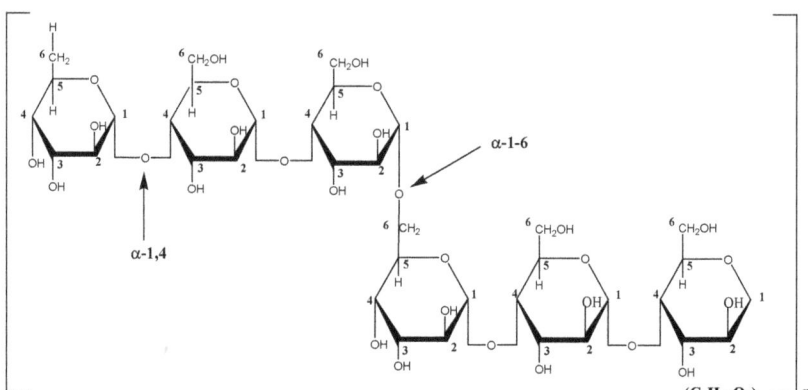

Figure 1. Structure of pullulan.

Pullulan is chosen as a biopolymer material to improve the reduction process without accelerating, reducing, or complexing agents. In addition, the structure of the polysaccharides in pullulan is reported to enhance the antibacterial activity by encapsulating the Ag-NPs, thus producing more stable nanoparticles [17]. This, in turn, will result in uniform and monodisperse nanoparticles being made [18]. Moreover, the particle size can also be tailored to the desired size with the incorporation of capping agents such as polymers, plant extracts, and, in this case, pullulan. It is reported that pullulan has a number average molecular weight (M_n) of 90,000–150,000 daltons (Da) and a weight average molecular weight (M_w) of 380,000–480,000 Da. The polydispersity index (PI) of pullulan ranges from 4.22 to 3.2 [13]. The capping agent molecules are believed to collide with the metal particles, which later induce the particles' aggregation, shaping the particle size into the desired shape [19].

The fabrication of silver nanoparticles with the right shape and uniform size distribution inside the matrix, on the other hand, remains challenging. A radiation-induced process is ideal for generating metal particles in a solution, notably those of silver. Metallic ions are reduced at each interaction. Previous studies have shown that radiolytically produced species, solvated electrons, and secondary radicals have substantial reducing potentials [19]. In this study, pullulan was selected for generating silver nanoparticles (Ag-NPs) as it is non-toxic, renewable, and biocompatible [20–26]. The gamma irradiation synthesis method was introduced because it reduces the reaction time, produces a higher yield, and is green compared to the chemical reduction method [6]. The radiation method offers many advantages for the preparation of metal nanoparticles. Hydrated electrons, resulting from the aqueous solution's gamma radiolysis, can reduce metal ions to zero-valent metal particles, avoiding the use of additional reducing agents and the consequent side reactions.

Traditional chemical methods for Ag-NP synthesis involve toxic chemicals (such as potent reducing agents, hydrazine, ethylenediaminetetraacetic acid, and, above all, sodium borohydride) and exposure to high-temperature conditions (such as in the citrate-based method). Therefore, they are not compatible with the green chemistry principle [27,28], which is why it is necessary to develop facile and green technologies in nanomaterial synthesis [29]. Therefore, this work addresses the need to synthesize silver nanoparticles in a green and clean process without the need for a chemical reducing agent. The gamma irradiation technique was chosen due to its sterile and inert process. In addition, this method also provides faster reaction times, higher yields, and improved material properties [14]. Hence, based on a previous study, Ag-NPs on pullulan were synthesized by the gamma irradiation method.

2. Materials and Methods

2.1. Materials

All materials and reagents used in this work were of analytical grade and used as received without further purification. Silver nitrate ($AgNO_3$-99.85%) was used as the silver precursor and was obtained from Acros Organic (Carlsbad, CA, USA). Pullulan powder (R&M Chemicals, London, UK) was applied as a solid support for Ag-NPs. All these aqueous solutions were used with double-distilled water.

2.2. Methods

2.2.1. Synthesis of Ag-NPs on Pullulan by Using γ-Irradiation

For the synthesis of Ag-NP/PL nanocomposites, 3.0 g of pullulan was dispersed in 100 mL double-distilled water and consistently stirred for 1 h at 90 °C until a clear solution was obtained. The solution of PL was left to cool at the ambient temperature; thereafter, 100 mL of the aqueous solution of $AgNO_3$ (0.1 mol/L) was added, and the mixture was further stirred for 1 h. The mixture was then divided into six equal part (20 mL) sample bottles, purged by N_2 for 30 min, and sealed. Finally, the suspension, which contained $AgNO_3$/PL (A0), was irradiated under γ-irradiation with absorbed doses of 5, 10, 15, 20, 25, and 50 kGy (A1–A6). The γ-irradiation process was carried out in a ^{60}Co Gammacell irradiator at room temperature (the dose rate was 35.7 kGy/min) provided by the Malaysia Nuclear Agency, Bangi.

2.2.2. Characterization Methods and Instruments

The formation of Ag-NPs was confirmed using UV–Vis spectroscopy analysis. At a medium scan rate, the produced Ag-NPs were scanned from 300 to 1000 nm with a UV–Vis spectrophotometer (UV-2600 Shimadzu, Kyoto, Japan). The spectra of the sample were measured using a designed holder for a sample of 2 cm × 2 cm in dimension. It is known that UV–Vis analysis can be used to determine the peak of the metal atoms. The peak detection is due to the colored metal ions in response to the absorption of visible light. In addition, the electrons within the metal atoms exit one electronic state to another, leading to a strong peak presence at a specific range.

The crystallinity of the nanomaterials was determined by XRD analysis. In this analysis, the XRD pattern was scanned from 10° to 80° at a 2θ angle. The identity of the silver nanoparticles can be confirmed by comparing the XRD pattern with the library. The XRD samples were dropped until they reached a certain thickness in the thin film by being repeatedly dropped and dried at 60 °C. The structure of the produced Ag-NPs was examined using XRD-Empyrean (PANanalytical, Malvern, UK).

The TEM sample was prepared by dropping the Ag-NPs on the surface of a carbon-coated copper grid. This technique determined the morphology and size of the nanoparticles. The dried sample was scanned using a JEM-2100F transmission electron microscope (JEOL, Akishima, Japan).

The stability of A-NP/PL was determined by measuring the zeta potential using a particle analyzer (Nano-Plus Zeta Sizer, Tokyo, Japan). The instrument was attached to a He-Ne laser lamp (0.4 mW) at a wavelength of 633 nm. Measurements were performed at 25 °C in an insulated chamber with 10 runs for each measurement. Prior to the measurement, the colloidal suspension was sonicated for 10 min.

The microbial activity assay of the biofilm embedded with Ag-NP/PL against Gram-positive *Staphylococcus aureus* was carried out using the culture medium toxicity method. The biofilm sample with a dimension of 1 cm × 1 cm was placed on the surface of an agar plate and seeded with 100 μL of test culture consisting of microorganisms. Petri dish covers were sealed by wax to avoid any type of contamination. The Petri dish was incubated for 24 h, and the microbial growth was followed by visual observation. Natrium Agar (NA) was used as a medium with a temperature of 37 °C. The clear zone that formed around the film sample in the medium was recorded as an indication of inhibition against the microbial species.

3. Results and Discussion

3.1. Synthesis and Characterization of Silver Nanoparticles from Pullulan

The formation of Ag-NPs by reducing silver ions to silver metal species was achieved in the presence of aqueous solutions of pullulan or oxidized pullulan at room temperature over a period of 2 h with different gamma radiation doses. This process was monitored closely using UV–Vis spectroscopy. From the prepared samples, it can be seen that a different color intensity was observed depending on the gamma ray dosage, as shown in Figure 2. Ag-NP/PL suspensions without γ-rays (A0) showed a light brown solution and exhibited no Ag^+ ion formation. As the γ-rays increased (A1–A6 with 5–50 kGy, respectively), the suspension marked a dark brown solution. It should be noted that the pullulan solution is transparent without color. The solution of Ag-NP/PL changed from a pale yellow to a dark brown solution after the gamma irradiation process was adopted.

Figure 2. Photograph of Ag-NP/PL at different different γ-irradiation doses: 0, 5, 10, 15, 20, 25, and 50 kGy (**A0–A6**).

3.2. Formation Mechanism of Silver Nanoparticles

In general, the radiolysis process involves producing a large number of homogenously distributed hydrated radicals [21,30–33]. Hydrated electrons and primary radicals and molecules appeared when the AgNO$_3$/PL aqueous suspensions were exposed to γ-rays, as shown in Equation (1). Ag$^+$ and NO$_3^-$ were created during the separation reaction of the AgNO$_3$ aqueous solution, as described in Equation (2). It is known that the e_{aq}^- and H atoms are vital reducing agents; thus, they can be effectively reduced to the zero-valent state (Equations (3) and (4)) [34]. Larger clusters of silver atoms (Equation (6)) were produced due to the coalescence of silver atoms created during the irradiation process. Then, a stabilized Ag cluster was formed after the aqueous electrons reacted with the Ag+ groups (Equation (7)) [35,36]. Finally, a large number of aqueous electrons (e_{aq}^-) were provided after γ-irradiation, resulting in silver ions being reduced to Ag-NPs (Equation (8)). In addition, the incorporation of pullulan also created more hydrated electrons (OH-). The hydrated electrons can reduce all silver ions to silver atoms, ending up as a nucleus for the successively formed atoms. The proposed mechanism of the formation of the silver nanoparticles is illustrated in the graphic in Figure 3.

The pullulan matrix is expected to better interact with the nanoparticles by preventing particle agglomeration between molecules [1,37]. As shown in Figure 3, the silver nanoparticles were coated with pullulan, indicative of interaction between multiple hydroxyl groups brought by pullulan. These, in turn, will arrange the silver by encapsulating the nanoparticles to form stable colloidal suspensions.

$$H_2O \xrightarrow{\gamma-Ray} e_{aq}^- + H + OH + H_3O^+ + H_2 + H_2O_2 + \ldots \tag{1}$$

$$AgNO_3 \xrightarrow{yields} Ag^+ + NO_3^- \tag{2}$$

$$Ag^+ + e_{aq}^- \xrightarrow{reduction} Ag^0 \tag{3}$$

$$Ag^+ + H \xrightarrow{reduction} Ag^0 + H^+ \tag{4}$$

$$Ag^0 + Ag^+ \rightarrow Ag_2^+ \tag{5}$$

$$nAg^+ + Ag_2^+ \rightarrow (Ag)_n^+ \tag{6}$$

$$(Ag)_n^+ + ne_{aq} \rightarrow (Ag)_n \tag{7}$$

$$(Ag)_n \xrightarrow{\gamma-Ray} (Ag)_n^+ + ne_{aq}^- \tag{8}$$

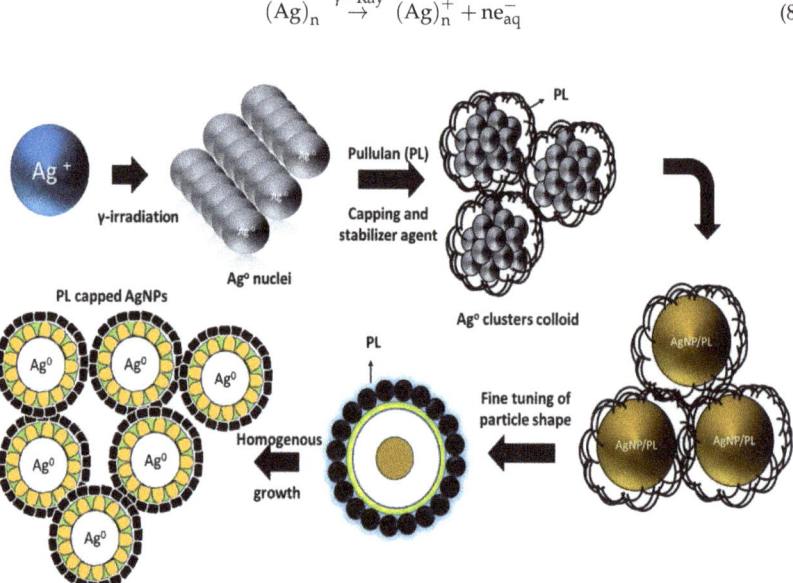

Figure 3. Proposed schematic mechanism of the synthesis and growth of silver nanoparticles on pullulan.

3.3. UV–Vis Analysis

UV–Vis analysis is a top screening analysis in confirming the formation and strength of silver nanoparticles. In this study, the appearance of Ag-NPs with different gamma radiation doses was monitored using UV–Vis spectroscopy. Figure 4 shows the UV–Vis absorption spectra of Ag-NP/PL at different gamma ray doses. The figures show that the highest-intensity absorption spectra exhibited a similar pattern for all radiation doses, known as surface plasmon resonance (SPR), at 410–420 nm. This result reveals that Ag-NP/PL prepared at 50 kGy, 25 kGy, 20 kGy, 15 kGy, 10 kGy, and 5 kGy produced absorption spectra with sharp peaks (surface plasmon resonance, SPR) at 410–420 nm. On the other hand, in Ag-NP/PL prepared without gamma radiation, the SPR peaks did not appear at 420 nm, indicating that no Ag-NPs were produced.

At a higher intensity, the gamma radiation provided a more spherical shape with a smaller size of produced NPs. For instance, at 50 kGy, the absorbance peak was at 0.9 compared that at 25 kGy, which was at 0.4. It can be depicted that the highest absorbance reflects a highly uniform particle size [38]. For reference, Ag-NP/PL was also prepared without the gamma radiation method. There was no absorption peak recorded at 420 nm, implying that Ag-NPs were not successfully produced. Despite this, the UV–Vis analysis also revealed that the shape and size of nanoparticles, which are dependent on the wavelength of absorption in the UV spectrum, were indicatively affected by the plasmon resonance. As shown in Figure 4, the spectra of silver nanoparticles are consistent with the spherical shape of different size Ag-NPs, which can later be proved in TEM analysis [39].

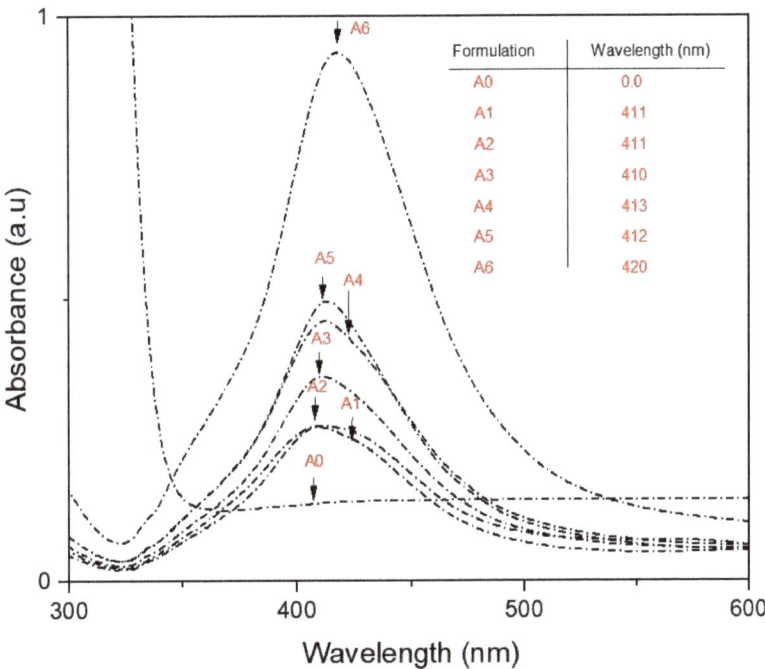

Figure 4. UV–Vis spectrum of Ag-NP/PL prepared at 0, 5, 10, 15, 20, 25, and 50 kGy gamma doses.

3.4. XRD Analysis

In order to further investigate the reaction mechanism that occurred during the synthesis, XRD analysis was carried out to determine the purity and crystallinity of Ag-NP/PL, as presented in Figure 4. It is known that pullulan (PL) exists in an amorphous state [40]. The incorporation of Ag-NPs and irradiation techniques was previously shown to transform the solution to exhibit crystallization characteristics [41]. Four (4) diffraction peaks can confirm the purity of the silver formation at 2θ of 38.54°, 44.91°, 65.04°, and 78.11°, corresponding to the (111), (200), (220), and (311) planes of the face-centered cubic (FCC) structure [42]. These four notable diffraction peaks are in good agreement with the theoretical standard figures (JCPDS file no. 01-087-0718) [42]. The formulation samples of A1–A6 in Figure 4 indicate that only silver metal is present in their crystalline phase without any significant traces of other substances. In addition, the occurrence of a nanosized particle of Ag-NP/PL promotes the considerable traces of silver formation in the XRD pattern peaks, as shown in A6 of Figure 5.

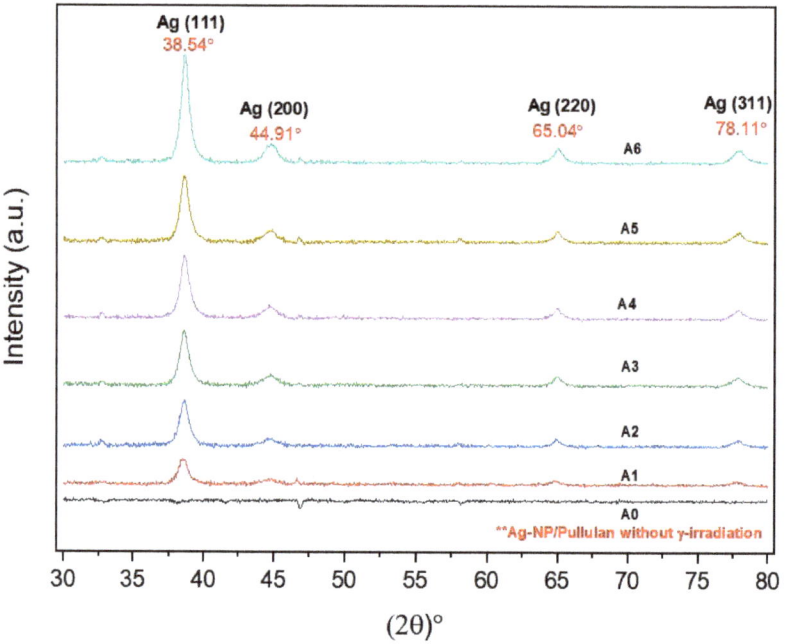

Figure 5. XRD patterns of Ag-NP/PL at different γ-irradiation doses: 0, 5, 10, 15, 20, 25, and 50 kGy (A0–A6).

3.5. TEM Analysis

Transmission electron microscopy (TEM) imaging was conducted to evaluate the morphology and particle size distribution of Ag-NPs. The TEM micrograph clearly reveals that the silver nanoparticles were well dispersed in the pullulan matrix, as shown in Figures 6–9, and this result is in line with that of previous studies [14]. In addition, the Ag-NPs were also segregated evenly with almost no agglomeration. Spherical and oval shapes were also observed in the TEM images [43]. Studies by Yoksan and Chirachancai (2010) showed that the particle size distribution (mean diameter and standard deviation) was drastically decreased to 3.98 ± 1.356 nm at 50 kGy doses, indicating an even Ag-NP segregation process [44]. As mentioned previously, in UV–Vis analysis, it can be stated that the γ-rays reduce the particle size distribution during the irradiation process. This condition confirms that the formation of the particle size distribution (mean diameter of particles) depends on the γ-ray doses. For low and medium γ-ray doses, i.e., 5, 10, and 25 kGy, the mean diameter and standard deviation of Ag-NPs were noted as 40.97 ± 31.64, 18.52 ± 13.906, 9.84 ± 3.595 nm, and 3.98 ± 1.356, respectively.

The combination of nanostructures at different sizes leads to the broadness of the size distribution peaks, as shown in Figures 6–9, which proves that the particles at high γ-doses were highly uniform and homogenous. In addition, 50 kGy γ-ray doses offer a more acceptable size (finer) of silver nanoparticles, indicating that hydrated radicals ($\bar{e}$ aq) are entirely produced; thus, this promotes the reduction of silver ions to silver atoms [30]. Moreover, the Ag-NPs had a narrow size distribution, implying that they were very uniform at 50 kGy (Figure 9). These results also reveal that as the gamma irradiation dosages were increased, the intermediate structure of the pullulan suspension significantly enhanced the interactions of (e_{aq}^-) electrons within the dissolved molecules, leading to an increased yield of nanoparticles. It is also found that pullulan also functions as a stabilizer during the synthesis of silver nanoparticles and as a capping agent by providing the template in the radiolysis process.

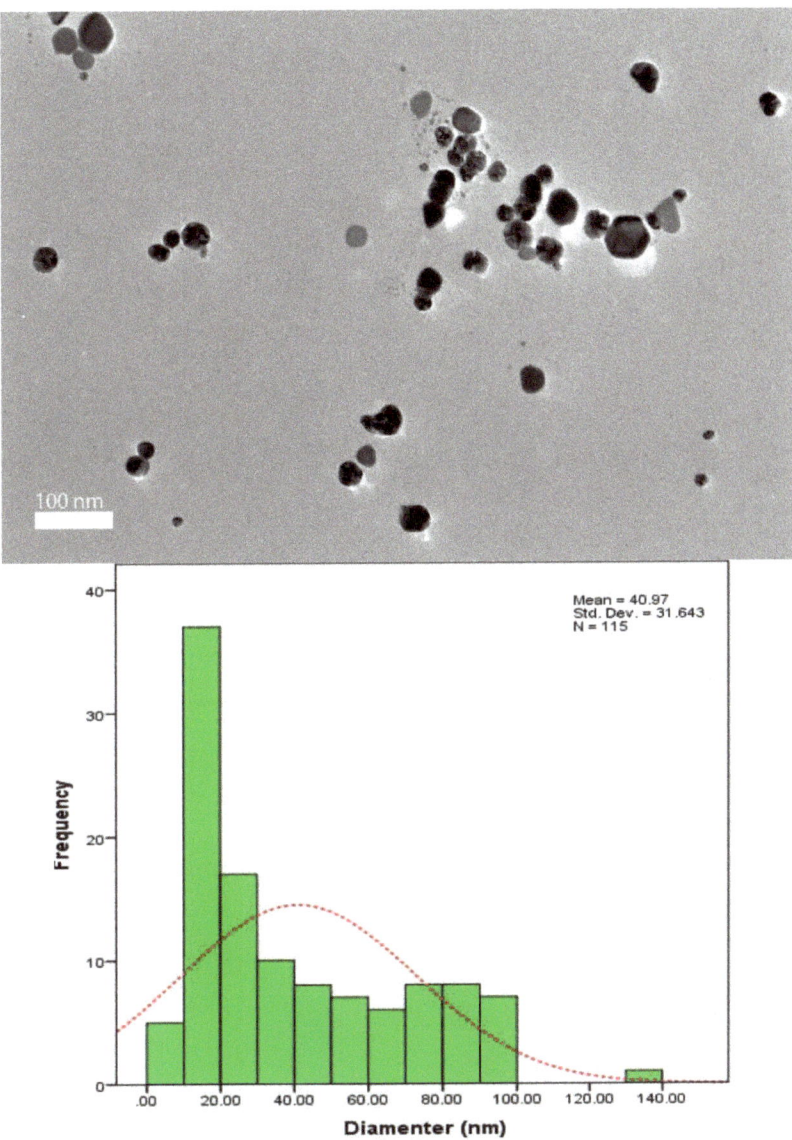

Figure 6. TEM image and corresponding particle size distribution of Ag-NP/PL at 5 kGy.

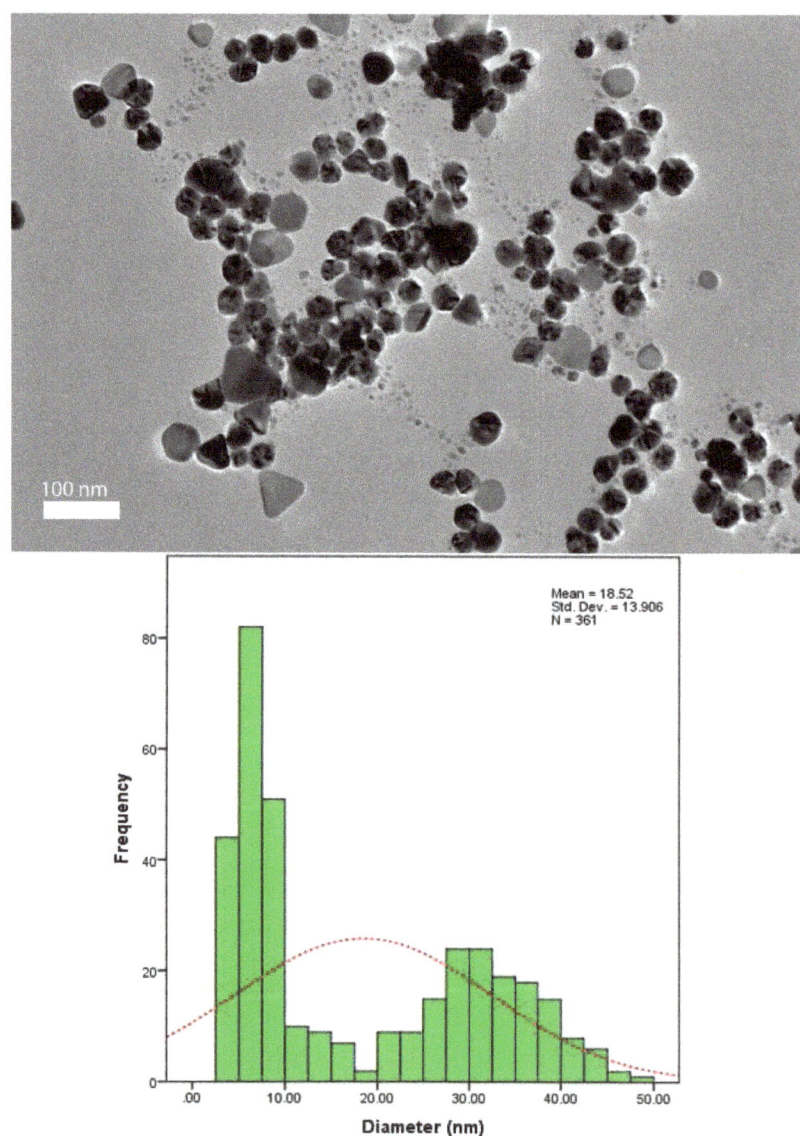

Figure 7. TEM image and corresponding particle size distribution of Ag-NP/PL at 10 kGy.

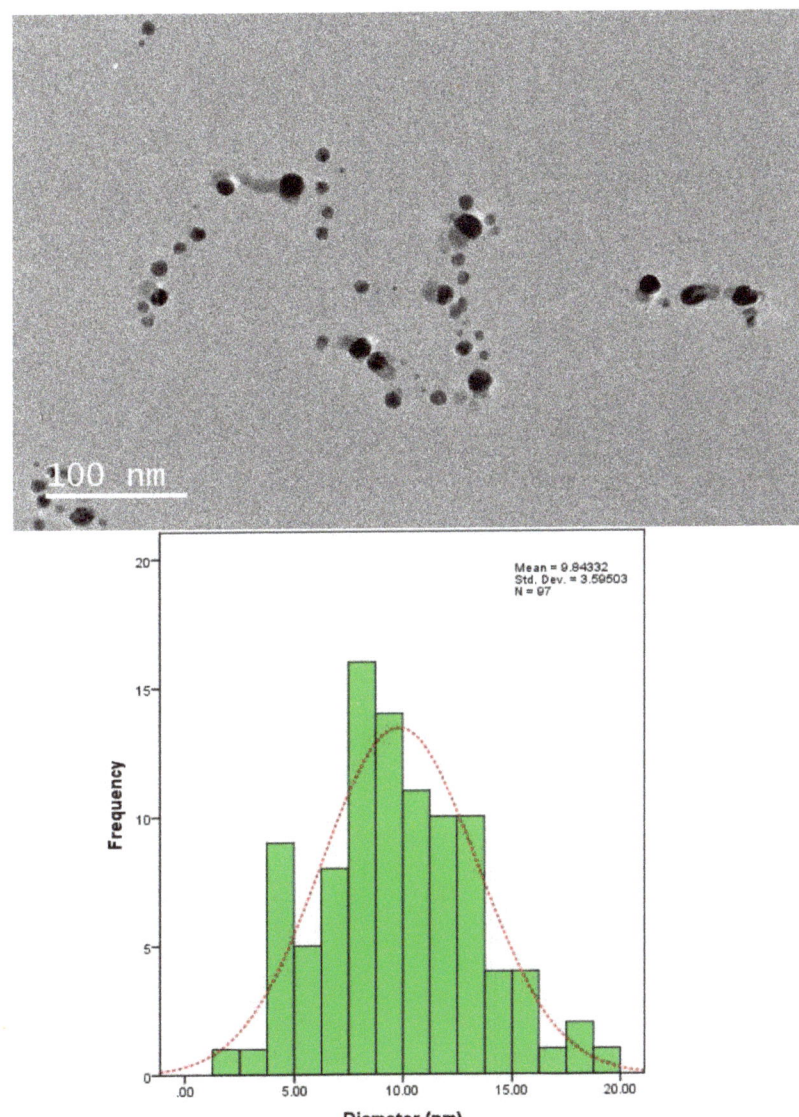

Figure 8. TEM image and corresponding particle size distribution of Ag-NP/PL at 25 kGy.

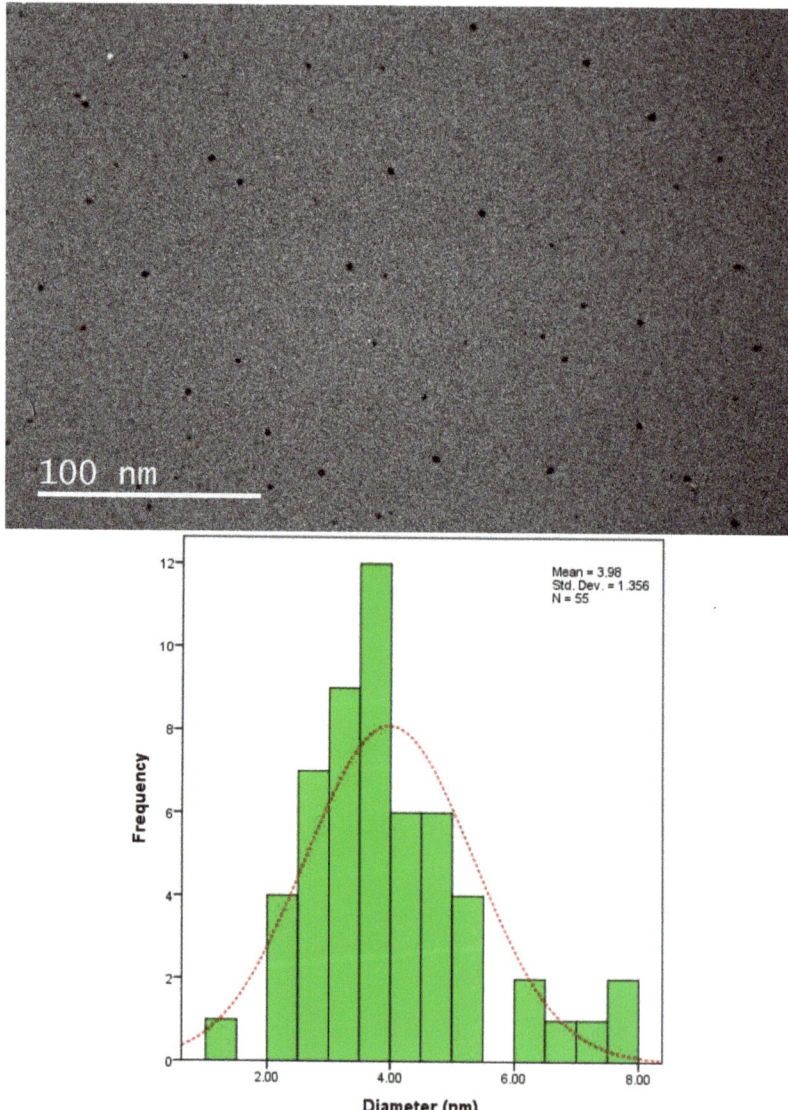

Figure 9. TEM image and corresponding particle size distribution of Ag-NP/PL at 50 kGy.

3.6. Antimicrobial Activity of Ag-NP/PL

It is vital to understand the stability of nanoparticles in aqueous solutions because the zeta potential is an indicator of the surface charge potential, which is a critical characteristic. The zeta potentials of Ag-NP/PL were found to be -72.05 ± 0.93 mV, as shown in Figure 10. It has been claimed that the generated nanoparticles have a negative charge on their surfaces. Having a positive or negative charge on the surface of nanoparticles makes them more stable and prevents them from aggregating together by pushing the same charges in the same direction [45]. Previous studies have confirmed that Ag-NPs can be considered stable if the zeta potential values are more than +30 mV or lower than −30 mV [45,46]. The zeta potential values recorded imply that the formation of Ag-NPs at the level of molecule activity led to stable forms of the nanoparticle solutions.

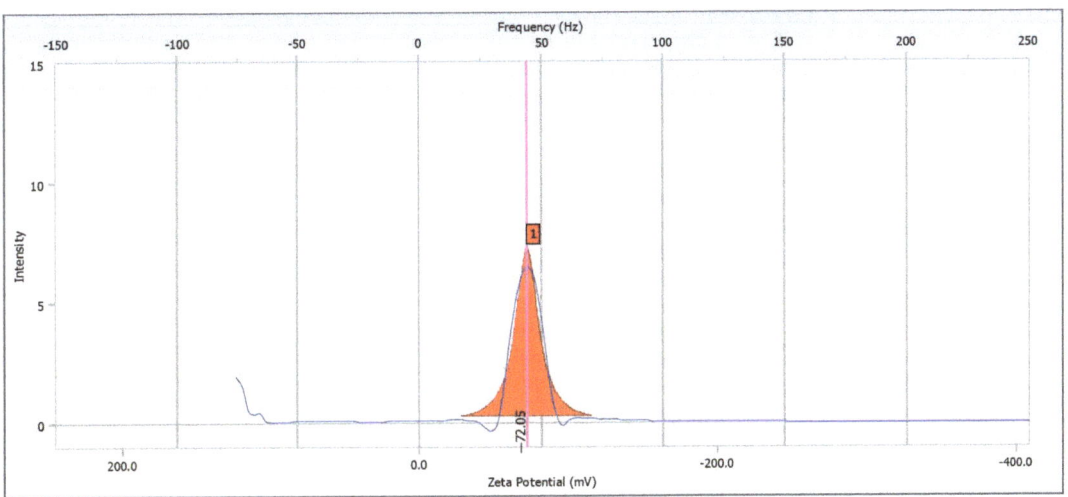

Figure 10. Zeta potential measurements for Ag-NP/PL at 50 kGy.

3.7. Antimicrobial Activity of Ag-NP/PL

The antimicrobial activity assay of the Ag-NP/pullulan biofilms against Gram-positive *Staphylococcus aureus* (*S.aureus*) was carried out by the culture medium toxicity method [47]. Typical broth and agar were used as media to grow the bacteria. The activity was evaluated after 24 h of incubation at 37 °C. The contact areas of all irradiation sample (0, 10, 25, and 50 kGy) biofilms were transparent, indicating inhibition of bacterial growth. The clear zone was prominent, as shown in Figure 11. At higher irradiation doses (50 kGy), the clear zone diameter was higher (16 mm) compared to that of the 25 kGy (12 mm) γ-doses.

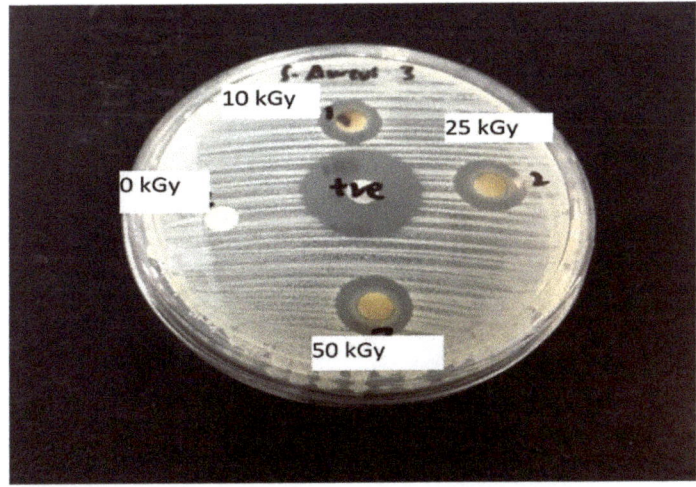

Figure 11. Photograph of antimicrobial test results against *Staphylococcus aureus*—After exposure to microbes.

The transparent zone was translated to the average diameter zone, which is tabulated in Table 1. Although the clear site was seen surrounding the biofilm embedded with Ag-NP/PL for both γ-ray doses, the microbe resistance was less effective for lower gamma rays. In short, the biofilm embedded with silver nanoparticles exhibited inhibitory activity against *S.aureus*. This antibacterial characteristic is expected to be created by the silver nanoparticles, which rupture the bacteria cell wall membrane. Disruption of the bacteria cell membrane instantly disturbs the respiration of the microbes, thus stopping the activity of the bacteria [48]. The outcomes of this analysis indicate that Ag-NP/PL embedded in the biofilm has good potential for antimicrobial food packaging applications.

Table 1. Average diameter of inhibition zone.

Sample	Average Diameter of Inhibition Zone (mm)
0 kGy	0 ± 0
10 kGy	10 ± 1.65
25 kGy	11 ± 2.34
50 kGy	13 ± 2.01

The antibacterial activity of Ag-NP/PL was primarily affected by the silver ions (Ag^+) that Ag-NPs released. As shown in Table 1, the higher the γ-ray doses, the better the antibacterial activity of the produced Ag-NPs. The release of more Ag^+ is influenced by the size of the particles [7]. In short, the smaller the particle size of Ag-NPs, the more they release Ag^+ due to the high surface area of Ag-NPs [49]. The antibacterial properties of Ag-NP/PL act by attachment of Ag-NP/PL on the cell membrane wall, leading to its rupture. Thus, Ag^+ is discharged inside the cell, which retards the cell's respiratory system by inducing the production of reactive oxygen species (ROS). An illustration of the mechanism of antibacterial activity that takes place with the aid of Ag^+ is shown in Figure 12. The stress placed on the cell wall membrane affected by the increase in ROS production is an effective mechanism of Ag-NP-induced silver ions. A previous study also revealed that the permeation of silver ions into bacterial cells changes DNA molecules into a condensed form and prevents the ability of the cells to replicate [50]. Moreover, bacterial sterilization is also caused by the silver ions due to the reaction with proteins, which leads to the direct binding with the sulfhydryl group (-SH) and causes the loss of activity of multiple enzymes [51].

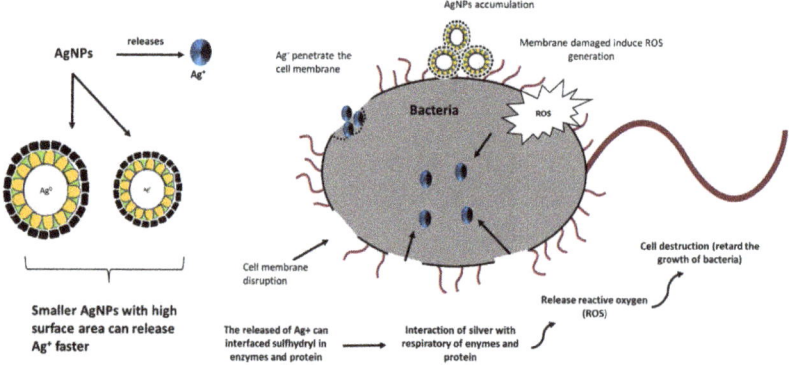

Figure 12. Antibacterial mechanism of Ag-NPs.

4. Conclusions

Pullulan-capped silver nanoparticles were successfully synthesized using the gamma irradiation technique. Stable Ag-NP/PL nanocomposites with an average size of 3.98 nm were prepared without any reducing agent. UV–visible spectroscopy confirmed the formation of silver nanoparticles by detecting a plasmonic band at 410–420 nm. The XRD pattern showed that the crystalline structure of the Ag-NPs for all samples was fcc. TEM imaging ascertained that the Ag-NPs were well dispersed in the pullulan matrix, with the particle diameter of the Ag-NPs gradually decreasing after higher doses at 50 kGy due to the γ-induced fragmentation in Ag-NPs. The zeta potential of Ag-NP/PL was negatively charged, and it was found to be a stable and good dispersion in a colloidal suspension. The biofilm embedded with Ag-NP/PL was found to have antibacterial activity against *Staphylococcus aureus*. In summary, pullulan-capped silver nanoparticles can be applied in a wide range of applications, particularly in antimicrobial biofilm packaging.

Author Contributions: Formal analysis, M.S.N.S.; investigation M.S.N.S. and R.R.A.; supervision R.R.A., K.S., M.Y.H., R.M.K. and M.M.N.; writing—original draft, M.S.N.S. and R.R.A. All authors have read and agreed to the published version of the manuscript.

Funding: This research was funded by Nippon Sheet Glass Foundation for Materials Science and Engineering (Grant: R.K130000.7343.4B635), Fundamental Research Grant Scheme (Grant: FRGS/1/2020/STG05/UTM/02/9), and a collaborative research grant between UTM, UMP, UMT, and UNIMAP (Grant: R.K130000.7343.4B539).

Institutional Review Board Statement: Not applicable.

Informed Consent Statement: Not applicable.

Data Availability Statement: Not applicable.

Acknowledgments: The authors are grateful for the financial support from Nippon Sheet Glass Foundation for Materials Science and Engineering (Grant: R.K130000.7343.4B635), Fundamental Research Grant Scheme (Grant: FRGS/1/2020/STG05/UTM/02/9) and a collaborative research grant between UTM, UMP, UMT, and UNIMAP (Grant: R.K130000.7343.4B539).

Conflicts of Interest: The authors declare that they have no known competing financial interests or personal relationships that could have appeared to influence the work reported in this paper. The authors also declare they have no conflict of interest.

References

1. Coseri, S.; Spatareanu, A.; Sacarescu, L.; Rimbu, C.M.; Suteu, D.; Spirk, S.; Harabagiu, V. Green synthesis of the silver nanoparticles mediated by pullulan and 6-carboxypullulan. *Carbohydr. Polym.* **2015**, *116*, 9–17. [CrossRef]
2. Khalil, K.A.; Fouad, H.; Elsarnagawy, T.; Almajhdi, F.N. Preparation and Characterization of Electrospun PLGA/silver Composite Nanofibers for Biomedical Applications. *Int. J. Electrochem. Sci.* **2013**, *8*, 3483–3493.
3. Ahmed, S.; Ahmad, M.; Swami, B.L.; Ikram, S. A review on plants extract mediated synthesis of silver nanoparticles for antimicrobial applications: A green expertise. *J. Adv. Res.* **2015**, *7*, 17–28. [CrossRef]
4. Alishah, H.; Pourseyedi, S.; Ebrahimipour, S.Y.; Mahani, S.E.; Rafiei, N. Green synthesis of starch-mediated CuO nanoparticles: Preparation, characterization, antimicrobial activities and in vitro MTT assay against MCF-7 cell line. *Rend. Lince-* **2016**, *28*, 65–71. [CrossRef]
5. Zhang, X.F.; Liu, Z.G.; Shen, W.; Gurunathan, S. Silver nanoparticles: Synthesis, characterization, properties, applications, and therapeutic approaches. *Int. J. Mol. Sci.* **2016**, *17*, 1534. [CrossRef]
6. Shameli, K.; Ahmad MBin Yunus, W.M.Z.W.; Ibrahim, N.A.; Gharayebi, Y.; Sedaghat, S. Synthesis of silver/montmorillonite nanocomposites using γ-irradiation. *Int. J. Nanomed.* **2010**, *5*, 1067–1077. [CrossRef]
7. Kaur, A.; Preet, S.; Kumar, V.; Kumar, R.; Kumar, R. Synergetic effect of vancomycin loaded silver nanoparticles for enhanced antibacterial activity. *Colloids Surf. B Biointerfaces* **2018**, *176*, 62–69. [CrossRef]
8. Sharma, V.; Yngard, R.A.; Lin, Y. Silver nanoparticles: Green synthesis and their antimicrobial activities. *Adv. Colloid Interface Sci.* **2009**, *145*, 83–96. [CrossRef] [PubMed]
9. Jayeoye, T.J.; Olatunde, O.O.; Benjakul, S.; Rujiralai, T. Synthesis and characterization of novel poly(3-aminophenyl boronic acid-co-vinyl alcohol) nanocomposite polymer stabilized silver nanoparticles with antibacterial and antioxidant applications. *Colloids Surf. B Biointerfaces* **2020**, *193*, 111112. [CrossRef] [PubMed]
10. van Hengel, I.A.J.; Putra, N.E.; Tierolf, M.W.A.M.; Minneboo, M.; Fluit, A.C.; Fratila-Apachitei, L.E.; Apachitei, I.; Zadpoor, A.A. Biofunctionalization of selective laser melted porous titanium using silver and zinc nanoparticles to prevent infections by antibiotic-resistant bacteria. *Acta Biomater.* **2020**, *107*, 325–337. [CrossRef] [PubMed]

11. Gupta, A.; Briffa, S.M.; Swingler, S.; Gibson, H.; Kannappan, V.; Adamus, G.; Kowalczuk, M.M.; Martin, C.; Radecka, I. Synthesis of Silver Nanoparticles Using Curcumin-Cyclodextrins Loaded into Bacterial Cellulose-Based Hydrogels for Wound Dressing Applications. *Biomacromolecules* **2020**, *21*, 1802–1811. [CrossRef]
12. Veerasamy, R.; Xin, T.Z.; Gunasagaran, S.; Xiang, T.F.W.; Yang, E.F.C.; Jeyakumar, N.; Dhanaraj, S.A. Biosynthesis of silver nanoparticles using mangosteen leaf extract and evaluation of their antimicrobial activities. *J. Saudi Chem. Soc.* **2011**, *15*, 113–120. [CrossRef]
13. Cheng, K.C.; Demirci, A.; Catchmark, J.M. Pullulan: Biosynthesis, production, and applications. *Appl. Microbiol. Biotechnol.* **2011**, *92*, 29–44. [CrossRef] [PubMed]
14. Zhang, K.; Ai, S.; Xie, J.; Xu, J. Comparison of direct synthesis of silver nanoparticles colloid using pullulan under conventional heating and microwave irradiation. *Inorg. Nano-Met. Chem.* **2017**, *47*, 938–945. [CrossRef]
15. Akter, M.; Sikder, T.; Rahman, M.; Ullah, A.K.M.A.; Hossain, K.F.B.; Banik, S.; Hosokawa, T.; Saito, T.; Kurasaki, M. A systematic review on silver nanoparticles-induced cytotoxicity: Physicochemical properties and perspectives. *J. Adv. Res.* **2018**, *9*, 1–16. [CrossRef]
16. Rahimi, M.; Noruzi, E.B.; Sheykhsaran, E.; Ebadi, B.; Kariminezhad, Z.; Molaparast, M.; Mehrabani, M.G.; Mehramouz, B.; Yousefi, M.; Ahmadi, R.; et al. Carbohydrate polymer-based silver nanocomposites: Recent progress in the antimicrobial wound dressings. *Carbohydr. Polym.* **2019**, *231*, 115696. [CrossRef]
17. Mohan, S.; Oluwafemi, O.S.; Songca, S.P.; Jayachandran, V.; Rouxel, D.; Joubert, O.; Kalarikkal, N.; Thomas, S. Synthesis, antibacterial, cytotoxicity and sensing properties of starch-capped silver nanoparticles. *J. Mol. Liq.* **2016**, *213*, 75–81. [CrossRef]
18. Yakout, S.M.; Mostafa, A.A. A novel green synthesis of silver nanoparticles using soluble starch and its antibacterial activity. *Int. J. Clin. Exp. Med.* **2015**, *8*, 3538–3544.
19. Krklješ, A.N.; Marinović-Cincović, M.T.; Kacarevic-Popovic, Z.M.; Nedeljković, J.M. Radiolytic synthesis and characterization of Ag-PVA nanocomposites. *Eur. Polym. J.* **2007**, *43*, 2171–2176. [CrossRef]
20. El-Batal, A.I.; El-Sayyad, G.; El-Ghamry, A.; Agaypi, K.; Elsayed, M.A.; Gobara, M. Melanin-gamma rays assistants for bismuth oxide nanoparticles synthesis at room temperature for enhancing antimicrobial, and photocatalytic activity. *J. Photochem. Photobiol. B Biol.* **2017**, *173*, 120–139. [CrossRef]
21. Morris, M.A.; Padmanabhan, S.C.; Cruz-Romero, M.C.; Cummins, E.; Kerry, J.P. Development of active, nanoparticle, antimicrobial technologies for muscle-based packaging applications. *Meat Sci.* **2017**, *132*, 163–178. [CrossRef]
22. Guan, Y. *Syntheses of Novel Water-Soluble Antimicrobial Polymers and Their Application in Papermaking*; The University of New Brunswick: Fredericton, NB, Canada, 2007.
23. Liu, M. *Synthesis of Bio-Based Nanocomposites for Controlled Release of Antimicrobial Agents in Food Packaging*; Food Science; The Pennsylvania State University: State College, PA, USA, 2014.
24. Mauriello, G. Control of Microbial Activity Using Antimicrobial Packaging. In *Jorge Barros-Velázquez, editor. Antimicrobial Food Packaging*; Academic Press, Elsevier: Amsterdam, The Netherlands, 2015; p. 141.
25. Bourtoom, T.; Chinnan, M.S. Preparation and properties of rice starch–chitosan blend biodegradable film. *LWT* **2008**, *41*, 1633–1641. [CrossRef]
26. Elsabee, M.Z.; Abdou, E.S. Chitosan based edible films and coatings: A review. *Mater. Sci. Eng. C* **2013**, *33*, 1819–1841. [CrossRef] [PubMed]
27. Cheviron, P.; Gouanvé, F.; Espuche, E. Green synthesis of colloid silver nanoparticles and resulting biodegradable starch/silver nanocomposites. *Carbohydr. Polym.* **2014**, *108*, 291–298. [CrossRef]
28. Salari, Z.; Danafar, F.; Dabaghi, S.; Ataei, S.A. Sustainable synthesis of silver nanoparticles using macroalgae Spirogyra varians and analysis of their antibacterial activity Sustainable synthesis of silver nanoparticles using macroalgae Spirogyra varians. *J. Saudi Chem. Soc.* **2016**, *20*, 459–464. [CrossRef]
29. Rao, Y.; Banerjee, D.; Datta, A.; Das, S.; Guin, R.; Saha, A. Gamma irradiation route to synthesis of highly re-dispersible natural polymer capped silver nanoparticles. *Radiat. Phys. Chem.* **2010**, *79*, 1240–1246. [CrossRef]
30. Ghazy, O.A.; Nabih, S.; Abdel-Moneam, Y.K.; Senna, M.M. Synthesis and characterization of silver/zein nanocomposites and their application. *Polym. Compos.* **2017**, *38*, E9–E15.
31. Kumar, B.; Smita, K.; Cumbal, L.; Debut, A. Green synthesis of silver nanoparticles using Andean blackberry fruit extract. *Saudi J. Biol. Sci.* **2015**, *24*, 45–50. [CrossRef]
32. Desai, K.; Kit, K.; Li, J.; Zivanovic, S. Morphological and Surface Properties of Electrospun Chitosan Nanofibers. *Biomacromolecules* **2008**, *9*, 1000–1006. [CrossRef]
33. Chen, P.; Song, L.; Liu, Y.; Fang, Y.-E. Synthesis of silver nanoparticles by γ-ray irradiation in acetic water solution containing chitosan. *Radiat. Phys. Chem.* **2007**, *76*, 1165–1168. [CrossRef]
34. Sheikh, N.; Akhavan, A.; Kassaee, M. Synthesis of antibacterial silver nanoparticles by γ-irradiation. *Phys. E Low-Dimens. Syst. Nanostructures* **2009**, *42*, 132–135. [CrossRef]
35. Janata, E. Structure of the Trimer Silver Cluster Ag 3^{2+} †. *J. Phys. Chem. B* **2003**, *109*, 7334–7336. [CrossRef]
36. Janata, E.; Henglein, A.; Ershov, B.G. First Clusters of Ag+ Ion Reduction in Aqueous Solution. *J. Phys. Chem.* **1994**, *98*, 10888–10890. [CrossRef]
37. Goia D., V. Preparation and formation mechanisms of uniform metallic particles in homogeneous solutions. *J. Mater. Chem.* **2004**, *14*, 451–458. [CrossRef]

38. Hosny, A.M.S.; Kashef, M.T.; Rasmy, S.A.; Aboul-Magd, D.S.; El-Bazza, Z.E. Antimicrobial activity of silver nanoparticles synthesized using honey and gamma radiation against silver-resistant bacteria from wounds and burns. *Adv. Nat. Sci. Nanosci. Nanotechnol.* **2017**, *8*, 045009. [CrossRef]
39. Razo-Lazcano, T.A.; Solans, C.; González-Muñoz, M.P.; Rivera-Rangel, R.D.; Avila-Rodriguez, M. Green synthesis of silver nanoparticles in oil-in-water microemulsion and nano-emulsion using geranium leaf aqueous extract as a reducing agent. *Colloids Surf. A Phys. Eng Asp* **2017**, *536*, 60–67. [CrossRef]
40. Biliaderis, C.; Lazaridou, A.; Mavropoulos, A.; Barbayiannis, N. WATER PLASTICIZATION EFFECTS ON CRYSTALLIZATION BEHAVIOR OF LACTOSE IN A CO-LYOPHILIZED AMORPHOUS POLYSACCHARIDE MATRIX AND ITS RELEVANCE TO THE GLASS TRANSITION. *Int. J. Food Prop.* **2002**, *5*, 463–482. [CrossRef]
41. Sun, J.-T.; Li, J.-W.; Tsou, C.-H.; Pang, J.-C.; Chung, R.-J.; Chiu, C.-W. Polyurethane/Nanosilver-Doped Halloysite Nanocomposites: Thermal, Mechanical Properties, and Antibacterial Properties. *Polymers* **2020**, *12*, 2729. [CrossRef]
42. Afify, T.; Saleh, H.; Ali, Z. Structural and morphological study of gamma-irradiation synthesized silver nanoparticles. *Polym. Compos.* **2015**, *38*, 2687–2694. [CrossRef]
43. Yoksan, R.; Chirachanchai, S. Silver nanoparticles dispersing in chitosan solution: Preparation by γ-ray irradiation and their antimicrobial activities. *Mater. Chem. Phys.* **2009**, *115*, 296–302. [CrossRef]
44. Yoksan, R.; Chirachanchai, S. Silver nanoparticle-loaded chitosan-starch based films: Fabrication and evaluation of tensile, barrier and antimicrobial properties. *Mater. Sci. Eng. C* 2010. [CrossRef]
45. Patil, M.; Singh, R.; Koli, P.; Patil, K.T.; Jagdale, B.S.; Tipare, A.R.; Kim, G.-D. Antibacterial potential of silver nanoparticles synthesized using Madhuca longifolia flower extract as a green resource. *Microb. Pathog.* **2018**, *121*, 184–189. [CrossRef]
46. Price, M.; Reiners, J.J.; Santiago, A.M.; Kessel, D. Monitoring Singlet Oxygen and Hydroxyl Radical Formation with Fluorescent Probes During Photodynamic Therapy. *Photochem. Photobiol.* **2009**, *85*, 1177–1181. [CrossRef]
47. Wei, D.; Sun, W.; Qian, W.; Ye, Y.; Ma, X. The synthesis of chitosan-based silver nanoparticles and their antibacterial activity. *Carbohydr. Res.* **2009**, *344*, 2375–2382. [CrossRef] [PubMed]
48. Wang, X.; Du, Y.; Fan, L.; Liu, H.; Hu, Y. Chitosan-metal complexes as antimicrobial agent: Synthesis, characterization and Structure-activity study. *Polym. Bull.* **2005**, *55*, 105–113. [CrossRef]
49. Qing, Y.; Cheng, L.; Li, R.; Liu, G.; Zhang, Y.; Tang, X.; Wang, J.; Liu, H.; Qin, Y. Potential antibacterial mechanism of silver nanoparticles and the optimization of orthopedic implants by advanced modification technologies. *Int. J. Nanomed.* **2018**, *13*, 3311–3327. [CrossRef] [PubMed]
50. Yan, X.; He, B.; Liu, L.; Qu, G.; Shi, J.; Hu, L.; Jiang, G. Antibacterial mechanism of silver nanoparticles in Pseudomonas aeruginosa: Proteomics approach. *Metallomics* **2018**, *10*, 557–564. [CrossRef]
51. Liau, S.Y.; Read, D.C.; Pugh, W.J.; Furr, J.R.; Russell, A.D. Interaction of silver nitrate with readily identifiable groups: Relationship to the antibacterial action of silver ions. *Lett. Appl. Microbiol.* 1997. [CrossRef] [PubMed]

Article

Development of Polydiphenylamine@Electrochemically Reduced Graphene Oxide Electrode for the D-Penicillamine Sensor from Human Blood Serum Samples Using Amperometry

Deivasigamani Ranjith Kumar [1,*], Kuppusamy Rajesh [2,*], Mostafa Saad Sayed [1,3], Ahamed Milton [1] and Jae-Jin Shim [1,*]

1 School of Chemical Engineering, Yeungnam University, 280 Daehak-ro, Gyeongsan 38541, Gyeongbuk, Republic of Korea
2 Research Centre, Sri Sivasubramaniya Nadar College of Engineering, Tamil Nadu 603110, India
3 Analysis and Evaluation Department, Egyptian Petroleum Research Institute, Nasr City, Cairo 11727, Egypt
* Correspondence: ranjith@yu.ac.kr (D.R.K.); rajeshche05@gmail.com (K.R.); jjshim@yu.ac.kr (J.-J.S.)

Abstract: D-penicillamine (PA) is a sulfur group-containing drug prescribed for various health issues, but overdoses have adverse effects. Therefore, regular, selective, and sensitive sensing is essential to reduce the need for further treatment. In this study, diphenylamine (DPA) was electropolymerized in an aqueous acidic medium. The PA detection sensitivity, selectivity, and limit of detection were enhanced by electropolymerizing DPA on an electrochemically reduced graphene oxide (ERGO)/glassy carbon (GC) surface. The formation of p-DPA and ERGO was investigated using various techniques. The as-prepared p-DPA@ERGO/GC revealed the excellent redox-active (N–C to N=C) sites of p-DPA. The p-DPA@ERGO/GC electrode exhibited excellent electrochemical sensing ability towards PA determination because of the presence of the –NH–functional moiety and effective interactions with the –SH group of PA. The p-DPA@ERGO/GC exhibited a high surface coverage of 9.23×10^{-12} mol cm^{-2}. The polymer-modified p-DPA@ERGO/GC electrode revealed the amperometric determination of PA concentration from the 1.4 to 541 µM wide range and the detection limit of 0.10 µM. The real-time feasibility of the developed p-DPA@ERGO/GC electrode was tested with a realistic PA finding in human blood serum samples and yielded a good recovery of 97.5–101.0%, confirming the potential suitability in bio-clinical applications.

Keywords: polydiphenylamine; electrochemically reduced graphene oxide; D-penicillamine; electropolymerization; cyclic voltammetry

Citation: Ranjith Kumar, D.; Rajesh, K.; Sayed, M.S.; Milton, A.; Shim, J.-J. Development of Polydiphenylamine @Electrochemically Reduced Graphene Oxide Electrode for the D-Penicillamine Sensor from Human Blood Serum Samples Using Amperometry. *Polymers* **2023**, *15*, 577. https://doi.org/10.3390/ polym15030577

Academic Editors: Arunas Ramanavicius and Claudio Gerbaldi

Received: 30 November 2022
Revised: 11 January 2023
Accepted: 17 January 2023
Published: 22 January 2023

Copyright: © 2023 by the authors. Licensee MDPI, Basel, Switzerland. This article is an open access article distributed under the terms and conditions of the Creative Commons Attribution (CC BY) license (https:// creativecommons.org/licenses/by/ 4.0/).

1. Introduction

Carbon and its composites, such as carbon nanotubes/graphene, with metal nanoparticles, organic dies, and conducting polymer-based electrode materials, have been studied extensively for various analyte determinations. Carbon and its composite electrode materials showed a low limit of detection (LOD) and high sensitivity and selectivity toward the target analyte sensor [1]. Graphene-based electrode materials exhibit excellent electrochemical properties for batteries, supercapacitors, fuel cells, and electrochemical sensor applications [2]. Graphene has high conductivity and a theoretical surface area of 2630 m^2 g^{-1}, with low intrinsic mass, excellent mechanical strength, and rapid charge transport through π–electron conjugation. The reduced graphene oxide-based nanocomposites and hybrids are attractive for electrode modification [3]. Several studies have evaluated electrochemically reduced graphene oxide (ERGO)-coated electrodes for diverse applications. ERGO-modified electrodes retained some unreduced oxygen functional groups that require further surface modification, such as metal nanoparticles, metal oxides, metal sulfides, and polymers, offering excellent electrochemical performance [4–6]. ERGO with conducting polymers have high sensing ability because of its active redox center, rapid electron transfer rate, and high specific surface area, making them desirable for selective sensing [7]. Reduced graphene oxide surface electropolymerized with polyaniline, polypyrrole, polydopamine, polydiphenylamine (p-DPA), and polyortho-phenylenediamine revealed extraordinary electrochemical performance [7–11]. Among the above polymers, N,N-diphenylamine ((Figure 1), DPA)-based polymer and its composites have attracted considerable attention owing to their outstanding electrochemical redox couple activity. Obrezkov et al. used polydiphenylamine (p-DPA) as an organic cathode material for lithium and potassium-ion (dual) batteries and reported a high energy density [12]. Muthusankar et al. synthesized p-DPA with a phosphotungstic acid/graphene composite electrode electrochemically and achieved excellent selectivity towards a urea sensor [13]. Suganandam et al. synthesized a p-DPA/platinum/indium tin oxide electrode for use as an iron (III) ion sensor using spectroelectrochemical and electrochemical methods [14]. These studies emphasized that p-DPA is an auspicious electrode material for energy storage and sensor applications.

Penicillamine (PA, Cuprimine), called 2-amino-3-mercapto-3-methylbutanoic acid, is a physiological sulfur (–SH) containing amino acid (R–CH(NH$_2$)COOH) in the amino-thiols family. PA exists in the L and D enantiomeric forms. L-Penicillamine is an extensive toxin that inhibits the role of pyridoxine in the body and leads to certain metabolic disorders. On the other hand, D-penicillamine (PA) is used as a therapeutic agent prescribed for treating rheumatoid arthritis, heavy metal poisoning cystinuria, liver disease, and Wilson's disease (Figure 1) [15,16]. PA is taken directly by active rheumatoid arthritis patients in the form of an oral tablet and capsule to reduce joint pain, collagen cross-linkage, and swelling. In daily life, PA is injected at several grams per day, but only 50% of PA is consumed by the human system, which increases the PA concentration in the blood and urine [17]. An overdose of PA in human fluids may cause severe health issues, such as rashes or itchy skin, loss of taste, loss of appetite, nausea, dysentery, and serious kidney problems [18–20]. Despite the biological importance, the unwanted health impact of PA in pharmaceutical and biological samples needs to be determined [21].

Therefore, the researchers focused on developing simple, reliable, real-time, and accurate PA determination methods [22]. Several methods are available for PA detection, such as colorimetry [23], nuclear magnetic resonance [24], high performance liquid chromatography [25], and gas chromatography [26]. Among these methods, electrochemical methods have attracted considerable attention because of the rapid response, easy sample preparation, and no need for trained staff to operate the electrochemical device. Although the electroanalytical technique revealed user-friendly and excellent performance, it still has issues, such as narrow concentration range, high LOD, and low selectivity and electrode sensitivity. The major issue of PA determination is that the operating potential of coexisting interfering species, such as L-cysteine, uric acid, dopamine, ascorbic acid, leucine, and glucose, are close to the target analyte [27,28]. Chemically modified electrodes have resolved these problems, owing to their high conductivity, sensitivity, selectivity, rapid electron transfer rate, and low operating potential with negligible surface fouling. Modified electrodes have been designed based on nanocomposite [29], nanohybrids [30], conducting polymer [31], and metal nanoparticles [32] and applied to PA sensing in pharmaceutical and biological samples.

This paper reports the electrochemical formation of p-DPA on an ERGO/glassy carbon (GC) surface for use as an electrochemical PA sensor. The formation mechanism of the p-DPA@ERGO/GC surface involved oxidative polymerization of DPA monomer through a 4,4 coupling reaction in an aqueous acidic medium. The p-DPA@ERGO/GC electrode was used innovatively for PA determination using voltammetry and amperometric techniques. Finally, the proposed sensor examined PA in the biological human serum samples.

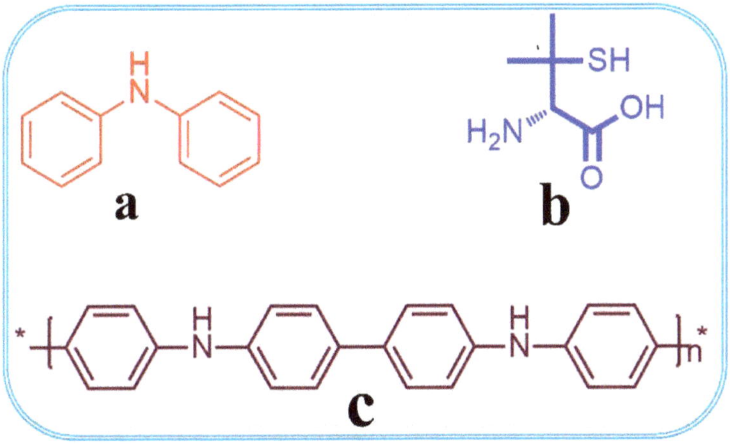

Figure 1. Chemical structure of (**a**) *N,N*-Diphenylamine; (**b**) D-Penicillamine; and (**c**) Poly(diphenylamine).

2. Experimental

2.1. Materials

Diphenylamine and D-penicillamine were purchased from Sigma-Aldrich (St. Louis, MO, USA). Type 1 GC plate (1 mm thickness), graphite, disodium hydrogen phosphate (Na_2HPO_4), monosodium hydrogen phosphate (NaH_2PO_4), potassium ferrocyanide ($K_4[Fe(CN)_6]$), and potassium ferricyanide ($K_3[Fe(CN)_6]$)] were obtained from Alfa Aesar (Shanghai, China). Potassium permanganate ($KMnO_4$) and sodium nitrate ($NaNO_3$) were obtained from SRL (Mumbai, India). All other regents and chemicals were of analytical grade and used as received. Deionized (DI) water was used in all solution preparations in this study. The phosphate buffer solution (PBS) pH 7.2 was prepared by mixing 0.1 M Na_2HPO_4 and NaH_2PO_4, which is used as a supporting electrolyte for PA detection.

2.2. Graphene Oxide Preparation

Graphene oxide was synthesized using a resemble procedure defined in former work [33]. The graphite (1 g) and $NaNO_3$ (1.5 g) were placed into a one-liter beaker containing 150 mL of H_2SO_4 with magnetic stirring. Subsequently, 3 g $KMnO_4$ was transferred carefully to the acid suspension at a reaction temperature of 5 °C. After $KMnO_4$ addition, the reaction temperature was increased to 90 °C and held at that temperature for one hour, resulting in a light brownish color product. The product was diluted slowly with 300 mL of H_2O and stirred for another two hours. Subsequently, 30 mL of 5% H_2O_2 were added to the above suspension, which turned brown. The product was washed with deionized water until the filtrate water was pH 7.0. The final product was freeze-dried.

2.3. p-DPA@ERGO/GC Electrode Assembly

Scheme 1 presents the comprehensive electrode coating steps. Before the GO coating on GC, the electrode was polished with 1.0, 0.3, and 0.05-micron alpha alumina powders and swept away thoroughly with DI water. The electrode probe was then exposed to an electrochemical activation process by successive CV cycling (10 cycles) in a 0.1 M H_2SO_4 electrolyte in the potential window from −0.2 V to 1.0 V [34,35]. The cleaned GC electrode was used for surface modification. The as-prepared GO (5 mg) was dispersed in 5 mL of ethanol in a 5% Nafion™ solution with ultrasonication for 20 min to acquire a homogenous suspension. Subsequently, 10 μL of the GO suspension spread onto a clean GC surface and dried at room temperature. The GO/GC coated electrode was subjected to electrochemical reduction in PBS (0.2 M, pH 7.2) by CV cycling over the potential range of 0.0 V to −1.5 V with fifteen sweeps to achieve the ERGO/GC. The ERGO/GC electrode underwent electropolymerization of DPA by cycling between the potential window of −0.4 V to +1.3 V in the presence of 1 mM DPA in 0.1 M H$_2$SO$_4$ supporting electrolyte, which yielded the p-DPA@ERGO/GC [36]. Similarly, the p-DPA@GC electrode was obtained using the same procedure on the GC surface. Similar to the above method, p-DPA@GC, ERGO/GC, and p-DPA@ERGO/GC electrodes were prepared on the GC-plate surface for X-ray photoelectron spectroscopy (XPS), Raman spectroscopy, and surface microstructure studies.

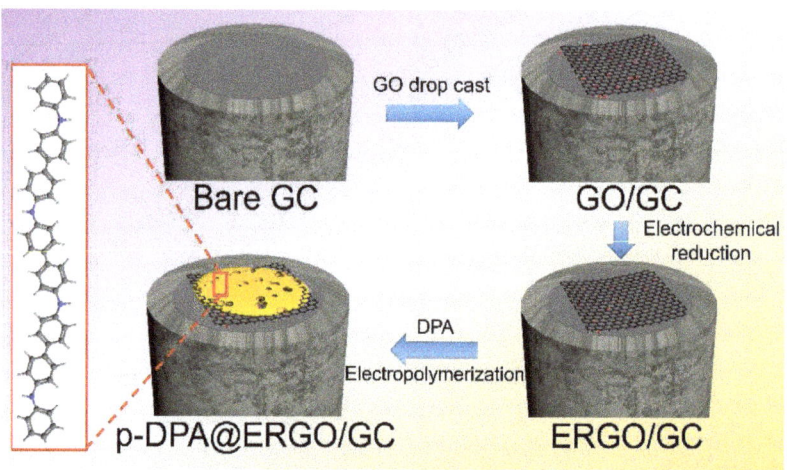

Scheme 1. Stepwise p-DPA@ERGO/GC electrode preparation.

2.4. Instrumentation and Analytical Procedure

The surface morphology of the as-prepared GO, ERGO, and polymer film was characterized by field-emission scanning electron microscopy (FESEM, S-4800, Hitachi, Tokyo, Japan). The Raman spectra (Horiba, XploRA PLUS, Kyoto, Japan) were obtained using an operating laser wavelength of 532 nm. The formation of GO and ERGO was investigated by X-ray diffraction (XRD, PANalytical, X'Pert-PRO MPD, PANalytical, Almelo, The Netherlands) using Cu K α1 radiation (1.5406 Å). The elemental composition of the as-prepared sample was examined XPS (Thermo Scientific, Boston, MA, USA). All the electrode fabrication and PA sensor experiments (electrochemical) were performed using a PGSTAT302N electrochemical (AUTOLAB, Utrecht, The Netherlands) workstation. A 25 mL volume electrochemical cell in a three-electrode system was used for all electrochemical tests. A glassy carbon (geometry of 0.0707 cm^2) electrode was used as the working electrode. The saturated (KCl) calomel electrode (SCE) was used as a reference electrochemical potential. A high surface geometrical area of Pt wire was used as the counter electrode. The 0.1 M PBS (pH 7.2) solution was used as a supporting electrolyte for PA determination, except for the p-DPA polymerization. A suitable volume of PA stock (0.01 M) solution was transferred to a 25 mL volume of an electrochemical cell (0.1 M PBS) in this solution immersed in three electrodes to evaluate the electrochemical results. The amperometry experiment was conducted at a constant potential of +0.62 V. In the case of cyclic voltammetry (CV) and differential pulse voltammetry (DPV) of the PA, electrochemical detection was measured over the potential of −0.20 V to +0.90 V range (0.1 M PBS).

2.5. Real Sample Analysis Procedure

A suitable standard aqueous PA solution was transferred to the 5% diluted human serum sample (Human serum received from Sigma-Aldrich, human male AB plasma, USA origin, sterile-filtered). The PA was determined using the standard addition method according to the amperometry technique. The PA concentration of the standard sample was yielded by extrapolating the respective standard calibration plot on the *x*-axis.

3. Results and Discussion
3.1. ERGO/GC and p-DPA@ERGO/GC Electrochemical Studies

Figure 2A displays the CV trace of the GO/GC electrochemical reduction process in the potential of 0.0 V to −1.50 V at a sweep rate of 50 mV s^{-1} in 0.2 M PBS (pH 7.2). The first cycle of the voltammogram showed a broad cathodic peak from −0.8 to −1.5 V. This cathodic peak corresponds to the epoxy, carbonyl, and hydroxyl group reduction [9]. The electrochemical reduction of GO to ERGO occurred by adding a proton from the supporting electrolyte and an electron from the electrode. The possible hydroxyl-group-reduction mechanism is given in Equation (1) [37,38]:

$$\text{GO-OH} + x\text{H}^+ + y\text{e}^- \rightarrow \text{ERGO} + z\text{H}_2\text{O} \tag{1}$$

In the subsequent cycles, the cathodic peak current was decreased gradually by increasing the cycle numbers, which described the removal of carbonyl, hydroxyl, and epoxy functional groups, resulting in restored sp^2 carbon [37,38]. After 10 CV cycles, the GO/GC electrode reduction peak current reached stability, which ensures that the GO was transferred electrochemically as ERGO on the GC surface and appeared as a black color.

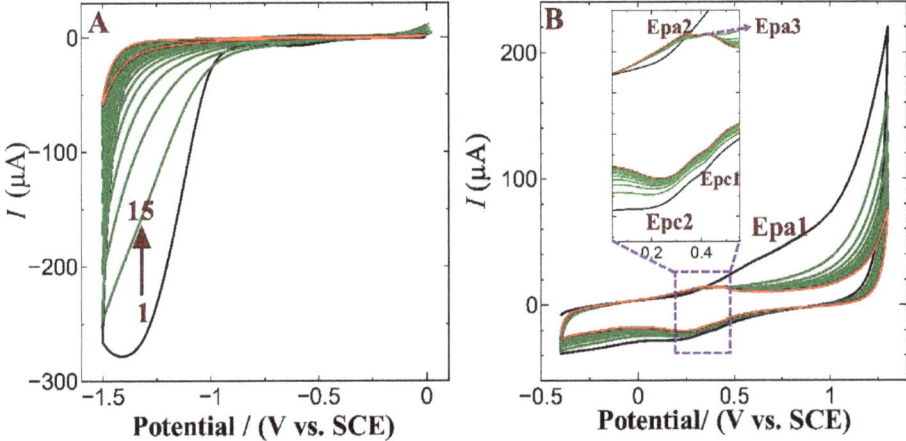

Figure 2. (**A**) GO electrochemical reduction CV curve trace in 0.2 M PBS (pH 7.2) at a sweep rate of 50 mV s^{-1}; (**B**) electropolymerization of DPA on ERGO/GC in a 0.1 M H$_2$SO$_4$ contains 1 mM DPA monomer at a 50 mV s^{-1} scan rate.

Figure 2B displayed a CV trace of DPA electropolymerization on the ERGO/GC electrode surface. The CV trace of 10 cycles was performed in the potential window from −0.40 V to +1.30 V in 1 mM DPA monomer with 0.1 M H$_2$SO$_4$ supporting electrolytes. In the first CV cycle, a broad anodic peak (Epa1) appeared at +0.55 V, corresponding to the positively charged nitrogen species of radical cation formation, followed by radical rearrangement [36]. These monomer radicals transform, dimerize, and join to form the p-DPA polymer (Figure 3). The polymer chain propagation occurred at the *para* site, such as the benzidine-type radical cation. In the reverse scan, cathodic peaks at +0.26 V to +0.4 V were observed, which correspond to the reduction of the dimeric and oligomeric products by electron/protonation. In the second cycle, the CV trace exhibited peaks at +0.32 V and +0.42 V ascribed to p-DPA oxidation and DPA monomer further deposition/conversion as a cation radical to oligomer on the ERGO/GC surface [36]. During the third and subsequent cycles, the anodic peak at +0.32/+0.23, with progressive increases in the peak current, suggests that the dimers undergo electrochemical oxidative polymerization resulting in the continued growth of p-DPA films on the ERGO/GC electrode surface. After eight cycles, the anodic (Epa2) and cathodic (Epc2) redox peak currents are stabilized, which confirmed the stable p-DPA@ERGO/GC film formation. Figure 3 shows the DPA electrochemical polymerization mechanism based on the CV trace result.

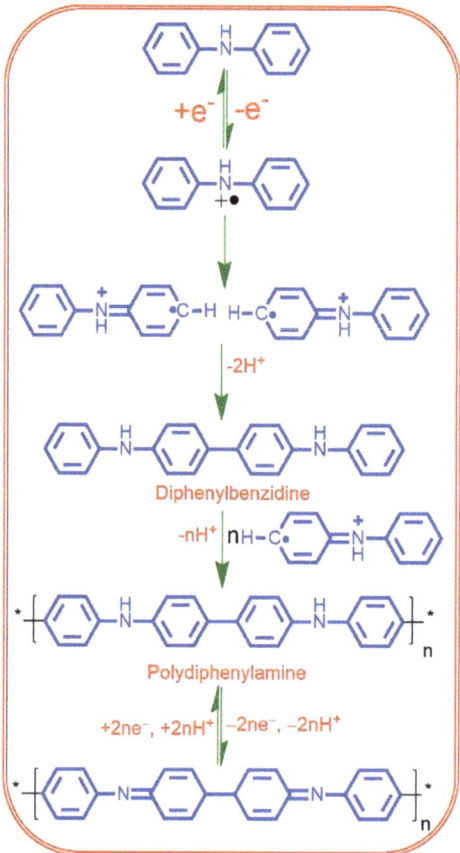

Figure 3. DPA stepwise electropolymerization reaction mechanism.

3.2. Structural and Surface Morphological Study

The chemical functional and surface properties of the GO and ERGO were characterized by Raman spectroscopy, as shown in Figure 4(Aa,Ab). The Raman spectrum of as-produced GO exhibited two intense peaks at 1355 cm^{-1} and 1606 cm^{-1} were attributed to the D and G bands of phonons A_{1g} (K-points) and E_{2g} symmetry carbon of graphene oxide [7]. The D and G bands confirmed the chemically oxidized form of the graphene basal plane and edge defects. The D band of ERGO showed nearly the same intensity with the GO due to the increased defect level. This shows that the sp^2 domains and edge defects confirm the electrochemical reduction of GO [39]. The calculated I_D/I_G ratios of GO and ERGO were 0.84 and 0.92, respectively, highlighting the significant reduction that occurred from GO to ERGO. The gained I_D/I_G value was used to calculate the average crystallite (L_a) size using Equation (2) [40]:

$$L_a(\text{nm}) = \left(2.4 \times 10^{-10}\right) \lambda_l^4 \left(\frac{I_D}{I_G}\right)^{-1} \quad (2)$$

where L_a is the average crystallite size (nm), and λ is the laser wavelength (532 nm) applied in the Raman spectroscopy. The calculated average crystallite size of a GO and ERGO were

22.80 and 20.80 nm, respectively. The decreasing crystallite size of GO to ERGO indicated the successful reduction of oxygen functional groups.

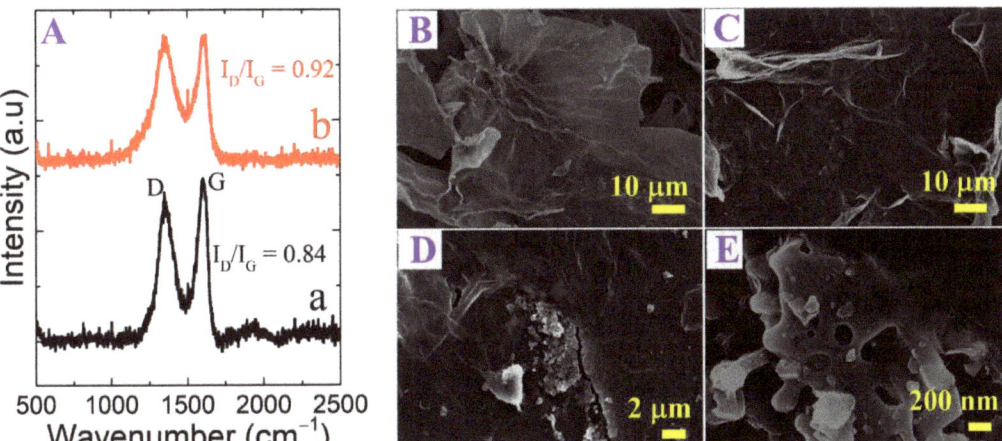

Figure 4. (**A**) Raman spectra of (a) GO and (b) ERGO; (**B**–**D**) FESEM of GO (**B**), ERGO (**C**), and p-DPA@ERGO (**D**); (**E**) high magnification image of p-DPA@ERGO.

FESEM was performed to understand the surface structure of the as-prepared GO/GC, ERGO/GC, and p-DPA@ERGO/GC film-coated electrodes (Figure 4B–E). The as-prepared GO showed a crumpled/wrinkled surface structure in Figure 4B, indicating the well-exfoliated graphene oxide layers. Figure 4C shows the oxygen functional groups removed ERGO, which revealed more crumpled/wrinkled-like surface morphology with substantial swelling. This might be due to the cathodic reduction of GO leading to the increased electrochemical active surface area or hydrogenation. The ERGO surface revealed the irregular bulk p-DPA particles, as shown in (Figure 4D,E). The amino (C-NH-C) functional group of p-DPA interacts with negatively charged (unreduced oxygen functional moiety on ERGO) groups, and the monomer aromatic π–π stacking interface is responsible for the p-DPA on the ERGO surface (Figure 5D). The high-magnified p-DPA@ERGO/GC surface morphology revealed an irregular polymer bulk particle-like structure.

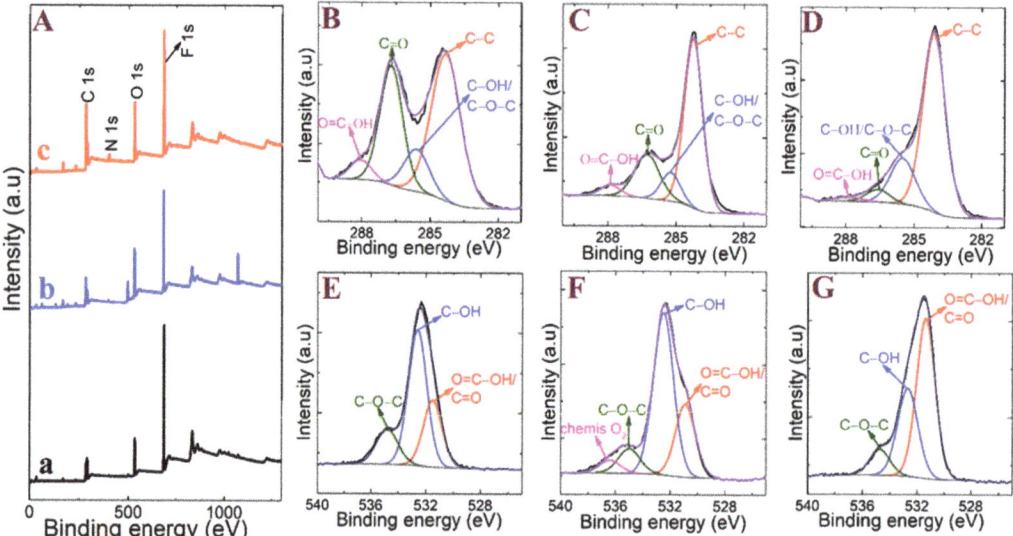

Figure 5. (**A**) XPS survey spectrum of (a) GO, (b) ERGO, and (c) p-DPA@ERGO; (**B,E**) GO C 1s and O 1s; (**C,F**) ERGO C 1s and O 1s; (**D,G**) p-DPA@ERGO/GC C 1s and O 1s region spectra.

3.3. XRD and XPS Analysis

Figure S1(Aa,Ab) (see Supplementary Materials) presents XRD patterns of the GO and ERGO on the FTO-coated electrode. The GO XRD peak appeared at 11.8° 2θ (001), indicating the 0.75 nm layer space of graphene sheets [41]. This might be because water and oxygen functional groups are dispersed within the oxidized graphene layers. After electrochemical reduction, the low angle peak diminished, and a broad shoulder peak appeared at 22° 2θ (002) explored the GO reduction, which is the removal of oxygen functional groups. The XPS spectra provide information on the sample surface elements. Figure 5(Aa–Ac) presents the GC plate surface-coated GO, ERGO, and p-DPA@ERGO samples' survey spectra investigated results. The GO and ERGO survey traces revealed C 1s, O 1s, and F 1s peaks (source from Nafion binder). By contrast, p-DPA showed an additional N 1s peak indicating that the DPA polymerized product was on the ERGO surface. The carbon/oxygen ratio increased from GO (0.5) to ERGO (0.88), and p-DPA@ERGO (1.0) emphasized the oxygen functional group reduction through an electrochemical approach. The comprehensive understanding of functional group changes on GO, ERGO, and p-DPA@ERGO elements, the C 1s, O 1s, and N 1s, the region was deconvoluted in Figures 5B–G and S1B. The high-resolution C 1s peak of GO and ERGO was fitted to the four envelopes: C–C, C–O–C/C–OH, C=O, and COOH, respectively (Figure 5B,C) [41,42]. The area under the envelope of oxygen functional groups decreased significantly, indicating the effective GO reduction (Table S1, see Supplementary Materials). Similarly, the ERGO O 1s peak area was lower than GO. Here, some of the –OH groups on the ERGO surface exploit the DPA monomer amine electrostatic interaction, leading to the facile microenvironment for polymer entrapment. In addition, p-DPA@ERGO revealed N 1s, indicating that the p-DPA polymer was successfully coated on the ERGO surface. The N 1s region is divided into two envelopes, such as imine (–N=) and amine (N–H) groups (Figure S1B, see Supplementary Materials). Table S1 (see Supplementary Materials) lists the peak positions, elements assignments, and area of the envelopes of the elements.

3.4. p-DPA@ERGO/GC Electrochemical Studies

The scan rate effects were examined from 5 to 500 mV s^{-1} in the 0.1 M H_2SO_4 supporting electrolyte to understand the electrochemical properties of the p-DPA@ERGO/GC-coated electrode (Figure 6A). CV traces of p-DPA@ERGO/GC showed well-defined anodic and cathodic peaks, corresponding to the conversion of p-DPA and N,N-diphenylbenzidine protonation of the N atom in the polymer backbone [36]. The anodic/cathodic peak currents increased with increasing scan rate. As the sweep rate was increased, the anodic and cathodic peak potentials moved toward the positive and negative sides, respectively. Figure 6B presents a plot of anodic/cathodic peak current versus sweep rate, showing that the peak current increased as the scan rate was increased. This indicates the rapid electron transfer kinetics of the p-DPA film. The surface coverage (τ) of the as-prepared polymer film (p-DPA@ERGO/GC)-coated electrode was calculated from the CV curve using Equation (3) [43]:

$$\tau = \frac{Q}{nFA} \quad (3)$$

where Q is a surface charge (C) obtained by integrating the CV area beneath the trace; n is the total number of electrons consumed in the present redox process (n = 2 in the p-DPA redox couple); A is the electrochemical surface area (p-DPA@ERGO/GC scan rate effect performed in the potassium ferrocynide solution (1 mM, in and 0.1 M KCl), using the I versus $v^{1/2}$ plot. The gained slope value substituted Randles–Sevcik equation for the electrochemical active surface area calculation. A = 0.0812 cm^2); F is the Faraday constant. The calculated surface concentration of electroactive species on the electrode surface was 9.23×10^{-12} mol cm^{-2}. The presence of electroactive p-DPA on the ERGO/GC surface enhanced surface coverage might be more feasible for determining a wide range of PA concentrations.

Electrochemical impedance spectroscopy (EIS) is a unique technique for modified electrode analysis. Table S2 (see Supplementary Materials) lists the fitted equivalent circuit model and corresponding component values. The EIS semicircle starting point of the high frequency on the x-axis intercept represents the electron-transfer limited process (electrolyte resistance, Rs). The Q1 and R_{th} parallel loop indicate the p-DPA and p-DPA@ERGO thin polymer materials impedance. The distance of the arch indicates the charge transfer resistance (Rct). The linear sector occurred at the low-frequency section because of a diffusion-controlled process, which dominates the electron transfer kinetics of the $[Fe(CN)_6]^{3-/4-}$ probe moiety on the p-DPA@ERGO/GC electrode interface [44]. Figure 6C shows the typical Nyquist plots of bare GC, p-DPA@GC, and p-DPA@ERGO/GC electrodes in $[Fe(CN)_6]^{3-/4-}$ (5 mM) with 0.1 M KCl medium. The bare GC and p-DPA@GC electrode revealed Rct of 156 Ω and 2790 Ω. The p-DPA@GC shows high Rct than the bare GC, which designates p-DPA alone sluggish electron transfer rate. In contrast, the p-DPA@ERGO/GC electrode revealed the Rct of 1158 Ω, suggesting effective contact between ERGO and p-DPA. The combination of ERGO and p-DPA acts as an excellent charge transfer material that allows the rapid acceleration of electrons on the $[Fe(CN)_6]^{3-/4-}$ redox probe with a p-DPA@ERGO/GC composite electrode.

Figure 6D shows a CV trace of bare GC, p-DPA/GC, and p-DPA@ERGO/GC-coated electrodes in the redox probe of $[Fe(CN)_6]^{3-/4-}$ (5 mM). The oxidation/reduction peak potential of the bare GC was observed at +0.29 V and +0.06 V, respectively. The p-DPA@GC-coated electrode showed a low peak current intensity and increased potential difference ΔE_p(anodic − cathodic) = 270 mV (E_{pa} = +0.38 V; E_{pc} = +0.02 V) because of the sluggish electron transfer rate compared to the bare carbon substrate. The bulk and irregular p-DPA growth on the GC surface leads to sluggish electron transfer. In contrast, the p-DPA@ERGO/GC electrode showed a low anodic and cathodic peak potential difference ΔE_p = 130 mV (E_{pa} = +0.25 V; E_{pc} = +0.12 V), which is lower than the p-DPA/GC and bare GC. The higher p-DPA@ERGO/GC current than p-DPA@GC indicates that composite electrodes fast electron transfer ability. The lower ΔE_p values of p-DPA@ERGO/GC compared to p-DPA/GC indicated that the ERGO and polymer composite offered a high electron transfer microenvironment.

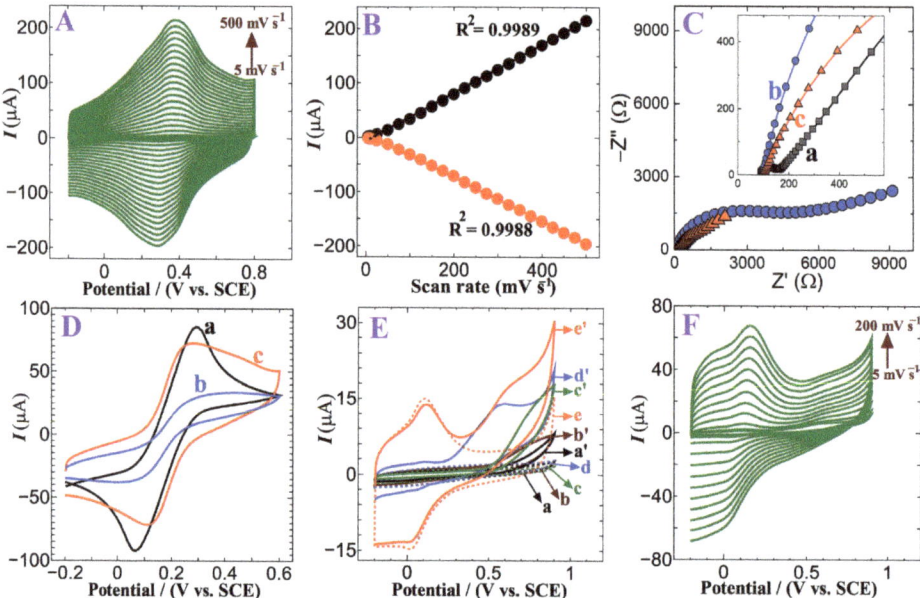

Figure 6. (**A**) CV scan rates of p-DPA@ERGO/GC in 0.1 M H_2SO_4 from 5 to 500 mV s^{-1}, (**B**) corresponding plot of peak current (I) versus scan rate; (**C,D**) electrochemical impedance spectroscopy and CV of (a) bare GC, (b) p-DPA/GC, and p-DPA@ERGO/GC in $[Fe(CN)_6]^{3-/4-}$ (5 mM) solution of 0.1 M KCl; (**E**) CV trace of (a) bare GC, (b) activated GC, (c) p-DPA/GC, and (d) ERGO/GC, and (e) p-DPA@ERGO/GC-modified electrodes in the absence of analyte and a′, b′, c′, d′, and e′ in the presence of 1 mM PA with 0.1 M PBS at a 50 mV s^{-1} scan rate; (**F**) scan rate effect of the p-DPA@ERGO/GC electrode in PA concentration of 1 mM.

3.5. Electrochemical Behavior of D-Penicillamine on Modified Electrodes

The different modified electrodes were evaluated for the electrochemical detection of PA. Figure 6(Ea–Ee) represents the CV response of the bare GC, activated GC, p-DPA/GC, ERGO/GC, and p-DPA@ERGO/GC-modified electrodes absence of PA (in a 0.1 M PBS supporting medium). The CV response of bare GC, activated GC and p-DPA/GC electrodes offered a low background current in a PBS solution, but the ERGO/GC electrode showed a high background current because of its high conductivity. In contrast, the p-DPA@ERGO/GC electrode shows well-defined polymer redox couple peaks at +0.10/+0.05 V. The polymer redox potential difference (ΔE_p) for p-DPA@ERGO a 50 mV, which is lower than the p-DPA/GC-coated electrode (70 mV). This is because the combination of electrochemically active p-DPA and conductive ERGO forms an effective electroactive platform on the p-DPA@ERGO/GC electrode surface. Figure 6(Ea′–Ee′) shows CV in the presence of 1 mM PA with bare GC, activated GC, p-DPA/GC, ERGO/GC, and p-DPA@ERGO/GC-modified electrodes (0.1 M PBS supporting medium). The bare GC, activated GC, and p-DPA/GC-modified electrode showed an anodic peak at approximately +0.65 V that was related to the oxidation of PA. The ERGO/GC had a PA oxidation peak current of 13 μA. In contrast, the p-DPA@ERGO/GC electrode showed an enhanced anodic peak current of 19 μA (+0.62 V), which was approximately two times higher current and lower oxidation potential than the bare GC, activated GC, and p-DPA/GC electrode. The PA sulfuryl (–SH) functional group and polymer nitrogen group electrostatic interaction confers outstanding electrochemical performance. This leads to better PA mass transport properties. Scheme 2 shows the proposed possible mechanism of PA detection on the p-DPA@ERGO/GC electrode. Figure S2A (see Supplementary Materials) shows the different concentrations of PA (from 50 to 500 μM) addition on the p-DPA@ERGO/GC-coated electrode. The CV curve exhibited a gradual current increase as the PA concentration was increased. The corresponding calibration curve of $Ipa = 0.0131[PA] + 3.98$ with $R^2 = 0.993$ (Figure S2B, see Supplementary Materials). Figure 6F presents the CV traces of various scan rate effects (5 to 200 mV s^{-1}) of PA in 0.1 M PBS (pH 7.2) using a p-DPA@ERGO/GC electrode. The anodic oxidation peak current increased linearly with a positive direction potential shift as the scan rate (v) was increased. A linear relationship existed with an anodic peak current against the square root of sweep rate (v)$^{1/2}$ (Figure S2C, see Supplementary Materials) with a correlation equation of $Ipa = 2.645v^{1/2} + 0.3480$, with ($R^2 = 0.998$). Furthermore, the double logarithm plot of log i_{pa} versus log scan rate (Figure S2D, see Supplementary Materials) showed the following regression equation: log $Ipa = 0.5148$ log (v) $- 0.3867$ and an R^2 value of 0.997. These results showed that the oxidation process at the p-DPA@ERGO/GC electrode was diffusion-controlled.

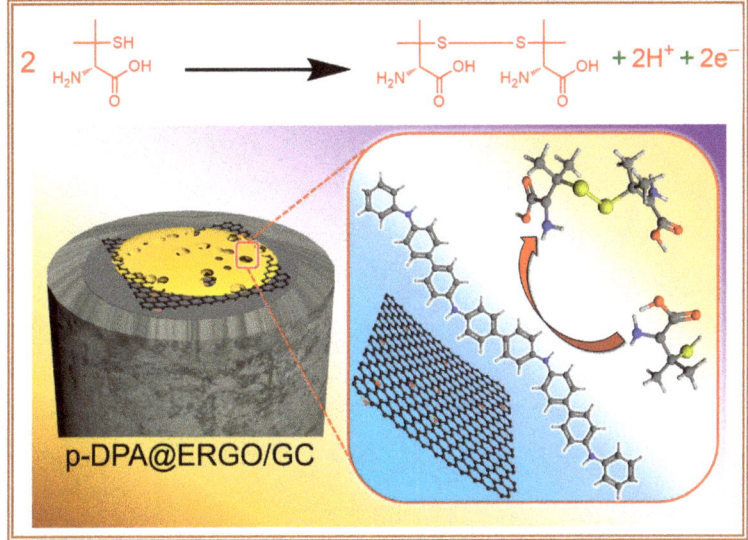

Scheme 2. Possible electrochemical oxidation mechanism of PA at the p-DPA@ERGO/GC electrode.

3.6. DPV and Amperometric Detection of PA

The DPV technique was performed to investigate the sensitivity of the p-DPA@ERGO/GC-coated electrode towards the wide concentration of PA determination ability. Figure 7A shows the DPV of PA detection on the p-DPA@ERGO/GC electrode; the PA oxidation peak potential was observed at +0.62 V. The modified electrode sensed PA concentrations from 10 to 4015.5 µM. Figure 7B shows the equivalent calibration plot of the PA oxidation current versus concentration. The two linear regression equations for the low and high concentration ranges were $Ipa = 0.0026[PA] + 5.593$ with $R^2 = 0.985$ and $Ipa = 0.0011[PA] + 6.638$ with $R^2 = 0.986$, respectively. The sensitivities of low and high concentration ranges were 0.0026 µAµM^{-1} and 0.0011 µAµM^{-1}, respectively. From the calibration plot, the limit of detection (LOD = 3s/S) was 0.30 µM. Figure 7C shows the successive 60 s interval PA additions performed with a fixed applied potential of +0.62 V. The amperometric current increased with every successive addition of PA with low and high concentration ranges (1.4–541 µM). The oxidation current increased considerably over a wide range as the PA concentration was increased, and Figure 7D shows the corresponding calibration curve. From the calibration curve, $Ipa = 0.0024[PA] + 0.0984$ with $R^2 = 0.981$. The calculated LOD of PA on the p-DPA@ERGO/GC electrode was 0.10 µM. The linear range and low LOD values from the present study technique were compared with the existing reports (Table 1). The present p-DPA@ERGO/GC result emphasizes the developed electrode ability to determine the wide PA sensor range and low LOD ability. The obtained high sensitivity and low LOD detection ability of the developed electrode might be useful for the bio-clinical analysis of PA in blood serum samples.

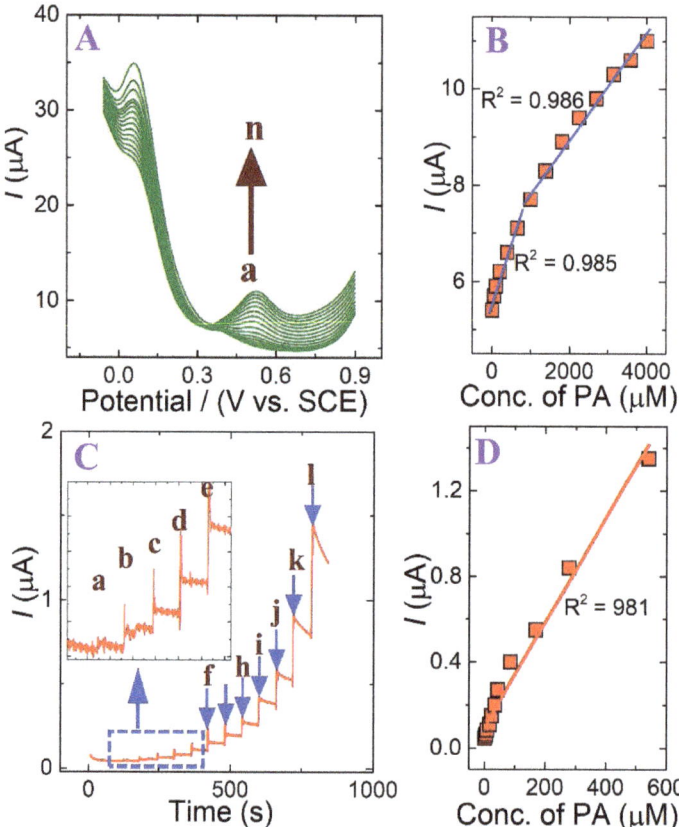

Figure 7. (**A**) DPV curves of different PA concentrations (a) 10.0, (b) 60.0, (c) 108.8, (d) 205.6, (e) 393.8, (f) 663.0, (g) 999.0, (h) 1386.7, (i) 810, (j) 2254.0, (k) 2706.5, (l) 3155.3, (m) 3593.8, and (n) 4015.5 µM; (**B**) calibration plot of the PA oxidation current versus concentration (pH 7.2); (**C**) amperometric (i–t) measurements for current steps upon the addition of PA from 1.4 to 541 µM on p-DPA@ERGO/GC-modified electrode with an applied potential of +0.62 V; (**D**) calibration plot of the current versus PA concentration.

Table 1. p-DPA@ERGO/GC electrode performance on PA detection compared to previous reports.

Electrode	Method	Linear Range (µM)	Limit of Detection (µM)	Ref.
Cu^{2+}/carbon paste electrode	CV	1–1000	0.1	[45]
β-CD@Ca-sacc/MeOH/GC [a]	DPV	100–500	0.79	[46]
CeO–ZnO/ILFCPE [b]	DPV	0.02–25.0	0.01	[47]
KI/GC	DPV	9.0–120	3.5	[48]
AuNPs/RGO [c]	DPV	5.0–110	3.9	[49]
t-Butylcatecho/GC	Amperometry	0.02–80	0.007	[50]
Quinizarine/TiO_2/CPE	SWV [d]	0.8–140	0.76	[51]
p-DPA@ERGO/GC	Amperometry	1.4–541	0.10	This work

[a] β-cyclodextrin/self-assembled chiral skeleton; [b] ionic liquid/ferrocene derivative n-hexyl-3-methylimidazolium hexafluoro phosphate; [c] reduced graphene oxide; [d] square wave voltammetry.

3.7. Selectivity, Repeatability, Reproducibility, and Storage Stability

The selectivity of the modified electrode was evaluated using the amperometric technique by determining 125 µM PA in the presence of common coexisting bioactive molecules. There is a two-fold high concentration of other interfering reagents, such as ascorbic acid (AA), uric acid (UA), dopamine (DA), glucose (GLU), cysteine (CY), tyrosine (TY), and leucine (Lu). In addition, inorganic ions of Ca^{2+}, Na^+, and K^+ were assessed (Figure S3 see Supplementary Materials). Among the foreign molecules investigated, cysteine showed a considerable influence of approximately 7% of the current responses, which can affect the determination of PA because cysteine has a similar chemical functional group. The proposed p-DPA@ERGO/GC electrode exhibits good tolerance against other common interfering bioactive molecules against PA determination. A repeatability test was performed using the same electrode set of PA detection and gained calibration plot slope values relative standard deviation (RSD, $n = 3$) to be 3.2%. The reproducibility of the developed sensor electrode was also studied in three replicate studies of the p-DPA@ERGO/GC-modified electrode, which was evaluated in 1 mM PA and resulted in an RSD ($n = 3$) of 3.85%. In addition, storage stability performed with a set of DPV measurements was carried with the p-DPA@ERGO/GC electrode after 20 days stored at 5 °C and obtained a similar current response for the determination of PA (RSD = 4.8). The observed selectivity, repeatability, reproducibility, and storage stability showed that the p-DPA@ERGO/GC-modified electrode is a potential tool for realistic PA detection.

3.8. Bio-Analytical Applications of p-DPA@ERGO/GC Electrode

The practical bio-analytical applicability of the as-prepared electrode was analyzed using real human serum samples. The determination of PA in serum samples performed by the standard addition method, and the obtained results are given in Figure S4A–D (see Supplementary Materials) and Table S3 (see Supplementary Materials). The results showed good recoveries of PA: 97.5 and 101.0%, with RSD values of 6.0 and 6.5%, respectively. These results confirmed that the p-DPA@ERGO/GC could be applied to the real-time analysis of PA in biological samples.

4. Conclusions

A p-DPA@ERGO/GC electrode was developed as a PA sensor. The detailed electrode fabrication through electrochemical reduction of GO and electropolymerization diphenylamine on the ERGO surface were discussed. The as-fabricated electrode was investigated by Raman spectroscopy, FESEM, EIS, and CV. The p-DPA formation mechanism on ERGO was given. The p-DPA@ERGO/GC-modified electrode was used for the oxidative determination of PA. The sulphuryl functional group of PA can adsorb on the electroactive amino functional surface of the p-DPA@ERGO/GC electrode. Amperometry revealed PA detection over a wide concentration range (1.4–541 µM), as well as high sensitivity, selectivity, and a low detection limit (100 nM). In addition, the designed sensor electrode was used in the real-time determination of PA in human serum samples, highlighting its bio-clinical applications.

Supplementary Materials: The following are available online at https://www.mdpi.com/article/10.3390/polym15030577/s1. Figure S1. (A,B) CV of concentration of PA addition and calibration plot, (C) plot of PA anodic oxidation peak current versus square root of scan rate, (D) double logarithmic plot of peak current versus scan rate; Figure S2. Amperometric selectivity curve; Figure S3. DPV standard addition human serum sample analysis; Figure S4. (A,C) Amperometry current response to the diluted human serum sample, by standard addition (R+S) method (B,D) corresponding analytical curve of the standard addition; Table S1. XPS region deconvolution evelopes peak positions, assignment, FWHM, and area; Table S2. The modified electrodes EIS data Randles circuit fitted values; Table S3. D-penicillamine determination in the human serum samples using p-DPA@ERGO/GC electrode.

Author Contributions: Conceptualization, Formal analysis, Visualization, and writing—review and editing, D.R.K.; Investigation and writing—original draft, K.R.; Investigation and visualization, M.S.S.; Formal analysis, A.M.; Project administration and supervision, J.-J.S. All authors have read and agreed to the published version of the manuscript.

Funding: This work was supported by the National Research Foundation (NRF) of the Republic of Korea, Priority Research Center's Program (NRF-2014R1A6A1031189); and the Regional University Superior Scientist Research Program (NRF-2020R1I1A3073981) funded by the National Research Foundation (NRF) of the Republic of Korea (Ministry of Education), and the author recognizes the SSN College of Engineering, Kalavakkam, Tamil Nadu, for financial assistance in the form of SSN PDF.

Institutional Review Board Statement: Not applicable.

Informed Consent Statement: Not applicable.

Data Availability Statement: The data presented in this study are available in the lab research notebook in SSN College and Yeungnam University.

Acknowledgments: The authors thank the National Research Foundation of Korea (NRF) for providing financial support.

Conflicts of Interest: The authors declare no conflict of interest.

References

1. Maziz, A.; Özgür, E.; Bergaud, C.; Uzun, L. Progress in conducting polymers for biointerfacing and biorecognition applications. *Sens. Actuators Rep.* **2021**, *3*, 100035. [CrossRef]
2. El-Kady, M.F.; Shao, Y.; Kaner, R.B. Graphene for batteries, supercapacitors and beyond. *Nat. Rev. Mater.* **2016**, *1*, 16033. [CrossRef]
3. Shamkhalichenar, H.; Choi, J.-W. Review—Non-Enzymatic Hydrogen Peroxide Electrochemical Sensors Based on Reduced Graphene Oxide. *J. Electrochem. Soc.* **2020**, *167*, 037531. [CrossRef]
4. He, B.; Yan, D. Au/ERGO nanoparticles supported on Cu-based metal-organic framework as a novel sensor for sensitive determination of nitrite. *Food Control* **2019**, *103*, 70–77. [CrossRef]
5. Zhang, Z.; Xiao, F.; Qian, L.; Xiao, J.; Wang, S.; Liu, Y. Facile Synthesis of 3D MnO_2–Graphene and Carbon Nanotube–Graphene Composite Networks for High-Performance, Flexible, All-Solid-State Asymmetric Supercapacitors. *Adv. Energy Mater.* **2014**, *4*, 1400064. [CrossRef]
6. Kim, J.; Yoon, D. A One-Pot Route for Uniform Deposition of Metal Oxide and Metal Sulfide Nanoparticles on Reduced Graphene Oxide Using Supercritical Alcohols. *ECS Meet. Abstr.* **2016**, *MA2016-03*, 992. [CrossRef]
7. Kumar, D.R.; Dhakal, G.; Nguyen, V.Q.; Shim, J.-J. Molecularly imprinted hornlike polymer@electrochemically reduced graphene oxide electrode for the highly selective determination of an antiemetic drug. *Anal. Chim. Acta* **2021**, *1141*, 71–82. [CrossRef]
8. Popov, A.; Aukstakojyte, R.; Gaidukevic, J.; Lisyte, V.; Kausaite-Minkstimiene, A.; Barkauskas, J.; Ramanaviciene, A. Reduced Graphene Oxide and Polyaniline Nanofibers Nanocomposite for the Development of an Amperometric Glucose Biosensor. *Sensors* **2021**, *21*, 948. [CrossRef]
9. Kumar, D.R.; Kesavan, S.; Nguyen, T.T.; Hwang, J.; Lamiel, C.; Shim, J.-J. Polydopamine@electrochemically reduced graphene oxide-modified electrode for electrochemical detection of free-chlorine. *Sens. Actuators B* **2017**, *240*, 818–828. [CrossRef]
10. Smaranda, I.; Benito, A.M.; Maser, W.K.; Baltog, I.; Baibarac, M. Electrochemical Grafting of Reduced Graphene Oxide with Polydiphenylamine Doped with Heteropolyanions and Its Optical Properties. *J. Phys. Chem. C* **2014**, *118*, 25704–25717. [CrossRef]
11. Mu, S. The electrocatalytic oxidative polymerization of o-phenylenediamine by reduced graphene oxide and properties of poly(o-phenylenediamine). *Electrochim. Acta* **2011**, *56*, 3764–3772. [CrossRef]
12. Obrezkov, F.A.; Shestakov, A.F.; Vasil'ev, S.G.; Stevenson, K.J.; Troshin, P.A. Polydiphenylamine as a promising high-energy cathode material for dual-ion batteries. *J. Mater. Chem. A* **2021**, *9*, 2864–2871. [CrossRef]
13. Muthusankar, E.; Ponnusamy, V.K.; Ragupathy, D. Electrochemically sandwiched poly(diphenylamine)/phosphotungstic acid/graphene nanohybrid as highly sensitive and selective urea biosensor. *Synth. Met.* **2019**, *254*, 134–140. [CrossRef]
14. Suganandam, K.; Santhosh, P.; Sankarasubramanian, M.; Gopalan, A.; Vasudevan, T.; Lee, K.-P. Fe^{3+} ion sensing characteristics of polydiphenylamine—Electrochemical and spectroelectrochemical analysis. *Sens. Actuators B* **2005**, *105*, 223–231. [CrossRef]
15. Pugliese, M.; Biondi, V.; Gugliandolo, E.; Licata, P.; Peritore, A.F.; Crupi, R.; Passantino, A. D-Penicillamine: The State of the Art in Humans and in Dogs from a Pharmacological and Regulatory Perspective. *Antibiotics* **2021**, *10*, 648. [CrossRef]
16. Kumar, V.; Singh, A.P.; Wheeler, N.; Galindo, C.L.; Kim, J.-J. Safety profile of D-penicillamine: A comprehensive pharmacovigilance analysis by FDA adverse event reporting system. *Expert Opin. Drug Saf.* **2021**, *20*, 1443–1450. [CrossRef]
17. Pitman, S.K.; Huynh, T.; Bjarnason, T.A.; An, J.; Malkhasyan, K.A. A case report and focused literature review of d-penicillamine and severe neutropenia: A serious toxicity from a seldom-used drug. *Clin. Case Rep.* **2019**, *7*, 990–994. [CrossRef]

18. Joly, D.; Rieu, P.; Méjean, A.; Gagnadoux, M.-F.; Daudon, M.; Jungers, P. Treatment of cystinuria. *Pediatr. Nephrol.* **1999**, *13*, 945–950. [CrossRef]
19. Dunea, G. Kidney Stones: Medical and Surgical Management. *JAMA* **1996**, *276*, 577. [CrossRef]
20. Kuśmierek, K.; Bald, E. Simultaneous determination of tiopronin and d-penicillamine in human urine by liquid chromatography with ultraviolet detection. *Anal. Chim. Acta* **2007**, *590*, 132–137. [CrossRef]
21. Ge, H.; Zhang, K.; Yu, H.; Yue, J.; Yu, L.; Chen, X.; Hou, T.; Alamry, K.A.; Marwani, H.M.; Wang, S. Sensitive and Selective Detection of Antibiotic D-Penicillamine Based on a Dual-Mode Probe of Fluorescent Carbon Dots and Gold Nanoparticles. *J. Fluoresc.* **2018**, *28*, 1405–1412. [CrossRef]
22. Saracino, M.A.; Cannistraci, C.; Bugamelli, F.; Morganti, E.; Neri, I.; Balestri, R.; Patrizi, A.; Raggi, M.A. A novel HPLC-electrochemical detection approach for the determination of d-penicillamine in skin specimens. *Talanta* **2013**, *103*, 355–360. [CrossRef]
23. Nazifi, M.; Ramezani, A.M.; Absalan, G.; Ahmadi, R. Colorimetric determination of D-penicillamine based on the peroxidase mimetic activity of hierarchical hollow MoS2 nanotubes. *Sens. Actuators B* **2021**, *332*, 129459. [CrossRef]
24. Ibrahim, S.E.; Al-badr, A.A. Application of PMR Spectrometry in Quantitative Analysis of Penicillamine. *Spectrosc. Lett.* **1980**, *13*, 471–478. [CrossRef]
25. Yusof, M.; Neal, R.; Aykin, N.; Ercal, N. High performance liquid chromatography analysis of D-penicillamine by derivatization with N-(1-pyrenyl)maleimide (NPM). *Biomed. Chromatogr.* **2000**, *14*, 535–540. [CrossRef]
26. Rushing, L.G.; Hansen, E.B., Jr.; Thompson, H.C., Jr. Analysis of D-penicillamine by gas chromatography utilizing nitrogen–phosphorus detection. *J. Chromatogr.* **1985**, *337*, 37–46. [CrossRef]
27. Yu, H.; Chen, X.; Yu, L.; Sun, M.; Alamry, K.A.; Asiri, A.M.; Zhang, K.; Zapien, J.A.; Wang, S. Fluorescent MUA-stabilized Au nanoclusters for sensitive and selective detection of penicillamine. *Anal. Bioanal. Chem.* **2018**, *410*, 2629–2636. [CrossRef]
28. Gunjal, D.B.; Gore, A.H.; Naik, V.M.; Pawar, S.P.; Anbhule, P.V.; Shejwal, R.V.; Kolekar, G.B. Carbon dots as a dual sensor for the selective determination of d-penicillamine and biological applications. *Opt. Mater.* **2019**, *88*, 134–142. [CrossRef]
29. Kumar, D.R.; Baynosa, M.L.; Dhakal, G.; Shim, J.-J. Sphere-like $Ni_3S_4/NiS_2/MoO_x$ composite modified glassy carbon electrode for the electrocatalytic determination of d-penicillamine. *J. Mol. Liq.* **2020**, *301*, 112447. [CrossRef]
30. Liu, Z.; Kuang, X.; Sun, X.; Zhang, Y.; Wei, Q. Electrochemical enantioselective recognition penicillamine isomers based on chiral C-dots/MOF hybrid arrays. *J. Electroanal. Chem.* **2019**, *846*, 113151. [CrossRef]
31. Zhang, Y.; Wang, H.-Y.; He, X.-W.; Li, W.-Y.; Zhang, Y.-K. Homochiral fluorescence responsive molecularly imprinted polymer: Highly chiral enantiomer resolution and quantitative detection of L-penicillamine. *J. Hazard. Mater.* **2021**, *412*, 125249. [CrossRef]
32. Alkahtani, S.A.; Mahmoud, A.M.; El-Wekil, M.M. Electrochemical sensing of copper-chelator D- penicillamine based on complexation with gold nanoparticles modified copper based-metal organic frameworks. *J. Electroanal. Chem.* **2022**, *908*, 116102. [CrossRef]
33. Hummers, W.S.; Offeman, R.E. Preparation of Graphitic Oxide. *J. Am. Chem. Soc.* **1958**, *80*, 1339. [CrossRef]
34. Kairy, P.; Hossain, M.M.; Khan, M.A.R.; Almahri, A.; Rahman, M.M.; Hasnat, M.A. Electrocatalytic oxidation of ascorbic acid in the basic medium over electrochemically functionalized glassy carbon surface. *Surf. Interfaces* **2022**, *33*, 102200. [CrossRef]
35. Islam, M.T.; Hasan, M.M.; Shabik, M.F.; Islam, F.; Nagao, Y.; Hasnat, M.A. Electroless deposition of gold nanoparticles on a glassy carbon surface to attain methylene blue degradation via oxygen reduction reactions. *Electrochim. Acta* **2020**, *360*, 136966. [CrossRef]
36. Tsai, T.-H.; Ku, S.-H.; Chen, S.-M.; Lou, B.-S.; Ali, M.A.; Al-Hemaid, F.M.A. Electropolymerized Diphenylamine on Functionalized Multiwalled Carbon Nanotube Composite Film and Its Application to Develop a Multifunctional Biosensor. *Electroanalysis* **2014**, *26*, 399–408. [CrossRef]
37. Wang, X.; Kholmanov, I.; Chou, H.; Ruoff, R.S. Simultaneous Electrochemical Reduction and Delamination of Graphene Oxide Films. *ACS Nano* **2015**, *9*, 8737–8743. [CrossRef]
38. Zhou, M.; Wang, Y.; Zhai, Y.; Zhai, J.; Ren, W.; Wang, F.; Dong, S. Controlled synthesis of large-area and patterned electrochemically reduced graphene oxide films. *Eur. J. Chem.* **2009**, *15*, 6116–6120. [CrossRef]
39. Toh, S.Y.; Loh, K.S.; Kamarudin, S.K.; Daud, W.R.W. Graphene production via electrochemical reduction of graphene oxide: Synthesis and characterisation. *Chem. Eng. J.* **2014**, *251*, 422–434. [CrossRef]
40. Cançado, L.G.; Takai, K.; Enoki, T.; Endo, M.; Kim, Y.A.; Mizusaki, H.; Jorio, A.; Coelho, L.N.; Magalhães-Paniago, R.; Pimenta, M.A. General equation for the determination of the crystallite size La of nanographite by Raman spectroscopy. *Appl. Phys. Lett.* **2006**, *88*, 163106. [CrossRef]
41. Stobinski, L.; Lesiak, B.; Malolepszy, A.; Mazurkiewicz, M.; Mierzwa, B.; Zemek, J.; Jiricek, P.; Bieloshapka, I. Graphene oxide and reduced graphene oxide studied by the XRD, TEM and electron spectroscopy methods. *J. Electron. Spectrosc. Relat. Phenom.* **2014**, *195*, 145–154. [CrossRef]
42. Oh, Y.J.; Yoo, J.J.; Kim, Y.I.; Yoon, J.K.; Yoon, H.N.; Kim, J.-H.; Park, S.B. Oxygen functional groups and electrochemical capacitive behavior of incompletely reduced graphene oxides as a thin-film electrode of supercapacitor. *Electrochim. Acta* **2014**, *116*, 118–128. [CrossRef]
43. Marinho, M.I.C.; Cabral, M.F.; Mazo, L.H. Is the poly (methylene blue)-modified glassy carbon electrode an adequate electrode for the simple detection of thiols and amino acid-based molecules? *J. Electroanal. Chem.* **2012**, *685*, 8–14. [CrossRef]

44. Hatamie, A.; Rahmati, R.; Rezvani, E.; Angizi, S.; Simchi, A. Yttrium Hexacyanoferrate Microflowers on Freestanding Three-Dimensional Graphene Substrates for Ascorbic Acid Detection. *ACS Appl. Nano Mater.* **2019**, *2*, 2212–2221. [CrossRef]
45. Ghaffarinejad, A.; Hashemi, F.; Nodehi, Z.; Salahandish, R. A simple method for determination of d-penicillamine on the carbon paste electrode using cupric ions. *Bioelectrochemistry* **2014**, *99*, 53–56. [CrossRef]
46. Hou, Y.; Liang, J.; Kuang, X.; Kuang, R. Simultaneous electrochemical recognition of tryptophan and penicillamine enantiomers based on MOF-modified β-CD. *Carbohydr. Polym.* **2022**, *290*, 119474. [CrossRef]
47. Parisa, B.; Shishehbore, M.R.; Beitollahi, H.; Sheibani, A. The Application of Ferrocene Derivative and CeO–ZnO Nanocomposite-Modified Carbon Paste Electrode for Simultaneous Detection of Penicillamine and Tryptophan. *Russ. J. Electrochem.* **2022**, *58*, 235–247. [CrossRef]
48. Raoof, J.-B.; Ojani, R.; Majidian, M.; Chekin, F. Voltammetric determination of D-penicillamine based on its homogeneous electrocatalytic oxidation with potassium iodide at the surface of glassy carbon electrode. *Russ. J. Electrochem.* **2010**, *46*, 1395–1401. [CrossRef]
49. Jafari, M.; Tashkhourian, J.; Absalan, G. Electrochemical sensing of D-penicillamine on modified glassy carbon electrode by using a nanocomposite of gold nanoparticles and reduced graphene oxide. *J. Iran. Chem. Soc.* **2017**, *14*, 1253–1262. [CrossRef]
50. Torriero, A.A.J.; Piola, H.D.; Martínez, N.A.; Panini, N.V.; Raba, J.; Silber, J.J. Enzymatic oxidation of tert-butylcatechol in the presence of sulfhydryl compounds: Application to the amperometric detection of penicillamine. *Talanta* **2007**, *71*, 1198–1204. [CrossRef]
51. Mazloum-Ardakani, M.; Beitollahi, H.; Taleat, Z.; Naeimi, H.; Taghavinia, N. Selective voltammetric determination of d-penicillamine in the presence of tryptophan at a modified carbon paste electrode incorporating TiO_2 nanoparticles and quinizarine. *J. Electroanal. Chem.* **2010**, *644*, 1–6. [CrossRef]

Disclaimer/Publisher's Note: The statements, opinions and data contained in all publications are solely those of the individual author(s) and contributor(s) and not of MDPI and/or the editor(s). MDPI and/or the editor(s) disclaim responsibility for any injury to people or property resulting from any ideas, methods, instructions or products referred to in the content.

MDPI
St. Alban-Anlage 66
4052 Basel
Switzerland
Tel. +41 61 683 77 34
Fax +41 61 302 89 18
www.mdpi.com

Polymers Editorial Office
E-mail: polymers@mdpi.com
www.mdpi.com/journal/polymers